Python Programming and Numerical Methods

A Guide for Engineers and Scientists

Python Programming and Numerical Methods

A Guide for Engineers and Scientists

Qingkai Kong

Timmy Siauw

Alexandre M. Bayen

ELSEVIER

ACADEMIC PRESS
An imprint of Elsevier

Academic Press is an imprint of Elsevier
125 London Wall, London EC2Y 5AS, United Kingdom
525 B Street, Suite 1650, San Diego, CA 92101, United States
50 Hampshire Street, 5th Floor, Cambridge, MA 02139, United States
The Boulevard, Langford Lane, Kidlington, Oxford OX5 1GB, United Kingdom

Library of Congress Cataloging-in-Publication Data
A catalog record for this book is available from the Library of Congress

British Library Cataloguing-in-Publication Data
A catalogue record for this book is available from the British Library

ISBN: 978-0-12-819549-9

For information on all Academic Press publications
visit our website at https://www.elsevier.com/books-and-journals

Publisher: Katey Birtcher
Acquisitions Editor: Steve Merken
Editorial Project Manager: Alice Grant
Production Project Manager: Kamesh Ramajogi
Designer: Miles Hitchen

Typeset by VTeX

To Fanyu, Fanqi, and Fan without whom this book would have been finished earlier ☺ To Xianzhong, and Ping!

To the students of UC Berkeley's E7 class: past, present, and future.

To my daughter Myriam,
who is discovering the world of programming.

Contents

List of Figures

Preface

PURPOSE

Because programming has become an essential component of engineering, science, medicine, media, business, finance, and many other fields, it is important for scientists and engineers to have a basic foundation in computer programming to be competitive. This book introduces programming to students from a wide range of backgrounds and gives them programming and mathematical tools that will be useful throughout their careers.

For the most part, this book follows the standard material taught at the University of California, Berkeley, in the class *E7: Introduction to computer programming for scientists and engineers*. This class is taken by most science and engineering freshmen in the College of Engineering, and by undergraduate students from other disciplines, including physics, biology, Earth, and cognitive sciences. The course was originally taught in Matlab, but with the recent trend of the data science movement at Berkeley, the Division of Data Sciences agreed on and supported the transform of this course into a Python-oriented course to prepare students from different fields for further data science courses. The course has two fundamental goals:

- Teach Python programming to science and engineering students who do not have prior exposure to programming.
- Introduce a variety of numerical analysis tools that are useful for solving science and engineering problems.

These two goals are reflected in the two parts of this book:

- Introduction to Programming for Scientists and Engineers
- Introduction to Numerical Methods

This book is written based on the book *An Introduction to MATLAB® Programming and Numerical Methods for Engineers* by Timmy Siauw and Alexandre Bayen. The current book was first written in `Jupyter Notebook` for interactive purposes, and then converted to LaTeX. Most of the codes showing in this book are from the Jupyter Notebook code cells, which can be run directly in the notebook cell. All the Jupyter Notebook codes can be found at pythonnumericalmethods.berkeley.edu.

Because this book covers such a wide range of topics, no topic is covered in great depth. Each chapter has been designed to be covered in at most two lecture hours, even though there are entire semester courses dedicated to these same chapters. Rather than an in-depth treatment, this book is intended to give students a wide breadth of programming knowledge and mathematical vocabulary on which they can expand.

We believe that just like learning a new foreign language, learning to program can be fun and illuminating. We hope that as you journey through this book, you will agree.

PREREQUISITES

This book is designed to introduce programming and numerical methods to students who have *absolutely no* prior experience with computer programming. We hope this underlying concept is reflected in the pace, tone, and content of the text. For the purpose of programming, we assume the reader has the following prerequisite knowledge:

- Understanding of the computer monitor and keyboard/mouse input devices
- Understanding of the folder structure used to store files in most operating systems

For the mathematical portions of the text, we assume the reader has the following prerequisite knowledge:

- High school level algebra and trigonometry
- Introductory, college-level calculus

That's it! Anything in the text that assumes more than this level of knowledge is our mistake, and we apologize in advance for any confusion of instances where concepts are unclear.

ORGANIZATION

Part 1 teaches the fundamental concepts of programming. Chapter 1 introduces the reader to Python and Jupyter Notebook. Chapters 2 through 7 teach the fundamentals of programming. Proficiency in the material from these chapters should provide enough background to enable you to program almost anything you imagine. Chapter 8 provides the theory that characterizes computer programs based on how fast they run, and Chapter 9 gives insights into how computers represent numbers and their effect on arithmetic. Chapter 10 provides useful tips on good programming practices to limit mistakes from popping up in computer code, and tells the user how to find them when they do. Chapter 11 explains how to store data over the long term and how to make results from Python useful outside of Python (i.e., for other programs). Chapter 12 introduces Python's graphical features that allow you to produce plots and charts, which is a really useful feature for engineers and scientists to visualize results. Finally, Chapter 13 introduces basics about the parallel programming in Python to take advantage of the multicore design of today's computers.

Part 2 gives an overview of a variety of numerical methods that are useful for engineers. Chapter 14 gives a crash course in linear algebra. Although theoretical in nature, linear algebra is the single most critical concept for understanding many advanced engineering topics. Chapter 15 discusses eigenvalues and eigenvectors, which are important tools in engineering and science, and the ways we can utilize them. Chapter 16 is about regression, a mathematical term that is simply a method of fitting theoretical models to observed data. Chapter 17 is about inferring the value of a function between data points, a framework known as "interpolation." Chapter 18 introduces the idea of approximating functions with polynomials, which can be useful for simplifying complicated functions. Chapter 19 teaches two algorithms for finding roots of functions, that is, finding an x such that $f(x) = 0$, where f is a function. Chapters 20 and 21 cover methods of approximating the derivative and integral of a function, respectively. Chapters 22 and Chapter 23 introduce a mathematical model type called "ordinary differential equations." These two chapters focus on different problems, i.e., initial value problems and boundary value problems, and present several methods for finding their solutions. Chapter 24 introduces the concepts of "discrete Fourier transform" and "fast Fourier transform" and their use in digital signal processing.

HOW TO READ THIS BOOK?

Learning to program is all about practice, practice, and practice. Just like learning a new language, there is no way to master computer programming without engaging with the material, internalizing it, and putting it into constant use.

We suggest that as you go through the text, you should ideally have Jupyter Notebook open or the interactive website in front of you, and run all of the numerous examples that are provided. Go slowly. Taking the time to understand what Python is doing in every example will pay large dividends compared to "powering through" the text like a novel.

In terms of the text itself, Chapters 1 through 5 should be read and understood first since they cover the fundamentals of programming. Chapters 6 through 11 can be covered in any order. Chapter 12 and 13 on plotting and parallel programming are must-read chapters if you wish to improve your problem-solving skills. In Part 2, Chapter 14 should be read first since subsequent chapters rely on linear algebraic concepts. The remaining chapters can be read in any order. We do suggest reading Chapters 17 and 18 before Chapters 19 and 20.

Throughout the text, there are words written in boldface. When you encounter one of them, you should take the time to commit the word to memory and understand its meaning in the context of the material being presented.

To keep the text from running on and overwhelming the reader, we have punctuated the material with smaller blocks of text. Different blocks of text have different functions. For example,

TRY IT! This is the most common block in the text. It will usually have a short description of a problem and/or an activity. We strongly recommend that you actually "try" all of these in Python.

TIP! This block gives some advice that we believe will make programming easier. Note that the blocks do not contain any new material that is essential for understanding the key concepts of the text.

EXAMPLE: These sections are concrete examples of new concepts. They are designed to help you think about new concepts; however, they do not necessarily need to be tried.

WARNING! Learning to program can have many pitfalls. These sections contain information that will help you avoid confusion, developing bad habits, or misunderstanding key concepts.

WHAT IS HAPPENING? These sections follow Python in scrutinizing detail to help you understand what goes on when Python executes programs.

CONSTRUCTION: In programming there are standard architectures that are reserved to perform common and important tasks. These sections outline these architectures and how to use them.

There are two sections that end every chapter. The Summary section lists the main take-aways of the chapter. These points should be intuitive to you by the end of the chapter. The Problems section gives exercises that will reinforce concepts from the chapter.

As one final note, there are many ways of doing the same thing in Python. Although at first this can seem like a useful feature, it can make learning Python confusing or overload the programming novice with possibilities when the task is actually straightforward. This book presents a single way of performing a task to provide structure for the new user's learning experience and to keep that user on track and from being inundated by extraneous information. You may discover solutions that differ from the text's solutions but solve the problem just the same or even better! We encourage you to find these alternative methods, and use your own judgment to given the tools we have provided herein to decide which way is better.

We hope you enjoy the book!

WHY PYTHON?

Python is a high-level and general-purpose computer language that lends itself to many applications. As it is beginner friendly, we hope that you will find it easy to learn and that it is fun to play with it. The language itself is very flexible, which means that there are no hard rules on how to build features, and you will find that there are several ways to solve the same problem. Perhaps its great strength is that it has a great user community that supports it, with lots of packages to essentially plug in and go with very little efforts. With the ongoing popular trend, Python suits the goals of data science today. Python is free (open source), and most of the packages are also free for use. The idea of an open source programming language makes a huge difference in the learning curve. Not only you can use these packages for free, but also you can learn many advanced skills from the source code of these packages developed by other users. We hope you can enjoy your learning of Python presented here and use it in your work and life.

PYTHON AND PACKAGE VERSIONS

This book was written using Python 3. Here is a list of packages with their versions that used in this book. As these packages are constantly under development, some features may be added, removed, or changed in the versions on your computer:

- `jupyter` – 1.0.0
- `IPython` – 7.5.0
- `NumPy` – 1.16.4
- `SciPy` – 1.2.1
- `h5py` – 2.9.0
- `matplotlib` – 3.1.0
- `cartopy` – 0.17.0
- `joblib` – 0.13.2

Acknowledgements

The first version of this book was written at a time when the standard generalist language taught in engineering (and beyond) at UC Berkeley was Matlab. Its genesis goes back to the mid-2000s, which precede the current era of data science, machine learning, in which Python emerged as a commonly used language across the engineering profession. The first version was thus written as part of the E7 class at UC Berkeley, which introduces many students to programming and numerical analysis. It would never have been written without the help of colleagues, teams of Graduate Student Instructors (GSI), graders, and administrative staff members who helped us through the challenging process of teaching E7 to several hundreds of students each semester at UC Berkeley. Furthermore, the first edition of this book would never have reached completion without the help of the students who had the patience to read the book and give us their feedback. In the process of teaching E7 numerous times, we have interacted with thousands of students, dozens of GSIs and graders, and a dozen colleagues and administrators, and we apologize to those we will inevitably forget given the number of people involved. We are extremely grateful for guidance from our colleagues Professors Panos Papadopoulos, Roberto Horowitz, Michael Frenklach, Andy Packard, Tad Patzek, Jamie Rector, Raja Sengupta, Mike Cassidy, and Samer Madanat. We owe thanks particularly to Professors Roberto Horowitz, Andy Packard, Sanjay Govindjee, and Tad Patzek for sharing the material they used for the class, which contributed to the material in this book. We also thank Professors Rob Harley and Sanjay Govindjee for using a draft of this book during the semesters they taught E7 and giving us feedback that helped improve the manuscript. The smooth running of the semester course gave the authors the time and energy to produce this book. Managing the course was greatly facilitated by numerous administrative staff members who bore much of the logistic load. We are particularly grateful to Joan Chamberlain, Shelley Okimoto, Jenna Tower, and Donna Craig. Civil and Environmental Engineering Vice Chair Bill Nazaroff deserves particular recognition for assigning the second author to teach the class in 2011. Without this assignment the two authors of this book would not have had an opportunity to work together and write this book. E7 is notoriously the hardest class to teach at UC Berkeley in the College of Engineering. However, it continued to run smoothly over the many semesters we learned to teach this class, mainly due to the help of the talented GSIs we had the pleasure of working with. During the years the coauthors taught the class, a series of legendary head GSIs have contributed to shaping the class and making it a meaningful experience for students. In particular, Scott Payne, James Lew, Claire Saint-Pierre, Kristen Parish, Brian McDonald, and Travis Walter have in their respective roles led a team of dedicated GSI to exceed expectations. The GSI and grader team during the Spring of 2011 greatly influenced the material of this book. For their contribution during that critical semester, we thank Jon Beard, Leah Anderson, Marc Lipoff, Sebastien Blandin, Sam Chiu, Rob Hansen, Jiangchuan Huang, Brad Adams, Ryan Swick, Pranthik Samal, Matthieu Lewandowski, and Romain Bourcier. We are also grateful to Claire Johnson and Katherine Mellis for finding errors in the text and helping us incorporate edits into the manuscript. We are indebted to the E7 students for their patience with us and their thorough reading of the material. Having seen thousands of them through the years, we are sorry to only be able to mention a few for their extraordinary feedback and performance: Gurshamnjot Singh, Sabrina Nicolle Atienza, Yi Lu, Nicole Schauser, Harrison Lee, Don Mai, Robin Parrish, and Mara Minner. In 2018, as the UC Berkeley campus was already deeply engaged in the transition leading to the birth of the *Division of Computing, Data Science, and Society*, numerous conversations started on the need for UC Berkeley students to learn Python, which in the mean time had become a commodity of choice for employment in most tech companies. Thus, this book started with the intention of preparing engineering and science students with basic data science tools. The UC Berkeley *Division of Computing, Data Science, and Society* played an active role in creating this book for a lower division course to prepare students for further study. We thank Cathryn Carson and David Culler for their support in writing this book and for the discussions on how to make it better. Their help happened in parallel to the herculean efforts they led to build the *Division of Computing, Data Science, and Society*. It is one of the many expressions of their scientific generosity and dedication to building a rich and innovative data science environment at UC Berkeley. Finally, we also appreciate the care and help from Eric Van Dusen and Keeley Takimoto. About two thirds of the book are adapted from the original Matlab version – *An Introduction to MATLAB® Programming and Numerical Methods for Engineers* by the two last authors. We thank Jennifer Grannen, Brian Mickel, Nick Bourlier, and Austin Chang for their help to convert some of the Matlab code to Python. We are (again!) grateful to Claire Johnson for her help with the second version of the book, and to Jennifer Taggart for finding errors in the text and helping us incorporate edits into the manuscript. We also thank the Berkeley Seismsology Lab for the support of writing this book and the Python training over the years.

Qingkai Kong
Timmy Siauw
Alexandre M. Bayen

June 2020

PART

INTRODUCTION TO PYTHON PROGRAMMING

1

CHAPTER

PYTHON BASICS

1

CONTENTS

1.1 GETTING STARTED WITH PYTHON

1.1.1 SETTING UP YOUR WORKING ENVIRONMENT

The first step in using Python is to set up the working environment on the computer. This section introduces the initial processes to get it started.

There are different ways to **install Python** and related packages, and we recommend using Anaconda[1] or Miniconda[2] to install and manage your packages. Depending on the *operating system* (OS) you are using (i.e., Windows, Mac OS X, or Linux), you will need to download a specific installer for your machine. Both Anaconda and Miniconda are aimed at providing easy ways to manage the Python work environment in scientific computing and data sciences.

[1] https://www.anaconda.com/download/.

[2] https://conda.io/miniconda.html.

Python Programming and Numerical Methods. https://doi.org/10.1016/B978-0-12-819549-9.00010-5

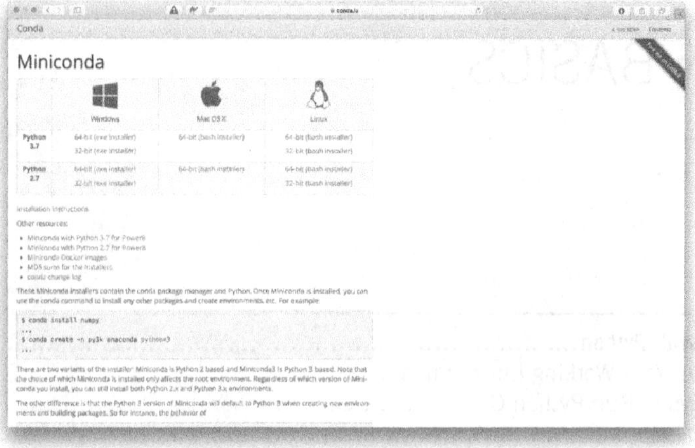

FIGURE 1.1

The Miniconda download page; choose the installer based on your operating system.

In this example, we will use Mac OS X to show you how to install Miniconda (the process of which is very similar to installing on Linux). For Windows users, please skip the rest of this section and read Appendix A on the installation instructions. The main differences between Anaconda and Miniconda are as follows:

- **Anaconda** is a complete distribution framework that includes the Python interpreter, package manager, and the commonly used packages in scientific computing.
- **Miniconda** is a "light" version of Anaconda that does not include the commonly used packages. You need to install all the different packages yourself, but it does include the Python interpreter and package manager.

The option we've chosen here is Miniconda, and we will install only those packages that we will need. The Miniconda install process is described below:

Step 1. Download the Miniconda installer from the website.[3] The download page is shown in Fig. 1.1. Here you can choose a different installer based on your OS. In this example, we choose Mac OS X and Python 3.7.

Step 2. Open a terminal (on a Mac, you can search for "terminal" in Spotlight search). Run the installer from the terminal using the commands showing in Fig. 1.2. After you run the installer, follow the guide to finish the installation.

Note that although you can change the installation location by giving it an alternative location on your machine, the default is your home directory (Fig. 1.3).

After installation, you can check the installed packages by typing the commands shown in Fig. 1.4.

[3] https://conda.io/miniconda.html.

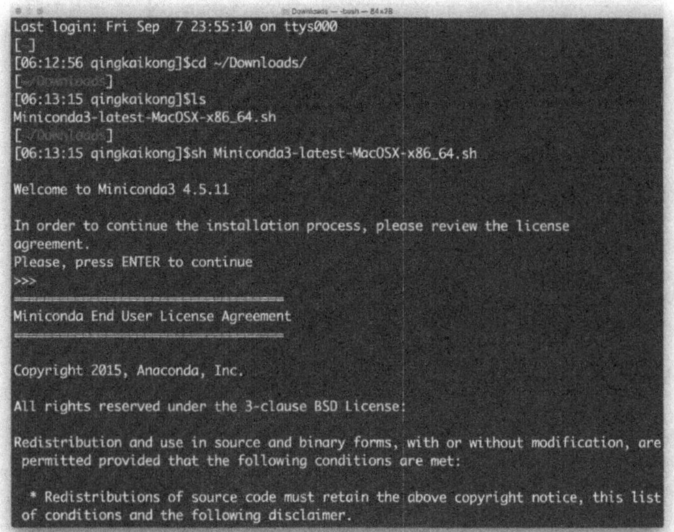

FIGURE 1.2

Screen shot of running the installer in a terminal.

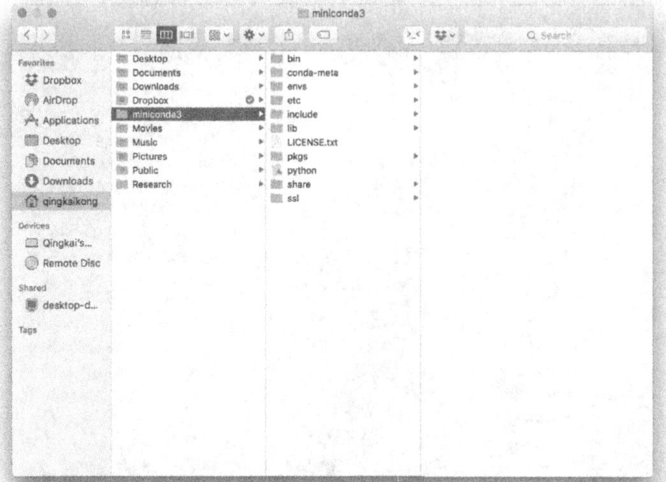

FIGURE 1.3

The default installation location of your file system.

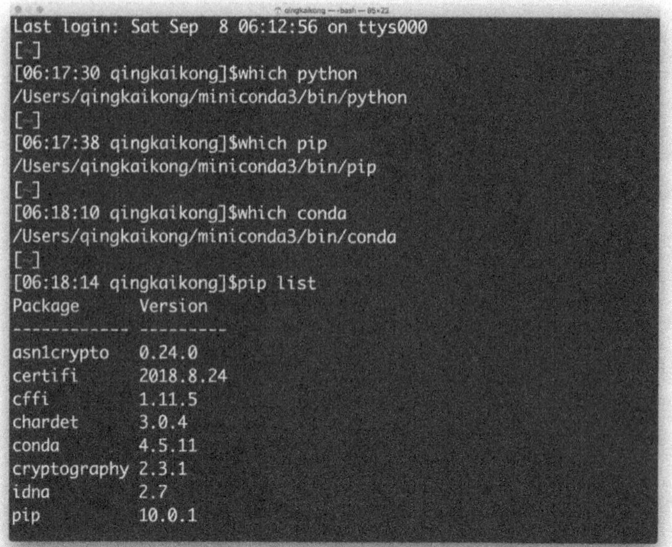

Last login: Sat Sep 8 06:12:56 on ttys000
[~]
[06:17:30 qingkaikong]$which python
/Users/qingkaikong/miniconda3/bin/python
[~]
[06:17:38 qingkaikong]$which pip
/Users/qingkaikong/miniconda3/bin/pip
[~]
[06:18:10 qingkaikong]$which conda
/Users/qingkaikong/miniconda3/bin/conda
[~]
[06:18:14 qingkaikong]$pip list
Package Version
------------ ---------
asn1crypto 0.24.0
certifi 2018.8.24
cffi 1.11.5
chardet 3.0.4
conda 4.5.11
cryptography 2.3.1
idna 2.7
pip 10.0.1

FIGURE 1.4

Quick way to check if Miniconda was installed successfully and the programs are run properly.

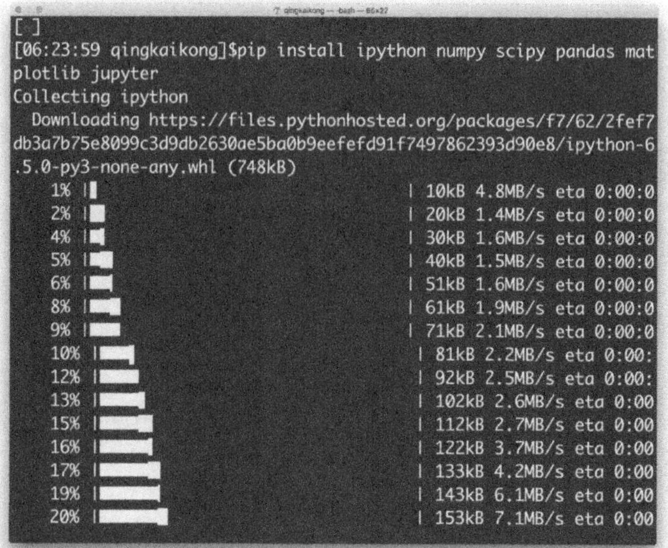

[~]
[06:23:59 qingkaikong]$pip install ipython numpy scipy pandas mat
plotlib jupyter
Collecting ipython
Downloading https://files.pythonhosted.org/packages/f7/62/2fef7
db3a7b75e8099c3d9db2630ae5ba0b9eefefd91f7497862393d90e8/ipython-6
.5.0-py3-none-any.whl (748kB)
1% | 10kB 4.8MB/s eta 0:00:0
2% | 20kB 1.4MB/s eta 0:00:0
4% | 30kB 1.6MB/s eta 0:00:0
5% | 40kB 1.5MB/s eta 0:00:0
6% | 51kB 1.6MB/s eta 0:00:0
8% | 61kB 1.9MB/s eta 0:00:0
9% | 71kB 2.1MB/s eta 0:00:0
10% | 81kB 2.2MB/s eta 0:00:
12% | 92kB 2.5MB/s eta 0:00:
13% | 102kB 2.6MB/s eta 0:00
15% | 112kB 2.7MB/s eta 0:00
16% | 122kB 3.7MB/s eta 0:00
17% | 133kB 4.2MB/s eta 0:00
19% | 143kB 6.1MB/s eta 0:00
20% | 153kB 7.1MB/s eta 0:00

FIGURE 1.5

Installation process for the packages that will be used in the rest of the book.

Step 3. As shown in Fig. 1.5, install the basic packages used in this book: `ipython`, `numpy`, `scipy`, `pandas`, `matplotlib`, and `jupyter notebook`. In a later section, we will talk about the management of the packages using `pip` and `conda`.

1.1.2 THREE WAYS TO RUN PYTHON CODE

There are different ways to run Python code and they all have different usages. This section will introduce the three different ways to get you started.

Using the Python shell or IPython shell. The easiest way to run Python code is through the **Python shell** or **IPython shell** (which stands for Interactive Python). The IPython shell is more powerful than the Python shell, including features such as Tab autocompletion, color-highlighted error messages, basic UNIX shell integration, and so on. Since we've just installed IPython, let us try to run the "Hello World" example using it. To launch the IPython shell, we begin by typing `ipython` in a terminal (see Fig. 1.6). Then we run a Python command by typing it into the shell and pressing `Enter`. We immediately see the results from the command. For example, we can print out "Hello World" by using the `print` function in Python, as shown in Fig. 1.6.

Run a Python script/file from the command line. The second way to run Python code is to put all the commands into a file and save it as a file with the extension `.py`; the extension of the file can be anything, but by convention, it is usually `.py`. For example, use your favorite text editor (here we used

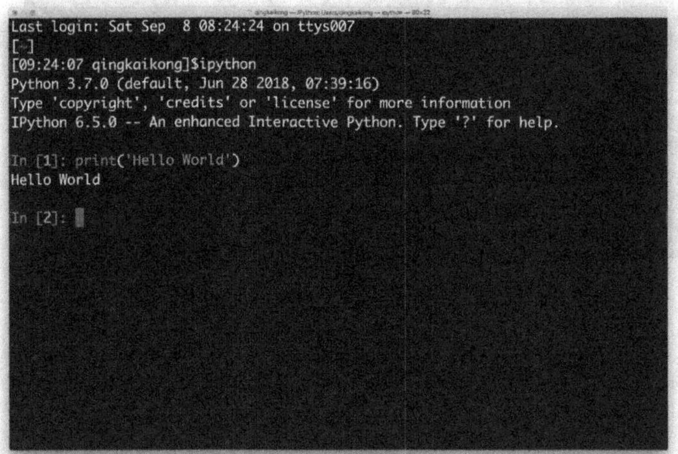

FIGURE 1.6

Run "Hello World" in IPython shell by typing the command. "print" is a function that is discussed later in the book that will print out anything within the parentheses.

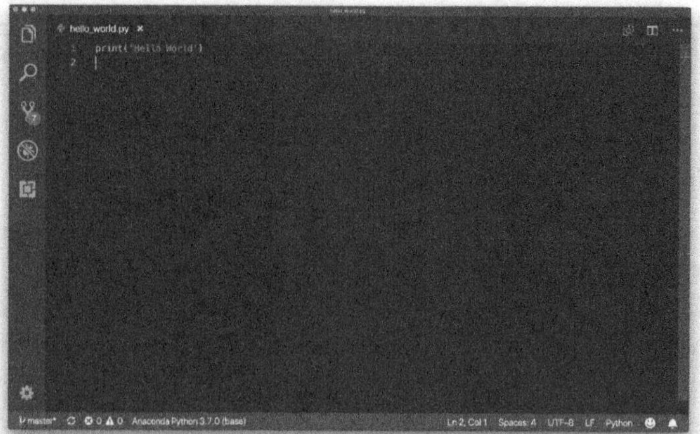

FIGURE 1.7

Example of a Python script file example using Visual Studio Code. Type in the commands you want to execute and save the file with a proper name.

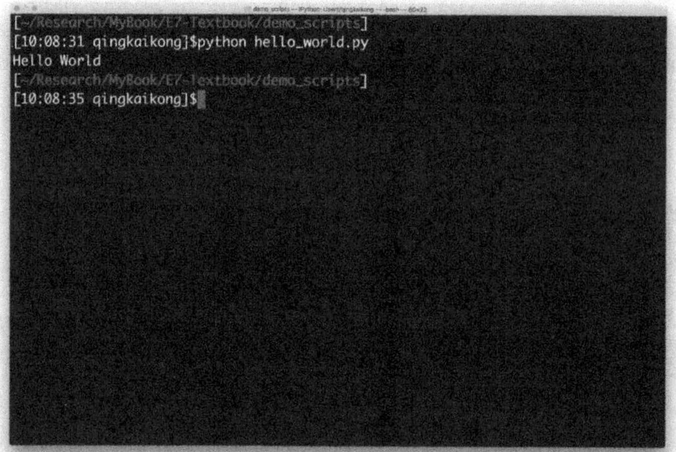

[~/Research/MyBook/E7-Textbook/demo_scripts]
[10:08:31 qingkaikong]$python hello_world.py
Hello World
[~/Research/MyBook/E7-Textbook/demo_scripts]
[10:08:35 qingkaikong]$

FIGURE 1.8

To run the Python script from command line, type "python hello_world.py". This line tells Python to execute the commands that saved in this file.

Visual Studio Code[4]) to type the commands you wish to execute in a file called *hello_world.py*, as shown in Fig. 1.7, which is then run from terminal (see Fig. 1.8).

[4] https://code.visualstudio.com.

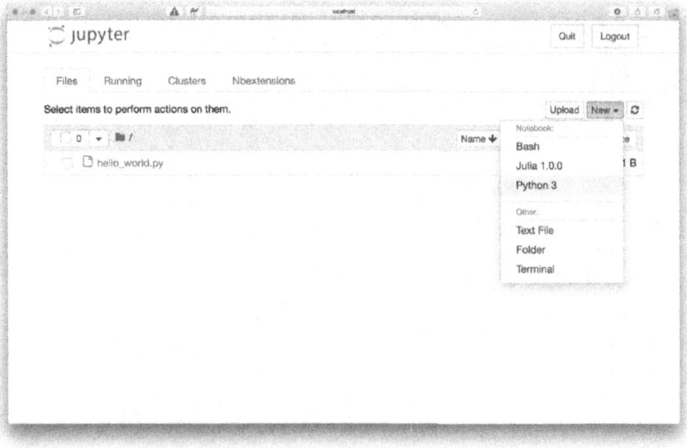

FIGURE 1.9

To launch a `Jupyter notebook` server, type `jupyter notebook` in the command line, which will open a browser page as shown here. Click the "New" button on the top right and choose "Python3". This will create a Python notebook from which to run Python code.

Using Jupyter Notebook. The third way to run Python is through **Jupyter Notebook**, which is a very powerful browser-based Python environment. We will discuss this in details later in this chapter. The example presented here is to demonstrate how quickly we can run the code using `Jupyter notebook`. If you type `jupyter notebook` in the terminal, a local web page will pop up; use the upper right button to create a new Python3 notebook, as shown in Fig. 1.9.

Running code in `Jupyter notebook` is easy. Type your code in the cell and press `Shift + Enter` to run the cell; the results will be shown below the code (Fig. 1.10).

1.2 PYTHON AS A CALCULATOR

Python contains functions found in any standard graphing calculator. An **arithmetic operation** is either addition, subtraction, multiplication, division, or powers between two numbers. An **arithmetic operator** is a symbol that Python has reserved to mean one of the aforementioned operations. These symbols are + for addition, - for subtraction, * for multiplication, / for division, and ** for exponentiation.

An instruction or operation is **executed** when it is resolved by the computer. An instruction is executed at the command prompt by typing it where you see the >>> symbol appears in the Python shell (or the `In [1]:` sign in IPython) and then pressing `Enter`. In the case of the `Jupyter notebook`, type the operation in the code cell and `Shift + Enter`. Since we will use `Jupyter notebook` for the rest of the book, to familiarize yourself with all the different options, all examples in this section will be shown in the IPython shell – see the previous section for how to begin working in IPython. For Windows users, you will use the Anaconda prompt shown in the Appendix A instead of using the terminal.

FIGURE 1.10

To run the Hello World example within `Jupyter notebook`, type the command in the code cell (the grey boxes) and press `Shift + Enter` to execute it.

> **TRY IT!** Compute the sum of 1 and 2.
>
> ```
> In [1]: 1 + 2
> Out[1]: 3
> ```

An **order of operations** is a standard order of precedence that different operations have in relationship to one another. Python utilizes the same order of operations you learned in grade school. Powers are executed before multiplication and division, which are executed before addition and subtraction. Parentheses () can also be used in Python to supersede the standard order of operations.

> **TRY IT!** Compute $\frac{3\times 4}{(2^2+4/2)}$.
>
> ```
> In [2]: (3*4)/(2**2 + 4/2)
> Out[2]: 2.0
> ```

> **TIP!** Note that `Out[2]` is the resulting value of the last operation executed. Use the underscore symbol _ to represent this result to break up complicated expressions into simpler commands.

TRY IT! Compute 3 divided by 4, then multiply the result by 2, and then raise the result to the 3rd power.

```
In [3]: 3/4
Out[3]: 0.75

In [4]: _*2
Out[4]: 1.5

In [5]: _**3
Out[5]: 3.375
```

Python has many basic arithmetic functions like `sin`, `cos`, `tan`, `asin`, `acos`, `atan`, `exp`, `log`, `log10`, and `sqrt` stored in a module (explained later in this chapter) called **math**. First, import this module to access to these functions.

```
In [6]: import math
```

TIP! In `Jupyter notebook` and IPython, you can have a quick view of what is in the module by typing the `module name + dot + TAB`. Furthermore, if you type the first few letters of the function and press TAB, it will automatically complete the function, which is known as "TAB completion" (an example is shown in Fig. 1.11).

These mathematical functions are executed via `module.function`. The inputs to them are always placed inside of parentheses that are connected to the function name. For trigonometric functions, it is useful to have the value of π available. You can call this value at any time by typing `math.pi` in the IPython shell.

TRY IT! Find the square root of 4.

```
In [7]: math.sqrt(4)
Out[7]: 2.0
```

TRY IT! Compute the $\sin(\frac{\pi}{2})$.

```
In [8]: math.sin(math.pi/2)
Out[8]: 1.0
```

Python composes functions as expected, with the innermost function being executed first. The same holds true for function calls that are composed with arithmetic operations.

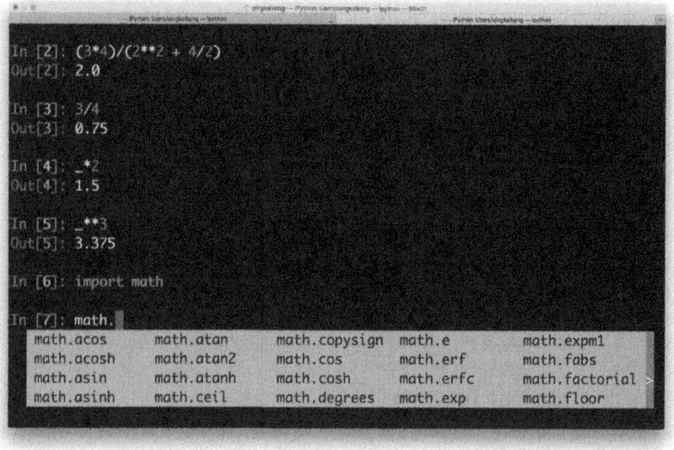

FIGURE 1.11

An example demonstrating the interactive search for functions within IPython by typing TAB after the dot. The grey box shown all available functions.

TRY IT! Compute $e^{\log 10}$.

```
In [9]: math.exp(math.log(10))
Out[9]: 10.000000000000002
```

Note that the `log` function in Python is $\log_e$, or the natural logarithm. It is not $\log_{10}$. To use $\log_{10}$, use the function `math.log10`.

TIP! You can see the result above should be 10, but in Python, it shows as 10.000000000000002. This is due to Python's number approximation, discussed in Chapter 9.

TRY IT! Compute $e^{\frac{3}{4}}$.

```
In [10]: math.exp(3/4)
Out[10]: 2.117000016612675
```

TIP! Using the UP ARROW in the command prompt recalls previously executed commands that were executed. If you accidentally type a command incorrectly, you can use the UP ARROW to recall it, and then edit it instead of retyping the entire line.

Often when using a function in Python, you need help specific to the context of the function. In IPython or `Jupyter notebook`, the description of any function is available by typing `function?`; the question mark is a shortcut for help. If you see a function you are unfamiliar with, it is good practice to use the question mark before asking your instructors what a specific function does.

TRY IT! Use the question mark to find the definition of the factorial function.

```
In [11]: math.factorial?


Signature: math.factorial(x, /)
Docstring:
Find x!.
Raise a ValueError if x is negative or non-integral.
Type: builtin_function_or_method
```

Python will raise an `ZeroDivisionError` when the expression 1/0 (which is infinity) appears, to remind you.

TRY IT! 1/0.

```
In [12]: 1/0


---------------------------------------------------------------------
ZeroDivisionError                 Traceback (most recent call last)
<ipython-input-12-9e1622b385b6> in <module>()
----> 1 1/0

ZeroDivisionError: division by zero
```

You can type `math.inf` at the command prompt to denote infinity or `math.nan` to denote something that is not a number that you wish to be handled as a number. If this is confusing, this distinction can be skipped for now; it will be explained in detail later. Finally, Python can also handle imaginary numbers.

TRY IT! Type $1/\infty$, and $\infty * 2$ to verify that Python handles infinity as you would expect.

```
In [13]: 1/math.inf
Out[13]: 0.0

In [14]: math.inf * 2
Out[14]: inf
```

TRY IT! Compute ∞/∞.

```
In [15]: math.inf/math.inf
Out[15]: nan
```

TRY IT! Compute sum 2 + 5i.

```
In [16]: 2 + 5j
Out[16]: (2+5j)
```

Note that in Python the imaginary part is represented by j instead of i.
Another way to represent a complex number in Python is to use the complex function.

```
In [17]: complex(2,5)
Out[17]: (2+5j)
```

Python can also handle scientific notation using the letter *e* between two numbers. For example, $1e6 = 1000000$ and $1e - 3 = 0.001$.

TRY IT! Compute the number of seconds in 3 years using scientific notation.

```
In [18]: 3e0*3.65e2*2.4e1*3.6e3
Out[18]: 94608000.0
```

TIP! Every time a function in `math` module is typed, it is always typed `math.function_name`. Alternatively, there is a simpler way. For example, if we want to use `sin` and `log` from `math` module, we can import them as follows: `from math import sin, log`. With this modified import statement, when using these functions, use them directly, e.g., `sin(20)` or `log(10)`.

The previous examples demonstrated how to use Python as a calculator to deal with different data values. In Python, there are additional data types needed for numerical values: `int`, `float`, and `complex` are the types associated with these values.

- **int**: Integers, such as 1, 2, 3, ...
- **float**: Floating-point numbers, such as 3.2, 6.4, ...
- **complex**: Complex numbers, such as $2 + 5j, 3 + 2j, \ldots$

Use function `type` to check the data type for different values.

TRY IT! Find out the data type for 1234.

```
In [19]: type(1234)
Out[19]: int
```

TRY IT! Find out the data type for 3.14.

```
In [20]: type(3.14)
Out[20]: float
```

TRY IT! Find out the data type for $2 + 5j$.

```
In [21]: type(2 + 5j)
Out[21]: complex
```

Of course, there are other data types, such as boolean, string, and so on; these are introduced in Chapter 2.

This section demonstrated how to use Python as a calculator by running commands in the IPython shell. Before we move on to more complex coding, let us go ahead to learn more about the managing packages, i.e., how to install, upgrade, and remove the packages.

1.3 MANAGING PACKAGES

One feature that makes Python really great is the various **packages/modules** developed by the user community. Most of the time, when you want to apply some functions or algorithms, often you will find multiple packages already available. All you need to do is to install the packages and use them in your code. Managing packages is one of the most important skills you need to learn to take fully advantage of Python. This section will show you how to manage packages in Python.

1.3.1 MANAGING PACKAGES USING PACKAGE MANAGERS

At the beginning of this book, we installed some packages using **pip** by typing `pip install package_name`. This is currently the most common and easy way to install Python packages. Pip is a package manager that automates the process of installing, updating, and removing the packages. It can install packages published on Python Package Index (*PyPI*).[5] If you install Miniconda, pip will be available for you to use as well.

Use `pip help` to get help for different commands, as shown in Fig. 1.12.

The most used commands usually include: installing, upgrading, and uninstalling a package.

Install a Package

To install the latest version of a package:

```
pip install package_name
```

To install a specific version, e.g., install version 1.5:

[5] https://pypi.org/.

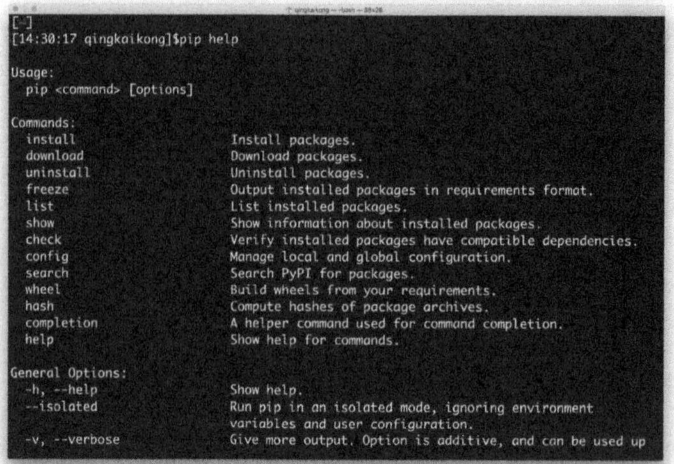

FIGURE 1.12

The help document of pip after executing `pip help`.

```
pip install package_name==1.5
```

`Pip` will install the package as well as the other dependent packages for you to use.

Upgrade a Package

To upgrade an installed package to the latest version from *PyPI*.

```
pip install --upgrade package_name
```

or simply

```
pip install -U package_name
```

Uninstall a Package

```
pip uninstall package_name
```

Other Useful Commands

Other useful commands that provide information about the installed packages are available. As shown in Fig. 1.13, if you want to get a list of all the installed packages, you can use the command:

```
pip list
```

If you want to know more about an installed package, such as the location of the package, the required other dependent packages, etc., you can use the following command as shown in Fig. 1.14:

```
pip show package_name
```

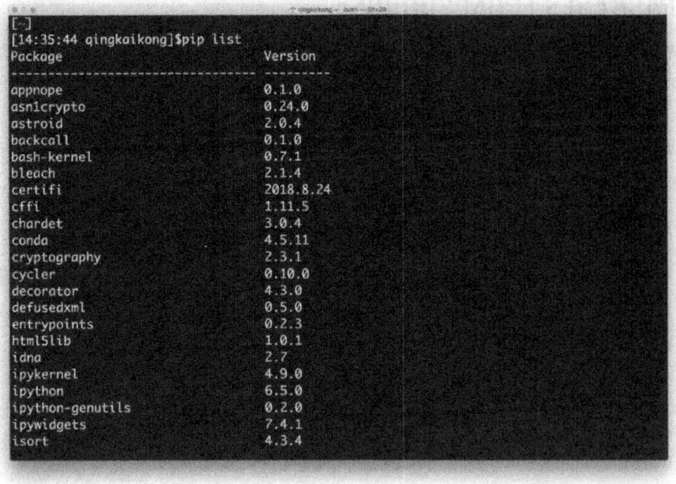

FIGURE 1.13

Using `pip list` to show all the packages installed on your machine.

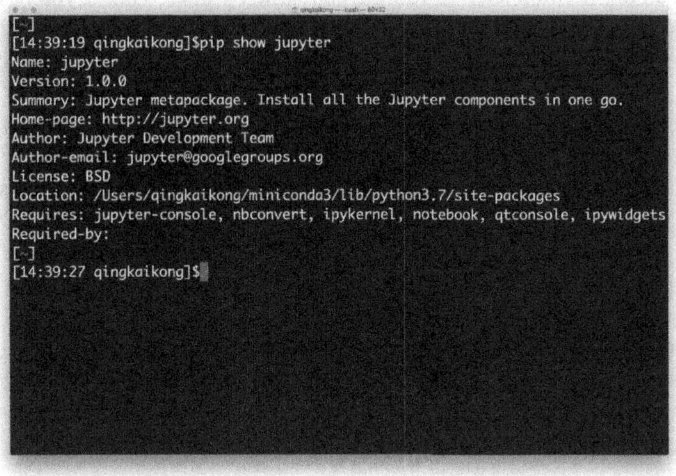

FIGURE 1.14

Using `pip show` to get detailed information about a installed package.

Other package managers exist, e.g., conda (which is included with the Anaconda distribution and is similar to pip in terms of its capability); therefore, it is not discussed further, and you can find more information by reading the documentation.[6]

[6] https://conda.io/docs/user-guide/getting-started.html.

1.3.2 INSTALL PACKAGES FROM SOURCE

Occasionally, you will need to download the source file for some project that is not in the PyPI. In that case, you need to install the package in a different way. In the standard install, after uncompressing the file you downloaded, usually you can see the folder contains a setup script `setup.py`, and a file named README, which documents how to build and install the module. For most cases, you just need to run one command from the terminal to install the package:

```
python setup.py install
```

Note that Windows users will need to run the following command from a command prompt window:

```
setup.py install
```

Now you know how to manage the packages in `Python`, which is a big step forward in using Python correctly. In the next section, we will talk more about the `Jupyter notebook` that we used for the rest of the book.

1.4 INTRODUCTION TO JUPYTER NOTEBOOK

You have already used the IPython shell to run the code line by line. What if you have more lines and want to run it block by block, and share it easily with others? In this case, the IPython shell is not a good option. This section will introduce you another option, **Jupyter notebook**, which we will use for the rest of the book. From the `Jupyter notebook` website[7]:

> *The* `Jupyter notebook` *is an open-source web application that allows you to create and share documents that contain live code, equations, visualizations and narrative text. Uses include: data cleaning and transformation, numerical simulation, statistical modeling, data visualization, machine learning, and much more.*

`Jupyter notebook` runs using your browser; it can run locally on your machine as a local server or remotely on a server. The reason it is called notebook is because it can contain live code, rich text elements such as equations, links, images, tables, and so on. Therefore, you can have a very nice notebook to describe your idea and the live code all in one document. Thus `Jupyter notebook` has become a popular way to test ideas, write blogs, papers, and even books. In fact, this book was written entirely within `Jupyter notebook` and converted to LaTeX afterwards. Although the `Jupyter notebook` has many other advantages, to get you started we will only cover the basics here.

1.4.1 STARTING THE JUPYTER NOTEBOOK

As seen earlier, we can start `Jupyter notebook` by typing the following command in our terminal (for Windows users, type this command in the Anaconda prompt) in the folder where you want to locate the notebooks:

```
jupyter notebook
```

[7] http://jupyter.org/.

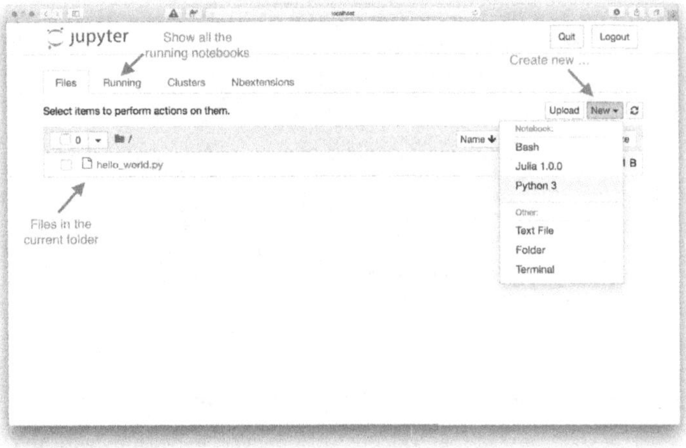

FIGURE 1.15

The `Jupyter notebook` dashboard after launching the server. Red arrows (light grey arrows in print version) are pointing to you the most common features in the dashboard.

The `Jupyter notebook` dashboard will appear in the browser as shown in Fig. 1.15. The default address is: http://localhost:8888, which is at the `localhost` with port `8888`, as shown in Fig. 1.15 (if the port `8888` is taken by other `Jupyter notebooks`, then it will automatically use another port). This is essentially creating a local server to run in your browser. When you navigate to the browser, you will see a dashboard. In this dashboard, you will see some important features labeled in red (light grey in print version). To create a new Python notebook, select "Python 3," which is usually called Python kernel. You can use Jupyter to run other kernels as well. For example, in Fig. 1.15, there are Bash and Julia kernels that you can run as a notebook, but you will need to install them first. We will use the Python kernel, so choose Python 3 kernel.

1.4.2 WITHIN THE NOTEBOOK

After you create a new Python notebook, it will look like the one shown in Fig. 1.16. The toolbar and menu are self-explanatory. If you hover the cursor over the toolbar, it will show you the function of the tool. When you press the menu, it will show you the drop down list. Some important things to know about the `Jupyter notebook` are as follows: a cell is the place you can write your code or text in. If you run this cell, it only executes code within this cell block. Two important cell types are code and markdown: (a) the code cell is where to type your code and where to run the code; (b) the markdown cell is where to type the description in rich text format (see Fig. 1.16 as an example). Search for the "Markdown cheatsheet" to get a quick start with markdown. To run the code or render the markdown in the notebook is simple: press `Shift + Enter`.

The notebook allows you to move the cell up or down, insert or delete the cell, etc. There are many other nice features available using Jupyter notebook, and we encourage you to access the many online tutorials available to broaden your knowledge.

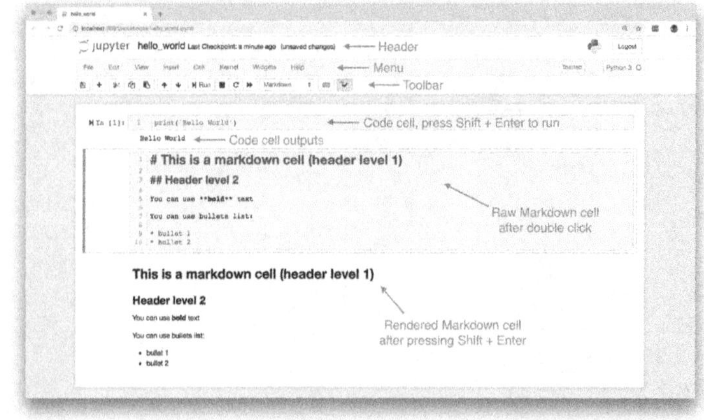

FIGURE 1.16

A quick view of a notebook. The Header of the notebook shows the name of the notebook. The menu has various drop-down lists that let you access to all the functionalities of the notebook. The tool bar provides you some shortcuts for the commonly used functionalities.

1.4.3 HOW DO I CLOSE A NOTEBOOK?

When you close the notebook browser tab, the notebook actually is not closed; it is still running in the background. To completely close a notebook, go to the dashboard, and check the box before the notebook. There is a shutdown option in the toolbar above; this is the correct way you close a notebook completely.

1.4.4 SHUTTING DOWN THE JUPYTER NOTEBOOK SERVER

Closing the browser will not shut down the notebook server since it is still running. You can reopen the previous address in a browser. To completely shut it down, we need to close the associated terminal that you used to launch the notebook.

Now that you have the basic knowledge to launch and run a Jupyter notebook, you are equipped to continue learning Python.

1.5 LOGICAL EXPRESSIONS AND OPERATORS

A **logical expression** is a statement that can either be *true* or *false*. For example, $a < b$ is a logical expression. It can be true or false depending on what values of a and b are given. Note that this differs from a **mathematical expression**, which denotes a truth statement. In the previous example, the mathematical expression $a < b$ means that a is less than b, and values of a and b where $a \geq b$ are not permitted. Logical expressions form the basis of computing. In this book all statements are assumed to be logical rather than mathematical unless otherwise indicated.

In Python, a logical expression that is true will compute to the value `True`. A false expression will compute to the value `False`. This is a new data type known as **boolean**, which has the built-in values `True` and `False`. In this book, "True" is equivalent to 1, and "False" is equivalent to 0. Logical expressions are used to pose questions to Python. For example, "3 < 4" is equivalent to, "Is 3 less than 4?" Since this statement is true, Python will compute it as 1; however, if we write 3 > 4, this is false, and Python will compute it as 0.

Comparison operators compare the value of two numbers, which are used to build logical expressions. Python reserves the symbols >, >=, <, <=, != , ==, to denote "greater than," "greater than or equal," "less than," "less than or equal," "not equal," and "equal," respectively; see and Table 1.1. Let us start with an example, $a = 4, b = 2$:

Table 1.1 Comparison operators.

Operator	Description	Example	Results
>	greather than	a > b	True
>=	greater than or equal	a >= b	True
<	less than	a < b	False
<=	less than or equal	a <= b	False
!=	not equal	a != b	True
==	equal	a == b	False

TRY IT! Compute the logical expression for "Is 5 equal to 4?" and "Is 2 smaller than 3?"

```
In [1]: 5 == 4

Out[1]: False

In [2]: 2 < 3

Out[2]: True
```

Logical operators, as shown in Table 1.2, are operations between two logical expressions that, for the sake of discussion, we will call P and Q. The fundamental logical operators we will use herein are **and**, **or**, and **not**.

Table 1.2 Logical operators.

Operator	Description	Example	Results
and	greater than	P and Q	True if both P and Q are True. False otherwise
or	greater than or equal	P or Q	True if either P or Q is True. False otherwise
not	less than	not P	True if P is False. False if P is True

The **truth table**, as shown in Fig. 1.17, of a logical operator or expression gives the result of every truth combination of P and Q. Fig. 1.17 shows the truth tables for "and" and "or".

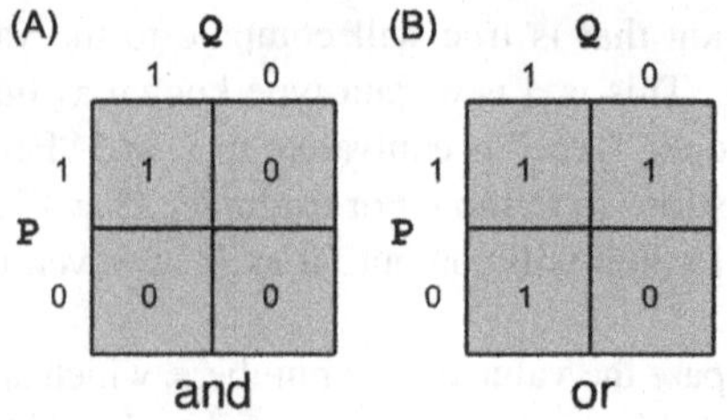

FIGURE 1.17

Truth tables for the logical and/or.

TRY IT! Assuming P is true, let us use Python to determine if the expression (P AND NOT(Q)) OR (P AND Q) is always true regardless of whether or not Q is true. Logically, can you see why this is the case? First assume Q is true:

```
In [3]: (1 and not 1) or (1 and 1)

Out[3]: 1
```

Now assume Q is false

```
In [4]: (1 and not 0) or (1 and 0)

Out[4]: True
```

Just as with arithmetic operators, logical operators have an order of operations relative to each other and in relation to arithmetic operators. All arithmetic operations will be executed before comparison operations, which will be executed before logical operations. Parentheses can be used to change the order of operations.

TRY IT! Compute $(1 + 3) > (2 + 5)$

```
In [5]: 1 + 3 > 2 + 5

Out[5]: False
```

TIP! Even when the order of operations is known, it is usually helpful for you and those reading your code to use parentheses to make your intentions clearer. In the preceding example $(1 + 3) > (2 + 5)$ is clearer.

WARNING! In Python's implementation of logic, 1 is used to denote true and 0 for false. But because 1 and 0 are still numbers, Python will allow abuses such as: (3 > 2) + (5 > 4), which will resolve to 2.

```
In [6]: (3 > 2) + (5 > 4)

Out[6]: 2
```

WARNING! Although in formal logic 1 is used to denote true and 0 to denote false, Python's notation system is different, and it will take any number not equal to 0 to mean true when used in a logical operation. For example, 3 and 1 will compute to true. Do not utilize this feature of Python. Always use 1 to denote a true statement.

TIP! A fortnight is a length of time consisting of 14 days. Use a logical expression to determine if there are more than 100,000 seconds in a fortnight.

```
In [7]: (14*24*60*60) > 100000

Out[7]: True
```

1.6 SUMMARY AND PROBLEMS

1.6.1 SUMMARY

1. You have now learned the basics of Python, which should enable you to set up the working environment and experiment with ways to run Python.
2. Python can be used as a calculator. It has all the functions and arithmetic operations commonly used with a scientific calculator.
3. You can manage the Python packages using package managers.
4. You learned how to interact with Jupyter notebook.
5. You can also use Python to perform logical operations.
6. You have now been introduced to int, float, complex, string, and boolean data types in Python.

1.6.2 PROBLEMS

1. Print "I love Python" using Python Shell.
2. Print "I love Python" by typing it into a `.py` file and run it from command line.
3. Type `import antigravity` in the IPython Shell, which will take you to xkcd and enable you to see the awesome Python.
4. Launch a new `Jupyter notebook` server in a folder called "exercise" and create a new Python notebook with the name "exercise_1." Put the rest of the problems within this notebook.

5. Compute the area of a triangle with base 10 and height 12. Recall that the area of a triangle is half the base times the height.

6. Compute the surface area and volume of a cylinder with radius 5 and height 3.

7. Compute the slope between the points $(3, 4)$ and $(5, 9)$. Recall that the slope between points (x_1, y_1) and (x_2, y_2) is $\frac{y_2-y_1}{x_2-x_1}$.

8. Compute the distance between the points $(3, 4)$ and $(5, 9)$. Recall that the distance between points in two dimensions is $\sqrt{(x_2-x_1)^2+(y_2-y_1)^2}$.

9. Use Python's `factorial` function to compute 6!

10. Although a year is considered to be 365 days long, a more exact figure is 365.24 days. As a consequence, if we held to the standard 365-day year, we would gradually lose that fraction of the day over time, and seasons and other astronomical events would not occur as expected. To keep the timescale on tract, a leap year is a year that includes an extra day, February 29, to keep the timescale on track. Leap years occur on years that are exactly divisible by 4, unless it is exactly divisible by 100, unless it is divisible by 400. For example, the year 2004 is a leap year, the year 1900 is not a leap year, and the year 2000 is a leap year. Compute the number of leap years between the years 1500 and 2010.

11. A very powerful approximation for π was developed by a brilliant mathematician named Srinivasa Ramanujan. The approximation is the following:

$$\frac{1}{\pi} \approx \frac{2\sqrt{2}}{9801} \sum_{k=0}^{N} \frac{(4k)!(1103+26390k)}{(k!)^4 396^{4k}}.$$

Use Ramanujan's formula for $N = 0$ and $N = 1$ to approximate π. Compare your approximation with Python's stored value for π. Hint: $0! = 1$ by definition.

12. The hyperbolic sin or sinh is defined in terms of exponentials as $\sinh(x) = \frac{\exp(x)-\exp(-x)}{2}$. Compute sinh for $x = 2$ using exponentials. Verify that the result is indeed the hyperbolic sin using Python's function `sinh` in the `math` module.

13. Verify that $\sin^2(x) + \cos^2(x) = 1$ for $x = \pi, \frac{\pi}{2}, \frac{\pi}{4}, \frac{\pi}{6}$.

14. Compute the $\sin 87°$.

15. Write a Python statement that generates the following error: "AttributeError: module 'math' has no attribute 'sni'." Hint: sni is a misspelling of the function sin.

16. Write a Python statement that generates the following error: "TypeError: sin() takes exactly one argument (0 given)." Hint: Input arguments refers to the input of a function (any function); for example, the input in $\sin(\pi/2)$ is $\pi/2$.

17. If P is a logical expression, the law of noncontradiction states that P AND (NOT P) is always false. Verify this for P true and P false.

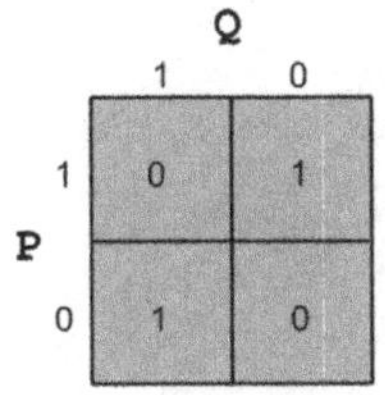

FIGURE 1.18

Truth tables for the logical XOR.

18. Let P and Q be logical expressions. De Morgan's rule states that NOT (P OR Q) = (NOT P) AND (NOT Q) and NOT (P AND Q) = (NOT P) OR (NOT Q). Generate the truth tables for each statement to show that De Morgan's rule is always true.

19. Under what conditions for P and Q is (P AND Q) OR (P AND (NOT Q)) false?

20. Construct an equivalent logical expression for OR using only AND and NOT.

21. Construct an equivalent logical expression for AND using only OR and NOT.

22. The logical operator XOR has the following truth table: Construct an equivalent logical expression for XOR using only AND, OR, and NOT that has the same truth table (see Fig. 1.18).

23. Do the following calculation at the Python command prompt:

$$e^2 \sin \pi/6 + \log_e(3) \cos \pi/9 - 5^3.$$

24. Do the following logical and comparison operations at the Python command prompt. You may assume that P and Q are logical expressions. For $P = 1$ and $Q = 1$, compute NOT(P) AND NOT(Q). For $a = 10$ and $b = 25$, compute ($a < b$) AND ($a = b$).

CHAPTER

VARIABLES AND BASIC DATA STRUCTURES

2

CONTENTS

2.1 VARIABLES AND ASSIGNMENT

When programming, it is useful to be able to store information in variables. A **variable** is a string of characters and numbers associated with a piece of information. The **assignment operator**, denoted by the "=" symbol, is the operator that is used to assign values to variables in Python. The line `x=1` takes the known value, 1, and **assigns** that value to the variable with name "x". After executing this line, this number will be stored into this variable. Until the value is changed or the variable deleted, the character `x` behaves like the value 1.

```
In [1]: x = 1
        x

Out[1]: 1
```

TRY IT! Assign the value 2 to the variable `y`. Multiply `y` by 3 to show that it behaves like the value 2.

```
In [2]: y = 2
        y

Out[2]: 2

In [3]: y*3
```


Python Programming and Numerical Methods. https://doi.org/10.1016/B978-0-12-819549-9.00011-7

```
Out[3]: 6
```

A variable is like a "container" used to store the data in the computer's memory. The name of the variable tells the computer where to find this value in the memory. For now, it is sufficient to know that the notebook has its own memory space to store all the variables in the notebook. As a result of the previous example, you will see the variables `x` and `y` in the memory. You can view a list of all the variables in the notebook using the magic command `%whos` (magic commands are a specialized set of commands host by the IPython kernel and require the prefix `%` to specify the commands).

TRY IT! List all the variables in this notebook.

```
In [4]: %whos

Variable   Type    Data/Info
----------------------------
x          int     1
y          int     2
```

Note! The equality sign in programming is *not* the same as a truth statement in mathematics. In math, the statement $x = 2$ declares the universal truth within the given framework, *x is 2*. In programming, the statement `x=2` means a known value is being associated with a variable name, *store* 2 in `x`. Although it is perfectly valid to say $1 = x$ in mathematics, assignments in Python always go *left*, meaning the value to the right of the equal sign is assigned to the variable on the left of the equal sign. Therefore, `1=x` will generate an error in Python. The assignment operator is always last in the order of operations relative to mathematical, logical, and comparison operators.

TRY IT! The mathematical statement $x = x + 1$ has no solution for any value of x. In programming, if we initialize the value of `x` to be 1, then the statement makes perfect sense. It means, "Add `x` and 1, which is 2, then assign that value to the variable `x`". Note that this operation overwrites the previous value stored in `x`.

```
In [5]: x = x + 1
        x

Out[5]: 2
```

There are some restrictions on the names variables can take. Variables can only contain alphanumeric characters (letters and numbers) as well as underscores; however, the first character of a variable name must be a letter or an underscore. Spaces within a variable name are not permitted, and the variable names are case sensitive (e.g., `x` and `X` are considered different variables).

TIP! Unlike in pure mathematics, variables in programming almost always represent *something* tangible. It may be the distance between two points in space or the number of rabbits in a population. Therefore, as your code becomes increasingly complicated, it is very important that your variables carry a name that can easily be associated with what they represent. For example, the distance between two points in space is better represented by the variable `dist` than `x`, and the number of rabbits in a population is better represented by `n_rabbits` than `y`.

Note that when a variable is assigned, it has no memory of how it was assigned. That is, if the value of a variable, `y`, is constructed from other variables, like `x`, reassigning the value of `x` will not change the value of `y`.

EXAMPLE: What value will `y` have after the following lines of code are executed?

```
In [7]: x = 1
        y = x + 1
        x = 2
        y

Out[7]: 2
```

WARNING! You can overwrite variables or functions that have been stored in Python. For example, the command `help = 2` will store the value 2 in the variable with name `help`. After this assignment `help` will behave like the value 2 instead of the function `help`. Therefore, you should always be careful not to give your variables the same name as built-in functions or values.

TIP! Now that you know how to assign variables, it is important that you remember to *never* leave unassigned commands. An **unassigned command** is an operation that has a result, but that result is not assigned to a variable. For example, you should never use `2+2`. You should instead assign it to some variable `x=2+2`. This allows you to "hold on" to the results of previous commands and will make your interaction with Python much less confusing.

You can clear a variable from the notebook using the `del` function. Typing `del x` will clear the variable `x` from the workspace. If you want to remove all the variables in the notebook, you can use the magic command `%reset`.

In mathematics, variables are usually associated with unknown numbers; in programming, variables are associated with a value of a certain type. There are many data types that can be assigned to variables. A **data type** is a classification of the type of information that is being stored in a variable. The basic data types that you will utilize throughout this book are boolean, int, float, string, list, tuple, dictionary, and set. A formal description of these data types is given in the following sections.

2.2 DATA STRUCTURE – STRING

We have introduced the different data types, such as int, float and boolean; these are all related to single value. The rest of this chapter will introduce you more data types so that we can store multiple values. The data structure related to these new types are strings, lists, tuples, sets, and dictionaries. We will start with the strings.

A **string** is a sequence of characters, such as "Hello World" we saw in Chapter 1. Strings are surrounded by either single or double quotation marks. We can use `print` function to output the strings to the screen.

TRY IT! Print "I love Python!" to the screen.

```
In [1]: print("I love Python!")
```

TRY IT! Assign the character "S" to the variable with name `s`. Assign the string "Hello World" to the variable `w`. Verify that `s` and `w` have the type string using the `type` function.

```
In [2]: s = "S"
        w = "Hello World"

In [3]: type(s)

Out[3]: str

In [4]: type(w)

Out[4]: str
```

Note! A blank space, " ", between "Hello" and "World" is also a type `str`. Any symbol can be a character, even those that have been reserved for operators. Note that as a `str`, they do not perform the same function. Although they look the same, Python interprets them completely differently.

TRY IT! Create an empty string. Verify that the empty string is an `str`.

```
In [5]: s = " "
        type(s)

Out[5]: str
```

Because a string is an array of characters, it has length to indicate the size of the string. For example, we can check the size of the string by using the built-in function `len`.

Character	H	e	l	l	o		W	o	r	l	d
Index	0	1	2	3	4	5	6	7	8	9	10

FIGURE 2.1

String index for the example of "Hello World".

```
In [6]: len(w)

Out[6]: 11
```

Strings also have indexes that enables us to find the location of each character, as shown in Fig. 2.1. The index of the position start with 0.

We can access a character by using a bracket and the index of the position. For example, if we want to access the character "W", then we type the following:

```
In [7]: w[6]

Out[7]: "W"
```

We can also select a sequence as well using string slicing. For example, if we want to access "World", we type the following command.

```
In [8]: w[6:11]

Out[8]: "World"
```

`[6:11]` means the start position is from index 6 and the end position is index 10. In the Python string slicing range, the upper-bound is exclusive; this means that `[6:11]` will "slice" the characters from index 6 to 10. The syntax for slicing in Python is `[start:end:step]`, the third argument, step, is optional. If you ignore the step argument, the default will be set to 1.

You can ignore the end position if you want to slice to the end of the string. For example, the following command is the same as the above one:

```
In [9]: w[6:]

Out[9]: "World"
```

TRY IT! Retrieve the word "Hello" from string `w`.

```
In [10]: w[:5]

Out[10]: "Hello"
```

You can also use a negative index when slicing the strings, which means counting from the end of the string. For example, -1 means the last character, -2 means the second to last and so on.

TRY IT! Slice the "Wor" within the word "World".

```
In [11]: w[6:-2]

Out[11]: "Wor"
```

TRY IT! Retrieve every other character in the variable `w`

```
In [12]: w[::2]

Out[12]: "HloWrd"
```

Strings cannot be used in the mathematical operations.

TRY IT! Use "+" to add two numbers. Verify that "+" does not behave like the addition operator, +.

```
In [13]: 1 "+" 2


      File "<ipython-input-13-46b54f731e00>", line 1
    1 "+" 2
          ^
  SyntaxError: invalid syntax
```

WARNING! Numbers can also be expressed as `str`. For example, `x = '123'` means that `x` is the string 123 not the number 123. However, strings represent words or text and so should not have addition defined on them.

TIP! You may find yourself in a situation where you would like to use an apostrophe as an `str`. This is problematic since an apostrophe is used to denote strings. Fortunately, an apostrophe can be used in a string in the following way. The backslash (\) is a way to tell Python this is part of the string, not to denote strings. The backslash character is used to escape characters that otherwise have a special meaning, such as newline, backslash itself, or the quote character. If either single or double quote is a part of the string itself, in Python, there is an easy way to do this, you can just place the string in double or single quotes respectively, as shown in the following example.

```
In [14]: "don't"

Out[14]: "don't"
```

One string can be concatenated to another string. For example:

```
In [15]: str_a = "I love Python! "
         str_b = "You too!"
         print(str_a + str_b)

I love Python! You too!
```

We can convert other data types to strings as well using the built-in function `str`. This is useful, for example, we have the variable `x` which has stored 1 as an integer type, if we want to print it out directly as a string, we will get an error saying we cannot concatenate string with an integer.

```
In [16]: x = 1
         print("x = " + x)


        ---------------------------------------------------------


        TypeError                 Traceback (most recent call last)

        <ipython-input-16-3e562ba0dd83> in <module>()
          1 x = 1
    ----> 2 print("x = " + x)


        TypeError: can only concatenate str (not "int") to str
```

The correct way to do it is to convert the integer to string first, and then print it out.

TRY IT! Print out `x = 1` to the screen.

```
In [17]: print("x = " + str(x))

x = 1


In [18]: type(str(x))

Out[18]: str
```

In Python, string is an object that has various methods that can be used to manipulate it (this is the so-called object-oriented programming and will be discussed later). To get access to these various methods, use this pattern `string.method_name`.

TRY IT! Turn the variable w to upper case.

```
In [19]: w.upper()

Out[19]: "HELLO WORLD"
```

TRY IT! Count the number of occurrences for letter "l" in w.

```
In [20]: w.count("l")

Out[20]: 3
```

TRY IT! Replace the "World" in variable w to "Berkeley".

```
In [21]: w.replace("World", "Berkeley")

Out[21]: "Hello Berkeley"
```

There are different ways to preformat a string. Here we introduce two ways to do it. For example, if we have two variables name and country, and we want to print them out in a sentence, but we do not want to use the string concatenation we used before since it will use many "+" signs in the string, we can do the following instead:

```
In [22]: name = "UC Berkeley"
         country = "USA"

         print("%s is a great school in %s!"%(name, country))

UC Berkeley is a great school in USA!
```

WHAT IS HAPPENING? In the previous example, the %s in the double quotation marks is telling Python that we want to insert some strings at this location (s stands for string in this case). The %(name, country) is the location where the two strings should be inserted.

NEW! There is a different way that only introduced in Python 3.6 and above, it is called **f-string**, which means formated-string. You can easily format a string with the following line:

```
In [23]: print(f"{name} is a great school in {country}.")

UC Berkeley is a great school in USA.
```

You can even print out a numerical expression without converting the data type as we did before.

TRY it! Print out the result of 3*4 directly using f-string.

```
In [24]: print(f"{3*4}")

12
```

By this point, we have learned about the string data structure; this is our first sequence data structure. Let us learn more now.

2.3 DATA STRUCTURE – LIST

In the previous section, we learned that "strings" can hold a sequence of characters. Now, we are introducing a more versatile sequential data structure in Python – **list**. The way to define it is to use a pair of brackets `[ ]`, and the elements within it are separated by commas. A list can hold any type of data: numerical, or strings, or other types. For example,

```
In [1]: list_1 = [1, 2, 3]
        list_1

Out[1]: [1, 2, 3]

In [2]: list_2 = ["Hello", "World"]
        list_2

Out[2]: ["Hello", "World"]
```

We can put mixed types in the list as well:

```
In [3]: list_3 = [1, 2, 3, "Apple", "orange"]
        list_3

Out[3]: [1, 2, 3, "Apple", "orange"]
```

We can also nest the lists, for example:

```
In [4]: list_4 = [list_1, list_2]
        list_4

Out[4]: [[1, 2, 3], ["Hello", "World"]]
```

The way to retrieve the element in the list is very similar to how it is done for strings, see Fig. 2.2 for the index of a string.

List	1	2	3	Apple	Orange
Index	0	1	2	3	4

FIGURE 2.2

Example of list index.

TRY IT! Get the 3rd element in list_3

```
In [5]: list_3[2]

Out[5]: 3
```

TRY IT! Get the first 3 elements in list_3

```
In [6]: list_3[:3]

Out[6]: [1, 2, 3]
```

TRY IT! Get the last element in list_3

```
In [7]: list_3[-1]

Out[7]: "orange"
```

TRY IT! Get the first list from list_4.

```
In [8]: list_4[0]

Out[8]: [1, 2, 3]
```

Similarly, we can obtain the length of the list by using the `len` function.

```
In [9]: len(list_3)

Out[9]: 5
```

We can also concatenate two lists by simply using a + sign.

TRY IT! Add list_1 and list_2 to one list.

```
In [10]: list_1 + list_2
```

```
Out[10]: [1, 2, 3, "Hello", "World"]
```

New items can be added to an existing list by using the `append` method from the list.

```
In [11]: list_1.append(4)
         list_1

Out[11]: [1, 2, 3, 4]
```

Note! The `append` function operate on the list itself as shown in the above example, 4 is added to the list. But in the `list_1 + list_2` example, `list_1` and `list_2` will not change. You can check `list_2` to verify this.

We can also insert or remove element from the list by using the methods `insert` and `remove`, but they are also operating on the list directly.

```
In [12]: list_1.insert(2,"center")
         list_1

Out[12]: [1, 2, "center", 3, 4]
```

Note! Using the `remove` method will only remove the first occurrence of the item (read the documentation of the method). There is another way to delete an item by using its index – function `del`.

```
In [13]: del list_1[2]
         list_1

Out[13]: [1, 2, 3, 4]
```

We can also define an empty list and add in new element later using the `append` method. It is used a lot in Python when you have to loop through a sequence of items; we will learn more about this method in Chapter 5.

TRY IT! Define an empty list and add values 5 and 6 to the list.

```
In [14]: list_5 = []
         list_5.append(5)
         list_5

Out[14]: [5]

In [15]: list_5.append(6)
         list_5
```

```
Out[15]: [5, 6]
```

We can also quickly check if an element is in the list using the operator `in`.

TRY IT! Check if number 5 is in the `list_5`.

```
In [16]: 5 in list_5

Out[16]: True
```

Using the `list` function, we can turn other sequence items into a list.

TRY IT! Turn the string "Hello World" into a list of characters.

```
In [17]: list("Hello World")

Out[17]: ["H", "e", "l", "l", "o", " ", "W", "o", "r", "l", "d"]
```

Lists are used frequently in Python when working with data, with many different possible applications as discussed in later sections.

2.4 DATA STRUCTURE – TUPLE

Let us learn one more different sequence data structure in Python – **tuple**. It is usually defined by using a pair of parentheses `( )`, and its elements are separated by commas. For example,

```
In [1]: tuple_1 = (1, 2, 3, 2)
        tuple_1

Out[1]: (1, 2, 3, 2)
```

As with strings and lists, there is a way to index tuples, slicing the elements, and even some methods are very similar to those we saw before.

TRY IT! Get the length of `tuple_1`.

```
In [2]: len(tuple_1)

Out[2]: 4
```

TRY IT! Get the elements from index 1 to 3 for `tuple_1`.

```
In [3]: tuple_1[1:4]

Out[3]: (2, 3, 2)
```

TRY IT! Count the occurrence for number 2 in `tuple_1`.

```
In [4]: tuple_1.count(2)

Out[4]: 2
```

You may ask, what is the difference between lists and tuples? If they are similar to each other, why do we need another sequence data structure?

Tuples are created for a reason. From the Python documentation[1]:

Though tuples may seem similar to lists, they are often used in different situations and for different purposes. Tuples are ***immutable****, and usually contain a* ***heterogeneous*** *sequence of elements that are accessed via unpacking (see later in this section) or indexing (or even by attribute in the case of named tuples). Lists are* ***mutable****, and their elements are usually* ***homogeneous*** *and are accessed by iterating over the list.*

What does it mean by immutable? It means the elements in the tuple, once defined, cannot be changed. In contrast, elements in a list can be changed without any problem. For example,

```
In [5]: list_1 = [1, 2, 3]
        list_1[2] = 1
        list_1

Out[5]: [1, 2, 1]

In [6]: tuple_1[2] = 1


    ---------------------------------------------------------------------------

    TypeError                 Traceback (most recent call last)

    <ipython-input-6-76fb6b169c14> in <module>()
    ----> 1 tuple_1[2] = 1
```

[1] https://docs.python.org/3/tutorial/datastructures.html#tuples-and-sequences.

```
TypeError: "tuple" object does not support item assignment
```

What does heterogeneous mean? Tuples usually contain a heterogeneous sequence of elements, while lists usually contain a homogeneous sequence. For example, we have a list that contains different fruits. Usually, the names of the fruits can be stored in a list, since they are homogeneous. Now we want to have a data structure to store how many pieces of fruit we have of each type. This is usually where the tuples comes in, since the name of the fruit and the number are heterogeneous. Such as ("apple", 3) which means we have 3 apples.

```
In [7]: # a fruit list
        ["apple", "banana", "orange", "pear"]

Out[7]: ["apple", "banana", "orange", "pear"]

In [8]: # a list of (fruit, number) pairs
        [("apple",3), ("banana",4) , ("orange",1), ("pear",4)]

Out[8]: [("apple",3), ("banana",4), ("orange",1), ("pear",4)]
```

Tuples or lists can be accessed by unpacking as shown in the following example, which requires that the number of variables on the left-hand side of the equality sign be equal to the number of elements in the sequence.

```
In [9]: a, b, c = list_1
        print(a, b, c)

1 2 1
```

Note! The opposite operation to unpacking is packing, as shown in the following example. We can see that we do not need the parentheses to define a tuple, but it is considered good practice to do so.

```
In [10]: list_2 = 2, 4, 5
         list_2

Out[10]: (2, 4, 5)
```

2.5 DATA STRUCTURE – SET

Another data type in Python is a **set**. It is a type that can store an unordered collection with no duplicate elements. It can also support mathematical operations like union, intersection, difference, and symmetric difference. It is defined by using a pair of braces, { }, and its elements are separated by commas.

```
In [1]: {3, 3, 2, 3, 1, 4, 5, 6, 4, 2}

Out[1]: {1, 2, 3, 4, 5, 6}
```

Using "sets" is a quick way to determine the unique elements in a string, list, or tuple.

TRY IT! Find the unique elements in list `[1, 2, 2, 3, 2, 1, 2]`.

```
In [2]: set_1 = set([1, 2, 2, 3, 2, 1, 2])
        set_1

Out[2]: {1, 2, 3}
```

TRY IT! Find the unique elements in tuple `(2, 4, 6, 5, 2)`.

```
In [3]: set_2 = set((2, 4, 6, 5, 2))
        set_2

Out[3]: {2, 4, 5, 6}
```

TRY IT! Find the unique character in string "Banana".

```
In [4]: set("Banana")

Out[4]: {"B", "a", "n"}
```

We mentioned earlier that sets support the mathematical operations like union, intersection, difference, and symmetric difference.

TRY IT! Get the union of `set_1` and `set_2`.

```
In [5]: print(set_1)
        print(set_2)

{1, 2, 3}
{2, 4, 5, 6}


In [6]: set_1.union(set_2)
```

```
Out[6]: {1, 2, 3, 4, 5, 6}
```

TRY IT! Get the intersection of `set_1` and `set_2`.

```
In [7]: set_1.intersection(set_2)

Out[7]: {2}
```

TRY IT! Is `set_1` a subset of `{1, 2, 3, 3, 4, 5}`?

```
In [8]: set_1.issubset({1, 2, 3, 3, 4, 5})

Out[8]: True
```

2.6 DATA STRUCTURE – DICTIONARY

We introduced several sequential data types in the previous sections. Now we will introduce to you a new and useful type – **dictionary**, which is a totally different data type than those we introduced earlier. Instead of using a sequence of numbers to index the elements (such as lists or tuples), dictionaries are indexed by keys, which can be a string, number, or even tuple (but not list). A dictionary comprises key-value pairs, and each key maps to a corresponding value. It is defined by using a pair of braces `{ }`, while the elements are a list of comma-separated key:value pairs (note that the key:value pair is separated by the colon, with key at front and value at the end).

```
In [1]: dict_1 = {"apple":3, "orange":4, "pear":2}
        dict_1

Out[1]: {"apple": 3, "orange": 4, "pear": 2}
```

Within a dictionary, because elements are stored without order, you cannot access a dictionary based on a sequence of index numbers. To access to a dictionary, we need to use the key of the element – `dictionary[key]`.

TRY IT! Get the element "apple" from `dict_1`.

```
In [2]: dict_1["apple"]

Out[2]: 3
```

We can get all the keys in a dictionary by using the `keys` method, or all the values by using the method `values`.

TRY IT! Get all the keys and values from `dict_1`.

```
In [3]: dict_1.keys()

Out[3]: dict_keys(["apple", "orange", "pear"])

In [4]: dict_1.values()

Out[4]: dict_values([3, 4, 2])
```

We can also get the size of a dictionary by using the `len` function.

```
In [5]: len(dict_1)

Out[5]: 3
```

We can define an empty dictionary and then fill in the element later. Or we can turn a list of tuples with (key, value) pairs to a dictionary.

TRY IT! Define an empty dictionary named `school_dict` and add value "UC Berkeley":"USA".

```
In [6]: school_dict = {}
        school_dict["UC Berkeley"] = "USA"
        school_dict

Out[6]: {"UC Berkeley": "USA"}
```

TRY IT! Add another element "Oxford":"UK" to `school_dict`.

```
In [7]: school_dict["Oxford"] = "UK"
        school_dict

Out[7]: {"UC Berkeley": "USA", "Oxford": "UK"}
```

TRY IT! Turn the list of tuples `[("UC Berkeley", "USA"), ("Oxford", "UK")]` into a dictionary.

```
In [8]: dict([("UC Berkeley", "USA"), ("Oxford", "UK")])

Out[8]: {"UC Berkeley": "USA", "Oxford": "UK"}
```

We can also check if an element belongs to a dictionary using the operator `in`.

TRY IT! Determine if "UC Berkeley" is in `school_dict`.

```
In [9]: "UC Berkeley" in school_dict

Out[9]: True
```

TRY IT! Determine whether "Harvard" is not in `school_dict`.

```
In [10]: "Harvard" not in school_dict

Out[10]: True
```

We can also use the `list` function to turn a dictionary with a list of keys. For example,

```
In [11]: list(school_dict)

Out[11]: ["UC Berkeley", "Oxford"]
```

2.7 INTRODUCING NUMPY ARRAYS

The second part of this book introduced numerical methods by using Python. We will use the array/-matrix construct a lot later in the book. So that you are prepared for this section, here we are going to introduce the most common way to handle arrays in Python using the `NumPy` module.[2] `NumPy` is probably the most fundamental numerical computing module in Python.

`NumPy` is coded both in Python and C (for speed). On its website, a few important features for `NumPy` are listed as follows:

- A powerful N-dimensional array object
- Sophisticated (broadcasting) functions
- Tools for integrating C/C++ and Fortran code
- Useful linear algebra, Fourier transform, and random number capabilities

Here, we will only introduce you the part of the `NumPy` array that is related to the data structure. Gradually, we will touch on other aspects of `NumPy` in later chapters.

In order to use `NumPy` module, we need to import it first. A conventional way to import it is to use `np` as a shortened name.

```
In [1]: import numpy as np
```

[2] http://www.numpy.org.

WARNING! Of course, you can call it any name, but "np" is considered convention and is accepted by the entire community, and it is a good practice to use it.

To define an array in Python, you can use the `np.array` function to convert a list.

TRY IT! Create the following arrays:

$$x = \begin{pmatrix} 1 & 4 & 3 \end{pmatrix}$$

$$y = \begin{pmatrix} 1 & 4 & 3 \\ 9 & 2 & 7 \end{pmatrix}$$

```
In [2]: x = np.array([1, 4, 3])
        x

Out[2]: array([1, 4, 3])

In [3]: y = np.array([[1, 4, 3], [9, 2, 7]])
        y

Out[3]: array([[1, 4, 3],
               [9, 2, 7]])
```

Note! A 2D array can use nested lists to represent, with the inner list representing each row.

Knowing the size or length of an array is often helpful. The array `shape` attribute is called on an array `M` and returns a 2 × 3 array where the first element is the number of rows in the matrix `M`; and the second element is the number of columns in `M`. Note that the output of the `shape` attribute is a tuple. The `size` attribute is called on an array `M` and returns the total number of elements in matrix `M`.

TRY IT! Find the rows, columns, and the total size for array `y`.

```
In [4]: y.shape

Out[4]: (2, 3)

In [5]: y.size

Out[5]: 6
```

Note! You may notice the difference that we only use `y.shape` instead of `y.shape()`; this is because `shape` is an attribute rather than a method in this array object. We will introduce more of the object-oriented programming in a later chapter. For now, just remember that when we call a method in an object, we need to use the parentheses, while with an attribute we do not.

Very often we would like to generate arrays that have a structure or pattern. For instance, we may wish to create the array `z = [1 2 3 ... 2000]`. It would be very cumbersome to type the entire description of `z` into Python. For generating arrays that are in order and evenly spaced, it is useful to use the `arange` function in `NumPy`.

TRY IT! Create an array z from 1 to 2000 with an increment 1.

```
In [6]: z = np.arange(1, 2000, 1)
        z

Out[6]: array([   1,    2,    3, ... , 1997, 1998, 1999])
```

Using the `np.arange`, we can create `z` easily. The first two numbers are the start and end of the sequence, and the last one is the increment. Since it is very common to have an increment of 1, if an increment is not specified, Python will use a default value of 1. Therefore `np.arange(1, 2000)` will have the same result as `np.arange(1, 2000, 1)`. Negative or noninteger increments can also be used. If the increment "misses" the last value, it will only extend until the value just before the ending value. For example, `x = np.arange(1,8,2)` would be [1, 3, 5, 7].

TRY IT! Generate an array with [0.5, 1, 1.5, 2, 2.5].

```
In [7]: np.arange(0.5, 3, 0.5)

Out[7]: array([0.5, 1. , 1.5, 2. , 2.5])
```

Sometimes we want to guarantee a start and end point for an array but still have evenly spaced elements. For instance, we may want an array that starts at 1, ends at 8, and has exactly 10 elements. To do this, use the function `np.linspace`. The function `linspace` takes three input values separated by commas; therefore, `A = linspace(a,b,n)` generates an array of `n` equally spaced elements starting from `a` and ending at `b`.

TRY IT! Use `linspace` to generate an array starting at 3, ending at 9, and containing 10 elements.

```
In [8]: np.linspace(3, 9, 10)

Out[8]: array([3., 3.66666667, 4.33333333, 5., 5.66666667,
          6.33333333, 7., 7.66666667, 8.33333333, 9.])
```

Getting access to the 1D `NumPy` array is similar to what we described for lists or tuples: it has an index to indicate the location. For example,

```
In [9]: # get the 2nd element of x
         x[1]

Out[9]: 4

In [10]: # get all the element after the 2nd element of x
         x[1:]

Out[10]: array([4, 3])
```

```
In [11]: # get the last element of x
         x[-1]

Out[11]: 3
```

For 2D arrays, it is slightly different, since we have rows and columns. To get access to the data in a 2D array `M`, we need to use `M[r, c]`, whereby the row `r` and column `c` are separated by a comma. This is referred to as "array indexing." The `r` and `c` can be single number, a list, etc. If you only think about the row index or the column index, then it is similar to the 1D array. Let us use the $y = \begin{pmatrix} 1 & 4 & 3 \\ 9 & 2 & 7 \end{pmatrix}$ as an example.

TRY IT! Obtain the element at first row and second column of array *y*.

```
In [12]: y[0,1]

Out[12]: 4
```

TRY IT! Obtain the first row of array `y`.

```
In [13]: y[0, :]

Out[13]: array([1, 4, 3])
```

TRY IT! Obtain the last column of array `y`.

```
In [14]: y[:, -1]

Out[14]: array([3, 7])
```

TRY IT! Obtain the first and third column of array `y`.

```
In [15]: y[:, [0, 2]]

Out[15]: array([[1, 3],
                [9, 7]])
```

Here are some predefined arrays that are really useful: the `np.zeros`, `np.ones`, and `np.empty` are three useful functions. See examples of these predefined arrays below:

TRY IT! Generate a 3 × 5 array with all the elements as 0.

```
In [16]: np.zeros((3, 5))

Out[16]: array([[0., 0., 0., 0., 0.],
                [0., 0., 0., 0., 0.],
                [0., 0., 0., 0., 0.]])
```

TRY IT! Generate a 5 × 3 array with all the elements as 1.

```
In [17]: np.ones((5, 3))

Out[17]: array([[1., 1., 1.],
                [1., 1., 1.],
                [1., 1., 1.],
                [1., 1., 1.],
                [1., 1., 1.]])
```

Note! The shape of the array is defined in a tuple with the number of rows as the first item, and the number of columns as the second. If you only need a 1D array, then use only one number as the input: `np.ones(5)`.

TRY IT! Generate a 1D empty array with 3 elements.

```
In [18]: np.empty(3)

Out[18]: array([-3.10503618e+231, -3.10503618e+231,
              -3.10503618e+231])
```

Note! The empty array is not really empty; it is filled with random very small numbers.

You can reassign a value of an array by using array indexing and the assignment operator. You can reassign multiple elements to a single number using array indexing on the left-hand side. You can also reassign multiple elements of an array as long as both the number of elements being assigned and the number of elements assigned are the same. You can create an array using array indexing.

TRY IT! Let `a = [1, 2, 3, 4, 5, 6]`. Reassign the fourth element of A to 7. Reassign the first, second, and third elements to 1. Reassign the second, third, and fourth elements to 9, 8, and 7.

```
In [19]: a = np.arange(1, 7)
         a

Out[19]: array([1, 2, 3, 4, 5, 6])
```

```
In [20]: a[3] = 7
         a

Out[20]: array([1, 2, 3, 7, 5, 6])

In [21]: a[:3] = 1
         a

Out[21]: array([1, 1, 1, 7, 5, 6])

In [22]: a[1:4] = [9, 8, 7]
         a

Out[22]: array([1, 9, 8, 7, 5, 6])
```

TRY IT! Create a 2 × 2 zero array b, and set $b = \begin{pmatrix} 1 & 2 \\ 3 & 4 \end{pmatrix}$ using array indexing.

```
In [23]: b = np.zeros((2, 2))
         b[0, 0] = 1
         b[0, 1] = 2
         b[1, 0] = 3
         b[1, 1] = 4
         b

Out[23]: array([[1., 2.],
                [3., 4.]])
```

Arrays are defined using basic arithmetic; however, there are operations between a scalar (a single number) and an array and operations between two arrays. We will start with operations between a scalar and an array. To illustrate, let c be a scalar, and b be a matrix. Then b + c, b - c, b * c and b / c adds a to every element of b, subtracts c from every element of b, multiplies every element of b by c, and divides every element of b by c, respectively.

TRY IT! Let $b = \begin{pmatrix} 1 & 2 \\ 3 & 4 \end{pmatrix}$. Add and subtract 2 from b. Multiply and divide b by 2. Square every element of b. Let c be a scalar. On your own, verify the reflexivity of scalar addition and multiplication: b + c = c + b and cb = bc.

```
In [24]: b + 2

Out[24]: array([[3., 4.],
                [5., 6.]])
```

```
In [25]: b - 2

Out[25]: array([[-1.,  0.],
                [ 1.,  2.]])

In [26]: 2 * b

Out[26]: array([[2., 4.],
                [6., 8.]])

In [27]: b / 2

Out[27]: array([[0.5, 1. ],
                [1.5, 2. ]])

In [28]: b**2

Out[28]: array([[ 1.,  4.],
                [ 9., 16.]])
```

Describing operations between two matrices is more complicated. Let b and d be two matrices of the same size. Then b - d takes every element of b and subtracts the corresponding element of d. Similarly, b + d adds every element of d to the corresponding element of b.

TRY IT! Let $b = \begin{pmatrix} 1 & 2 \\ 3 & 4 \end{pmatrix}$ and $d = \begin{pmatrix} 3 & 4 \\ 5 & 6 \end{pmatrix}$. Compute b + d and b - d.

```
In [29]: b = np.array([[1, 2], [3, 4]])
         d = np.array([[3, 4], [5, 6]])

In [30]: b + d

Out[30]: array([[ 4,  6],
                [ 8, 10]])

In [31]: b - d

Out[31]: array([[-2, -2],
                [-2, -2]])
```

There are two different kinds of multiplication (and division) for matrices. There is element-by-element matrix multiplication and standard matrix multiplication. This section will only demonstrate how element-by-element matrix multiplication and division works. Standard matrix multiplication will be described in the later chapter on Linear Algebra. Python takes the * symbol to mean element-by-element multiplication. For matrices b and d of the same size, b * d takes every element of b and multiplies it by the corresponding element of d. The same is true for / and **.

TRY IT! Compute `b * d`, `b / d`, and `b**d`.

```
In [32]: b * d

Out[32]: array([[ 3,  8],
                [15, 24]])

In [33]: b / d

Out[33]: array([[0.33333333, 0.5       ],
                [0.6       , 0.66666667]])

In [34]: b**d

Out[34]: array([[   1,   16],
                [ 243, 4096]])
```

The transposition of an array, `b`, is an array, `d`, where `b[i, j] = d[j, i]`. In other words, the transposition switches the rows and the columns of `b`. You can transpose an array in Python using the array method `T`.

TRY IT! Compute the transpose of array `b`.

```
In [35]: b.T

Out[35]: array([[1, 3],
                [2, 4]])
```

`NumPy` has many arithmetic functions, such as `sin`, `cos`, etc., that can take arrays as input arguments. The output is the function evaluated for every element of the input array. A function that takes an array as input and performs the function on it is said to be **vectorized**.

TRY IT! Compute `np.sqrt` for x = [1, 4, 9, 16].

```
In [36]: x = [1, 4, 9, 16]
         np.sqrt(x)

Out[36]: array([1., 2., 3., 4.])
```

Logical operations are defined only between a scalar and an array and between two arrays of the same size. Between a scalar and an array, the logical operation is conducted between the scalar and each element of the array. Between two arrays, the logical operation is conducted element-by-element.

TRY IT! Check which elements of the array `x = [1, 2, 4, 5, 9, 3]` are larger than 3. Check which elements in x are larger than the corresponding element in `y = [0, 2, 3, 1, 2, 3]`.

```
In [37]: x = np.array([1, 2, 4, 5, 9, 3])
         y = np.array([0, 2, 3, 1, 2, 3])

In [38]: x > 3

Out[38]: array([False, False,  True,  True,  True, False])

In [39]: x > y

Out[39]: array([ True, False,  True,  True,  True, False])
```

Python can index elements of an array that satisfy a logical expression.

TRY IT! Let `x` be the same array as in the previous example. Create a variable `y` that contains all the elements of `x` that are strictly bigger than 3. Assign all the values of `x` that are bigger than 3, the value 0.

```
In [40]: y = x[x > 3]
         y

Out[40]: array([4, 5, 9])

In [41]: x[x > 3] = 0
         x

Out[41]: array([1, 2, 0, 0, 0, 3])
```

2.8 SUMMARY AND PROBLEMS

2.8.1 SUMMARY

1. Storing, retrieving, and manipulating information and data is important in any scientific and engineering field.
2. Assigning variables is an important tool for handling data values.
3. There are different data types for storing information in Python: `int`, `float`, and `boolean` for single values, and `strings`, `lists`, `tuples`, `sets`, and `dictionaries` for sequential data.
4. The `NumPy` array is a powerful data structure that used a lot in scientific computing.

2.8.2 PROBLEMS

1. Assign the value 2 to the variable `x` and the value 3 to the variable `y`. Clear just the variable `x`.

2. Write a line of code that generates the following error:

```
NameError: name "x" is not defined
```

3. Let `x = 10` and `y = 3`. Write a line of code that will make each of the following assignments.

```
u = x + y
v = xy
w = x/y
z = sin(x)
r = 8sin(x)
s = 5sin(xy)
p = x**y
```

4. Show all the variables in the `Jupyter notebook` after you finish Problem 3.
5. Assign string "123" to the variable `S`. Convert the string into a float type and assign the output to the variable `N`. Verify that `S` is a string and `N` is a float using the `type` function.
6. Assign the string "HELLO" to the variable `s1` and the string "hello" to the variable `s2`. Use the == operator to show that they are not equal. Use the == operator to show that `s1` and `s2` are equal if the `lower` method is used on s1. Use the == operator to show that `s1` and `s2` are equal if `upper` method is used on `s2`.
7. Use the `print` function to generate the following strings:
   - The world "Engineering" has 11 letters.
   - The word "Book" has 4 letters.
8. Check if "Python" is in "Python is great!".
9. Get the last word "great" from "Python is great!"
10. Assign list `[1, 8, 9, 15]` to a variable `list_a` and insert 2 at index 1 using the `insert` method. Append 4 to the `list_a` using the `append` method.
11. Sort the `list_a` in problem 10 in ascending order.
12. Turn "Python is great!" into a list.
13. Create one tuple with element "One", 1 and assign it to `tuple_a`.
14. Get the second element in the `tuple_a` in Problem 13.
15. Get the unique element from (2, 3, 2, 3, 1, 2, 5).
16. Assign (2, 3, 2) to `set_a`, and (1, 2, 3) to `set_b`. Obtain the following:
    - union of `set_a` and `set_b`
    - intersection of `set_a` and `set_b`
    - difference of `set_a` to `set_b` using `difference` method

17. Create a dictionary that has the keys "A", "B", "C" with values "a", "b", "c" individually. Print all the keys in the dictionary.

18. Check if key "B" is in the dictionary defined in Problem 17.

19. Create array `x` and `y`, where `x = [1, 4, 3, 2, 9, 4]` and `y=[2, 3, 4, 1, 2, 3]`. Compute the assignments from Problem 3.

20. Generate an array with size 100 evenly spaced between −10 to 10 using `linspace` function in `NumPy`.

21. Let `array_a` be an array `[-1, 0, 1, 2, 0, 3]`. Write a command that will return an array consisting of all the elements of `array_a` that are larger than zero. Hint: Use logical expression as the index of the array.

22. Create an array $y = \begin{pmatrix} 3 & 5 & 3 \\ 2 & 2 & 5 \\ 3 & 8 & 9 \end{pmatrix}$ and calculate its transpose.

23. Create a 2×4 zero array.

24. Change the second column in the above array to 1.

25. Write a magic command to clear all the variables in the `Jupyter notebook`.

CHAPTER

FUNCTIONS

3

CONTENTS

3.1 FUNCTION BASICS

In programming, a **function** is a sequence of instructions that performs a specific task. A function is a block of code that can run when it is called. A function can have **input arguments**, which are made available by the user (the entity calling the function). Functions also have **output parameters**. These are the results of the function once it has completed its task. For example, the function `math.sin` has one input argument—an angle in radians, and one output argument—an approximation to the sin function computed at the input angle. The sequence of instructions to compute this approximation constitutes the **body of the function**, which is being introduced here.

3.1.1 BUILT-IN FUNCTIONS IN PYTHON

Many built-in Python functions have been introduced already, such as `type`, `len`, etc. In addition, we have introduced various functions available from different packages, for example, `math.sin`, `np.array`, etc. Do you still remember how to call and use these functions?

TRY IT! Verify that `len` is a built-in function using the `type` function.

```
In [1]: type(len)

Out[1]: builtin_function_or_method
```


Python Programming and Numerical Methods. https://doi.org/10.1016/B978-0-12-819549-9.00012-9

TRY it! Verify that `np.linspace` is a function using the `type` function. Next, figure out how to use the function using the question mark.

```
In [2]: import numpy as np

        type(np.linspace)

Out[2]: function

In [3]: np.linspace?
```

3.1.2 DEFINE YOUR OWN FUNCTION

We can define our own functions. A function can be specified in several ways. The most common way to define a function is to call it using a keyword `def`, as shown below:

```
def function_name(parameter_1, parameter_2, ...):
    """
    Descriptive String
    """
    # comments about the statements
    function_statements

    return output_parameters (optional)
```

Defining a Python function requires the following two components:

1. **Function header** that starts with a keyword `def`, followed by a pair of parentheses with the input parameters inside, and ends with a colon (:);
2. **Function Body** which is an indented block (usually four white spaces) indicating the main body of the function. It consists of three parts:

   - *Descriptive string,* which is string that describes the function that can be accessed by the `help()` function or the question mark. The triple single or triple double quotes show where to put (or locate) your descriptive strings. You can write any strings inside the quotes, either in one line or multiple lines.
   - *Function statements,* which are the step-by-step instructions the function will execute when calling the function. Note that there is a line that starts with #; this is a single line comment, which means that it is *not* part of the function and cannot be executed.
   - *Return statements,* which may contain some parameters to be returned after the function is called. As discussed in more detail later, any data type can be returned, even a function.

NOTE! Input parameter vs argument. A parameter is a variable defined by a function that receives a value when the function is called. An argument is a value that is passed to a method when it is invoked. For example, if we define a function `hello(name)`, then `name` is an input parameter. When we call the function, and pass in a value 'Qingkai', then this value is an input argument.

Due to the very subtle difference, in the rest of the book, we will use parameters and arguments interchangeably.

TIP! When your code becomes longer and more complicated, comments can help you and those reading your code to navigate through the commands and provide a logical "road map" to understand what you are trying to do. Getting in the habit of commenting frequently will prevent coding mistakes, understand where your code is going when you write it, and assist you in finding errors when you make mistakes. Even though it is optional, it is also customary to put a description of the function, author, and creation date in the descriptive string under the function header (you can skip the descriptive string). We highly recommend that you comment heavily in your own code.

TRY IT! Define a function named `my_adder` that takes three numbers and sum them.

```
In [4]: def my_adder(a, b, c):
            """
            function to sum the 3 numbers
            Input: 3 numbers a, b, c
            Output: the sum of a, b, and c
            author:
            date:
            """

            # this is the summation
            out = a + b + c

            return out
```

WARNING! If you do not indent your code when defining a function, you will get an `IndentationError`.

```
In [5]: def my_adder(a, b, c):
        """
        function to sum the 3 numbers
        Input: 3 numbers a, b, c
        Output: the sum of a, b, and c
        author:
        date:
```

```
    """

    # this is the summation
    out = a + b + c

    return out


      File "<ipython-input-5-e6a61721f00e>", line 8
    """

    ^
IndentationError: expected an indented block
```

TIP! Manually typing four white spaces is one level of indentation. Deeper levels of indentation are required when you have nested functions or if-statements, which we will discuss in the next chapter. Note that sometimes you need to indent or unindent a block of code. You can do this by first selecting all the lines in the code block and then pressing `Tab` and `Shift+Tab` to increase or decrease one level of indentation.

TIP! Build good coding practices by giving variables and functions descriptive names, commenting often, and avoiding extraneous lines of code.

For contrast, consider the following function that performs the same task as `my_adder` but is not constructed using best practices. As you can see, it is extremely difficult to follow the logic of the code and the intentions of the author.

EXAMPLE: Poor representation of `my_adder`.

```
In [6]: def abc(a, s2, d):
            z = a + s2
            z = z + d
            x = z
            return x
```

Functions must conform to a naming scheme similar to variables. They can only contain alphanumeric characters and underscores, and the first character must be a letter.

TIP! As is the convention with variable names, function names should be lowercase, with words separated by underscores as necessary to improve readability.

TIP! It is good programming practice to save often while writing your function. In fact, many programmers save their code by using the shortcut `Ctrl+s` (PC) or `cmd+s` (Mac) every time they stop typing!

TRY IT! Use your function my_adder to compute the sum of a few numbers. Verify that the result is correct. Try calling the `help` function on `my_adder`.

```
In [7]: d = my_adder(1, 2, 3)
        d

Out[7]: 6

In [8]: d = my_adder(4, 5, 6)
        d

Out[8]: 15

In [9]: help(my_adder)

Help on function my_adder in module __main__:

my_adder(a, b, c)
    function to sum the 3 numbers
    Input: 3 numbers a, b, c
    Output: the sum of a, b, and c
    author:
    date:
```

WHAT IS HAPPENING? First recall that the assignment operator works from right to left. This means that `my_adder(1,2,3)` is resolved before the assignment to `d`.

1. Python finds the function `my_adder`.
2. `my_adder` takes the first input argument value 1 and assigns it to the variable with name `a` (first variable name in input argument list).
3. `my_adder` takes the second input argument value 2 and assigns it to the variable with name `b` (second variable name in input argument list).
4. `my_adder` takes the third input argument value 3 and assigns it to the variable with name `c` (third variable name in input argument list).
5. `my_adder` computes the sum of `a`, `b`, and `c`, which is $1 + 2 + 3 = 6$.
6. `my_adder` assigns the value 6 to the variable `out`.
7. `my_adder` outputs the value contained in the output variable `out`, which is 6.

8. `my_adder(1,2,3)` is equivalent to the value 6, and this value is assigned to the variable with name `d`.

Python gives the user tremendous freedom to assign variables to different data types. For example, it is possible to give the variable `x` a dictionary or a float value. In other programming languages, this is not always the case. In these programs, you must declare at the beginning of a session whether `x` will be a dictionary or a float type, and once you decide which type it is, you cannot change it. This can be both a benefit and a drawback. For instance, `my_adder` was built assuming that the input arguments were numerical types, either int or float; however, the user may accidentally input a list or string into `my_adder`, which is not correct. If you try to input a nonnumerical type input argument into `my_adder`, Python will continue to execute the function until something goes wrong.

TRY IT! Use the string "1" as one of the input arguments to `my_adder`; in addition, use a list as one of the input arguments to `my_adder`.

```
In [10]: d = my_adder("1", 2, 3)


        ---------------------------------------------------------


        TypeError                Traceback (most recent call last)

        <ipython-input-10-245d0f4254a9> in <module>
    ----> 1 d = my_adder("1", 2, 3)


        <ipython-input-4-72d064c3ba7a> in my_adder(a, b, c)
          9
         10      # this is the summation
    ---> 11      out = a + b + c
         12
         13      return out


        TypeError: must be str, not int

In [11]: d = my_adder(1, 2, [2, 3])


        ---------------------------------------------------------


        TypeError                Traceback (most recent call last)
```

```
    <ipython-input-11-04f0428ffc51> in <module>
  ----> 1 d = my_adder(1, 2, [2, 3])


    <ipython-input-4-72d064c3ba7a> in my_adder(a, b, c)
       9
      10      # this is the summation
  ---> 11     out = a + b + c
      12
      13      return out

    TypeError: unsupported operand type(s) for +: "int"
    and "list"
```

TIP! Remember to read the error messages that Python provides. They usually tell you exactly where the problem was. In this case, the error says `---> 11 out = a + b + c`, meaning there was an error in `my_adder` on the 11th line. The reason there was an error is **TypeError**, because `unsupported operand type(s) for +: "int" and "list"`, which means that we cannot add int and list.

At this point, you do not have any control over what the user assigns your function as input arguments and whether they correspond to what you intended those input arguments to be. So for the moment, write your functions assuming that they will be used correctly. You can help yourself and other users use your function correctly by providing comments detailing your code.

You can compose functions by assigning function calls as the input to other functions. In the order of operations, Python will execute the innermost function call first. You can also assign mathematical expressions as the input to functions. In this case, Python will execute the mathematical expressions first.

TRY IT! Use the function `my_adder` to compute the sum of $\sin(\pi)$, $\cos(\pi)$, and $\tan(\pi)$. Use mathematical expressions as the input to `my_adder` and verify that the function performs the operations correctly.

```
In [12]: d=my_adder(np.sin(np.pi), np.cos(np.pi), np.tan(np.pi))
         d

Out[12]: -1.0

In [13]: d = my_adder(5 + 2, 3 * 4, 12 / 6)
         d

Out[13]: 21.0
```

```
In [14]: d = (5 + 2) + 3 * 4 + 12 / 6
         d

Out[14]: 21.0
```

Python functions can have multiple output parameters. When calling a function with multiple output parameters, you can unpack the results with multiple variables, which you should separate by commas. The function essentially will return the multiple result parameters in a tuple, which then allows you to unpack the returned tuple. See the following example (note that it has multiple output parameters):

EXAMPLE: Compute the function `my_trig_sum` for `a=2` and `b=3`. Assign the first output parameter to the variable `c`, the second output parameter to the variable `d`, and the third parameter to the variable `e`.

```
In [15]: def my_trig_sum(a, b):
             """
             function to demo return multiple
             author
             date
             """
             out1 = np.sin(a) + np.cos(b)
             out2 = np.sin(b) + np.cos(a)
             return out1, out2, [out1, out2]

In [16]: c, d, e = my_trig_sum(2, 3)
         print(f"c ={c}, d={d}, e={e}")

c =-0.0806950697747637, d=-0.2750268284872752,
e=[-0.0806950697747637, -0.2750268284872752]
```

If you assign the results to one variable, you will get a tuple that includes all the output parameters.

TRY IT! Compute the function `my_trig_sum` for `a=2` and `b=3`. Verify the output is a tuple.

```
In [17]: c = my_trig_sum(2, 3)
         print(f"c={c}, and the returned type is {type(c)}")

c=(-0.0806950697747637, -0.2750268284872752,
[-0.0806950697747637, -0.2750268284872752]),
and the returned type is <class "tuple">
```

A function can be defined without an input argument and returning any value. For example,

EXAMPLE: Function with no inputs and outputs.

```
In [18]: def print_hello():
             print("Hello")

In [19]: print_hello()

Hello
```

Note! Even there is no input argument, when you call the function, you still need to include the parentheses.

For the input of the argument, we can include the default value as well. See the following example:

EXAMPLE: Run the following function with and without an input:

```
In [20]: def print_greeting(day = "Monday", name = "Qingkai"):
             print(f"Greetings, {name}, today is {day}")

In [21]: print_greeting()

Greetings, Qingkai, today is Monday


In [22]: print_greeting(name = "Timmy", day = "Friday")

Greetings, Timmy, today is Friday


In [23]: print_greeting(name = "Alex")

Greetings, Alex, today is Monday
```

We can see that if we assign a value to the argument when we define the function, this value will be the default value of the function. If the user does not provide an input to this argument, then this default value will be used during calling of the function. Note that the order of the argument is not important when calling the function if you provide the name of the argument.

3.2 LOCAL VARIABLES AND GLOBAL VARIABLES

Chapter 2 introduced the idea of a memory block associated with the notebook, where variables created in the notebook are stored. A function also has its own memory block that is reserved for variables created within that function. This block of memory is not shared with the whole notebook memory block. Therefore, a variable with a given name can be assigned within a function without changing a variable with the same name outside of the function. The memory block associated with the function is opened every time a function is used.

TRY IT! What will the value of `out` be after the following lines of code are executed? Note that it is not 6, which is the value `out` was assigned inside of `my_adder`.

```
In [1]: def my_adder(a, b, c):
            out = a + b + c
            print(f"The value out within the function is {out}")
            return out

        out = 1
        d = my_adder(1, 2, 3)
        print(f"The value out outside the function is {out}")

The value out within the function is 6
The value out outside the function is 1
```

In `my_adder`, the variable `out` is a **local variable**. That is, because it is only defined in the function of `my_adder`, it cannot affect variables outside of the function. Actions taken in the notebook outside the function cannot affect it, even if they have the same name. So in the previous example, there is the variable, `out`, which has been defined in the notebook cell. When `my_adder` is called on the next line, Python opens a new memory block for that function's variables. One of the variables created within the function is another variable, `out`. Because they are located in different memory blocks, the assignment to `out` inside `my_adder` does not change the value assigned to `out` outside the function.

Why have separate function memory blocks rather than a single memory block? Although it may not seem logical for Python to separate memory blocks, it is very efficient for large projects consisting of many functions working together. If one programmer is responsible for coding one function and another is responsible for coding a different function, having separate memory blocks allows each programmer to work independently and be confident that their coding will not produce errors when considering another programmer's code, and vice versa. Separate memory blocks protect a function from outside influences. The only things from outside the function's memory block that can affect what happens inside a function are the input arguments, and the only things that can escape to the outside world from a function's memory block when the function terminates are the output arguments.

The next examples are designed to be exercises to gain experience with concept of local variables. They are intentionally very confusing, but if you can untangle them, then you will have mastered the concept of local variables within a function. Focus on exactly what Python is doing and in the order Python does it.

EXAMPLE: Consider the following function:

```
In [2]: def my_test(a, b):
            x = a + b
```

```
    y = x * b
    z = a + b

    m = 2

    print(f"Within function: x={x}, y={y}, z={z}")
    return x, y
```

TRY IT! What will the values of a, b, x, y, and z be after the following code is run?

```
In [3]: a = 2
        b = 3
        z = 1
        y, x = my_test(b, a)

        print(f"Outside function: x={x}, y={y}, z={z}")

Within function: x=5, y=10, z=5
Outside function: x=10, y=5, z=1
```

TRY IT! What will the values of a, b, x, y, and z be after the following code is run?

```
In [4]: x = 5
        y = 3
        b, a = my_test(x, y)

        print(f"Outside function: x={x}, y={y}, z={z}")

Within function: x=8, y=24, z=8
Outside function: x=5, y=3, z=1
```

TRY IT! What will the value of m be if you print m outside of the function?

```
In [5]: m


        ---------------------------------------------------


        NameError             Traceback (most recent call last)


        <ipython-input-5-9a40b379906c> in <module>
    ----> 1 m
```

```
NameError: name "m" is not defined
```

Note that the value m is not defined outside of the function because it is defined within the function. The same is true if you define a variable outside a function, using it inside the function will change the value, and the same error message will occur.

EXAMPLE: Try to use and change the value n within the function.

```
In [6]: n = 42

        def func():
            print(f"Within function: n is {n}")
            n = 3
            print(f"Within function: change n to {n}")

        func()
        print(f"Outside function: Value of n is {n}")


        -------------------------------------------------------

        UnboundLocalError      Traceback (most recent call last)

        <ipython-input-6-85f3215553ae> in <module>
          6     print(f"Within function: change n to {n}")
          7
    ----> 8 func()
          9 print(f"Outside function: Value of n is {n}")


        <ipython-input-6-85f3215553ae> in func()
          2
          3 def func():
    ----> 4     print(f"Within function: n is {n}")
          5     n = 3
          6     print(f"Within function: change n to {n}")


        UnboundLocalError: local variable "n" referenced before
                           assignment
```

The solution is to use the keyword **global** to let Python know this variable is a **global variable** and can be used both outside and inside the function.

EXAMPLE: Define n as the global variable, and then use and change the value n within the function.

```
In [7]: n = 42

        def func():
            global n
            print(f"Within function: n is {n}")
            n = 3
            print(f"Within function: change n to {n}")

        func()
        print(f"Outside function: Value of n is {n}")

Within function: n is 42
Within function: change n to 3
Outside function: Value of n is 3
```

3.3 NESTED FUNCTIONS

Once you have created and saved a new function, it behaves just like any other Python built-in function. You can call the function from anywhere in the notebook, and any other function can call on the function as well. A **nested function** is a function that is defined within another function – **parent function**. Only the parent function is able to call the nested function. Remember that the nested function retains a separate memory block from its parent function.

TRY IT! Consider the following function and nested function:

```
In [1]: import numpy as np

        def my_dist_xyz(x, y, z):
            """
            x, y, z are 2D coordinates contained in a tuple
            output:
            d - list, where
                d[0] is the distance between x and y
                d[1] is the distance between x and z
                d[2] is the distance between y and z
            """

            def my_dist(x, y):
```

```
        """
        subfunction for my_dist_xyz
        Output is the distance between x and y,
        computed using the distance formula
        """
        out = np.sqrt((x[0]-y[0])**2+(x[1]-y[1])**2)
        return out

    d0 = my_dist(x, y)
    d1 = my_dist(x, z)
    d2 = my_dist(y, z)

    return [d0, d1, d2]
```

Note that the variables x and y appear in both my_dist_xyz and my_dist. This is permissible because a nested function has a separate memory block from its parent function. Nested functions are useful when a task must be performed many times within the function but not outside the function. In this way, nested functions help the parent function perform its task while hiding in the parent function.

TRY IT! Call the function my_dist_xyz for x = (0, 0), y = (0, 1), z = (1, 1). Try to call the nested function my_dist in the following cell:

```
In [2]: d = my_dist_xyz((0, 0), (0, 1), (1, 1))
        print(d)
        d = my_dist((0, 0), (0, 1))

[1.0, 1.4142135623730951, 1.0]


        ---------------------------------------------------------

        NameError              Traceback (most recent call last)

        <ipython-input-2-1bec838581d7> in <module>
          1 d = my_dist_xyz((0, 0), (0, 1), (1, 1))
          2 print(d)
    ----> 3 d = my_dist((0, 0), (0, 1))


        NameError: name "my_dist" is not defined
```

The following example is the code repeated without using nested function. Notice how much busier and cluttered the function looks and how much more difficult it is to understand what is going on. This version is much more prone to mistakes because you have three chances to mistype the distance formula. Note that this function can be written more compactly using vector operations. We leave this as an exercise.

```
In [ ]: import numpy as np

        def my_dist_xyz(x, y, z):
            """
            x, y, z are 2D coordinates contained in a tuple
            output:
            d - list, where
                d[0] is the distance between x and y
                d[1] is the distance between x and z
                d[2] is the distance between y and z
            """

            d0 = np.sqrt((x[0]-y[0])**2+(x[1]-y[1])**2)
            d1 = np.sqrt((x[0]-z[0])**2+(x[1]-z[1])**2)
            d2 = np.sqrt((y[0]-z[0])**2+(y[1]-z[1])**2)

            return [d0, d1, d2]
```

3.4 LAMBDA FUNCTIONS

Sometimes it is not optimal to define a function the usual way, especially if our function is just one line. In this case, we use the anonymous function in Python, which is a function that is defined without a name. These types of function are also called **lambda function**, since they are defined using the `lambda` keyword. A typical lambda function is defined as follows:

CONSTRUCTION:

```
lambda arguments: expression
```

It can have any number of arguments but has only one expression.

TRY IT! Define a lambda function, which squares the input number; call the function with input 2 and 5.

```
In [1]: square = lambda x: x**2

        print(square(2))
        print(square(5))
```

```
4
25
```

In the above lambda function, x is the argument and x**2 is the expression that gets evaluated and returned. The function itself has no name, and it returns a function object (discussed in later chapter) to square it. After it is defined, we can call it as a normal function. The lambda function is equivalent to:

```
def square(x):
    return x**2
```

TRY IT! Define a lambda function, which adds x and y.

```
In [2]: my_adder = lambda x, y: x + y

        print(my_adder(2, 4))

6
```

Lambda functions can be useful in many cases, and we will provide other examples in later chapters. Here we just show a common use case for the lambda function.

EXAMPLE: Sort [(1, 2), (2, 0), (4, 1)] based on the second item in the tuple.

```
In [3]: sorted([(1, 2), (2, 0), (4, 1)], key=lambda x: x[1])

Out[3]: [(2, 0), (4, 1), (1, 2)]
```

What happens? The function sorted has an argument key, where a custom key function can be supplied to customize the sort order. We use the lambda function as a shortcut for this custom key function.

3.5 FUNCTIONS AS ARGUMENTS TO FUNCTIONS

Up until now, you have assigned various data structures to variable names. Being able to assign a data structure to a variable allows for the passing of information to the various functions and retrieve information back from them in a neat and orderly way. Sometimes it is useful to be able to pass a function and have it act as a variable to another function. In other words, the input to some functions may be other functions. In last section, we saw the lambda function returning a function object to the variable. In this section, we will see additional examples of how the function object can be used as the input to another function.

TRY IT! Assign the function max to the variable f. Verify the type of f.

```
In [1]: f = max
        print(type(f))

<class "builtin_function_or_method">
```

In the previous example, f is now equivalent to the max function. Because x = 1 means that x and 1 are interchangeable, f and max function are now interchangeable.

TRY IT! Get the maximum value from list [2, 3, 5] using f. Verify that the result is the same as using max.

```
In [2]: print(f([2, 3, 5]))
        print(max([2, 3, 5]))

5
5
```

TRY IT! Write a function my_fun_plus_one that takes a function object, f, and a float number x as input arguments; my_fun_plus_one should return f evaluated at x, and the result added to the value 1. Verify that it works for various functions and values of x.

```
In [3]: import numpy as np

        def my_fun_plus_one(f, x):
            return f(x) + 1

        print(my_fun_plus_one(np.sin, np.pi/2))
        print(my_fun_plus_one(np.cos, np.pi/2))
        print(my_fun_plus_one(np.sqrt, 25))

2.0
1.0
6.0
```

In the above example, different functions are used as inputs into the function. Of course, we can use the lambda functions as well.

```
In [4]: print(my_fun_plus_one(lambda x: x + 2, 2))

5
```

3.6 SUMMARY AND PROBLEMS

3.6.1 SUMMARY

1. A function is a self-contained set of instructions designed to perform a specific task.
2. A function has its own memory block for its variables. Information can be added to a function's memory block only through a function's input variables. Information can leave the function's memory block only through a function's output variables.
3. A function can be defined within another function, which is called a nested function. This nested function can be only accessed by the parent function.
4. You can define an anonymous function using the keyword lambda, which is the so-called lambda function.
5. You can assign functions to variables using function handles.

3.6.2 PROBLEMS

1. Recall that the hyperbolic sine, denoted by sinh, is $\frac{\exp(x)-\exp(-x)}{2}$. Write a function `my_sinh(x)` where the output `y` is the hyperbolic sine computed on x. Assume that `x` is a 1 by 1 float.

   Test Cases:

```
In: my_sinh(0)
Out: 0

In: my_sinh(1)
Out: 1.1752

In: my_sinh(2)
Out: 3.6269
```

2. Write a function `my_checker_board(n)` where the output `m` is an $n \times n$ array with the following form:

$$m = \begin{matrix} 1 & 0 & 1 & 0 & 1 \\ 0 & 1 & 0 & 1 & 0 \\ 1 & 0 & 1 & 0 & 1 \\ 0 & 1 & 0 & 1 & 0 \\ 1 & 0 & 1 & 0 & 1 \end{matrix}$$

   Note that the upper-left element should always be 1. Assume that `n` is a strictly positive integer.
   Test Cases:

```
In: my_checker_board(1)
Out: 1

In: my_checker_board(2)
Out: array([[1, 0],
```

```
         [0, 1]])

In: y = my_sinh(3)
Out: array([[1, 0, 1],
            [0, 1, 0],
            [1, 0, 1]])

In: y = my_sinh(5)
Out: array([[1, 0, 1, 0, 1],
            [0, 1, 0, 1, 0],
            [1, 0, 1, 0, 1],
            [0, 1, 0, 1, 0],
            [1, 0, 1, 0, 1]])
```

3. Write a function `my_triangle(b,h)` where the output is the area of a triangle with base, `b`, and height, `h`. Recall that the area of a triangle is one-half the base times the height. Assume that `b` and `h` are just 1 by 1 float numbers.

   Test Cases:

```
In: my_triangle(1, 1)
Out: 0.5

In: my_triangle(2, 1)
Out: 1

In: my_triangle(12, 5)
Out: 30
```

4. Write a function `my_split_matrix(m)`, where `m` is an array, the output is a list `[m1, m2]` where `m1` is the left half of `m`, and `m2` is the right half of `m`. In the case where there is an odd number of columns, the middle column should go to `m1`. Assume that `m` has at least two columns.

   Test Cases:

```
In: m = np.array([[1, 2, 3], [4, 5, 6], [7, 8, 9]])
In: m1, m2 = my_split_matrix(m)
Out: m1 = array([[1, 2],
                 [4, 5],
                 [7, 8]])
Out: m2 = array([3, 6, 9])


In: m = np.ones((5, 5))
In: m1, m2 = my_split_matrix(m)
```

```
Out: m1 = array([[1., 1., 1.],
    [1., 1., 1.],
    [1., 1., 1.],
    [1., 1., 1.],
    [1., 1., 1.]])
Out: m2 =  array([[1., 1.],
    [1., 1.],
    [1., 1.],
    [1., 1.],
    [1., 1.]])
```

5. Write a function `my_cylinder(r,h)`, where `r` and `h` are the radius and height of a cylinder, respectively, and the output is a list `[s, v]` where `s` and `v` are the surface area and volume of the same cylinder, respectively. Recall that the surface area of a cylinder is $2\pi r^2 + 2\pi rh$, and the volume is $\pi r^2 h$. Assume that `r` and `h` are 1 by 1 floats.

   Test Cases:

```
In: my_cylinder(1,5)
Out: [37.6991, 15.7080]

In: my_cylinder(2,4)
Out: [62.8319, 37.6991]
```

6. Write a function `my_n_odds(a)`, where `a` is a one-dimensional array of floats and the output is the number of odd numbers in `a`.

   Test Cases:

```
In: my_n_odds(np.arange(100))
Out: 50

In: my_n_odds(np.arange(2, 100, 2))
Out: 0
```

7. Write a function `my_twos(m,n)` where the output is an $m \times n$ array of twos. Assume that `m` and `n` are strictly positive integers.

   Test Cases:

```
In: my_twos(3, 2)
Out: array([[2, 2],
   [2, 2],
   [2, 2]])

In: my_twos(1, 4)
Out: array([2, 2, 2, 2])
```

8. Write a lambda function that takes in input x and y, and the output is the value of x - y.

9. Write a function add_string(s1, s2) where the output is the concatenation of the strings s1 and s2.

   Test Cases:

```
In: s1 = add_string("Programming", " ")
In: s2 = add_string("is ", "fun!")
In: add_string(s1, s2)
Out: "Programming is fun!"
```

10. Generate the following errors:

   - TypeError: fun() missing 1 required positional argument: "a"
   - IndentationError: expected an indented block

11. Write a function greeting(name, age) where name is a string, age is a float, and the output is a string "Hi, my name is XXX and I am XXX years old." where XXX are the input name and age, respectively.

   Test Cases:

```
In: greeting("John", 26)
Out: "Hi, my name is John and I am 26 years old."

In: greeting("Kate", 19)
Out: "Hi, my name is Kate and I am 19 years old."
```

12. Let r1 and r2 be the radius of circles with the same center and let r2>r1. Write a function my_donut_area(r1, r2) where the output is the area outside of the circle with radius r1 and inside the circle with radius r2. Make sure that the function is vectorized. Assume that r1 and r2 are one-dimensional arrays of the same size.

   Test Cases:

```
In: my_donut_area(np.arange(1, 4), np.arange(2, 7, 2))
Out: array([9.4248, 37.6991, 84.8230])
```

13. Write a function my_within_tolerance(A, a, tol) where the output is an array or list of the indices in A such that $|A - a| < \text{tol}$. Assume that A is a one-dimensional float list or array, and that a and tol are 1 by 1 floats.

   Test Cases:

```
In: my_within_tolerance([0, 1, 2, 3], 1.5, 0.75)
Out: [1, 2]

In: my_within_tolerance(np.arange(0, 1.01, 0.01), 0.5, 0.03)
Out: [47, 48, 49, 50, 51, 52]
```

14. Write a function `bounding_array(A, top, bottom)` where the output is equal to the array `A` wherever `bottom < A < top`, the output is equal to `bottom` wherever `A <= bottom`, and the output is equal to `top` wherever `A >= top`. Assume that `A` is a one-dimensional float array and that `top` and `bottom` are 1 by 1 floats.

Test Cases:

```
In: bounding_array(np.arange(-5, 6, 1), 3, -3)
Out: [-3, -3, -3, -2, -1, 0, 1, 2, 3, 3, 3]
```

CHAPTER

BRANCHING STATEMENTS

4

CONTENTS

4.1 IF-ELSE STATEMENTS

A **branching statement**, **If-Else Statement**, or **If-Statement** for short, is a code construct that executes blocks of code only if certain conditions are met. These conditions are represented as logical expressions.

CONSTRUCTION: Simple If Statement Syntax

```
if logical expression:
    code block
```

CONSTRUCTION: Simple If-Else Statement Syntax

```
if logical expression:
    code block 1
else:
    code block 2
```

The word "if" is a keyword. When Python sees an if-statement, it will determine if the associated logical expression is true. If it is true, then the code in `code block` will be executed. If it is false, then the code in the if-statement will not be executed. The way to read this is "If the logical expression is true then do code block." Similarly, if the logical expression in the if-else-statement is true, then the code in `code block1` will be executed. Otherwise, `code block2` will be executed.

When there are several conditions to consider, you can include elif statements; if you want a condition that covers any other case, then you may use an else statement. Let `P`, `Q`, and `R` be three logical expressions in Python. The following example shows multiple branches.

Python Programming and Numerical Methods. https://doi.org/10.1016/B978-0-12-819549-9.00013-0

Note! Python gives the same level of indentation to every line of code within a conditional statement.

CONSTRUCTION: Extended If-Else Statement Syntax

```
if logical expression P:
    code block 1
elif logical expression Q:
    code block 2
elif logical expression R:
    code block 3
else:
    code block 4
```

In the previous code, Python will first check if `P` is true. If `P` is true, then code block 1 will be executed, and then the `if-statement` will end. In other words, Python will *not* check the rest of the statements once it reaches a true statement. If `P` is false, then Python will check if `Q` is true. If `Q` is true, then code block 2 will be executed, and the `if-statement` will end. If it is false, then `R` will be executed, and so forth. If `P`, `Q`, and `R` are all false, then code block 4 will be executed. You can have any number of `elif statements` (or none) as long as there is at least one `if-statement` (the first statement). You do not need an `else statement`, but you can have, at most, one else statement. The logical expressions after the `if` and `elif` (i.e., such as `P`, `Q`, and `R`) will be referred to as conditional statements.

TRY IT! Write a function `my_thermo_stat(temp, desired_temp)`. The return value of the function should be the string `"Heat"` if temp is less than `desired_temp` minus 5 degrees, `"AC"` if temp is more than the `desired_temp` plus 5, and `"off"` otherwise.

```
In [1]: def my_thermo_stat(temp, desired_temp):
            """
            Changes the status of the thermostat based on
            temperature and desired temperature
            author
            date
            :type temp: Int
            :type desiredTemp: Int
            :rtype: String
            """
            if temp < desired_temp - 5:
                status = "Heat"
            elif temp > desired_temp + 5:
                status = "AC"
            else:
```

```
            status = "off"
        return status

In [2]: status = my_thermo_stat(65,75)
        print(status)

Heat

In [3]: status = my_thermo_stat(75,65)
        print(status)

AC

In [4]: status = my_thermo_stat(65,63)
        print(status)

off
```

EXAMPLE: What will be the value of y after the following script is executed?

```
In [5]: x = 3
        if x > 1:
            y = 2
        elif x > 2:
            y = 4
        else:
            y = 0
        print(y)

2
```

We can also insert more complicated conditional statements using logical operators.

EXAMPLE: What will be the value of y after the following code is executed?

```
In [6]: x = 3
        if x > 1 and x < 2:
            y = 2
        elif x > 2 and x < 4:
            y = 4
        else:
            y = 0
        print(y)
```

```
4
```

Note that if you want the logical statement $a < x < b$, this is considered as two conditional statements, $a < x$ and $x < b$. Python allows you to type `a < x < b` as well. For example,

```
In [7]: x = 3
        if 1 < x < 2:
            y = 2
        elif 2 < x < 4:
            y = 4
        else:
            y = 0
        print(y)

4
```

A statement is called **nested** if it is entirely contained within another statement of the same type as itself. For example, a **nested if-statement** is an if-statement that is entirely contained within a clause of another if-statement.

EXAMPLE: Think about what will happen when the following code is executed. What are all the possible outcomes based on the input values of `x` and `y`?

```
In [8]: def my_nested_branching(x,y):
            """
            Nested Branching Statement Example
            author
            date
            :type x: Int
            :type y: Int
            :rtype: Int
            """
            if x > 2:
                if y < 2:
                    out = x + y
                else:
                    out = x - y
            else:
                if y > 2:
                    out = x*y
                else:
```

```
            out = 0
        return out
```

Note! As before, Python gives the same level of indentation to every line of code within a conditional statement. The nested if-statement should be indented by an additional four white spaces. You will get an `IndentationError` if the indentation is not correct, as we saw earlier when discussing how to define functions.

```
In [9]: import numpy as np

In [10]: all([1, 1, 0])

Out[10]: False
```

There are many logical functions that are designed to help you build branching statements. For example, you can ask if a variable has a certain data type with function `isinstance`. There are also functions that can tell you information about arrays of logicals like `any`, which computes to true if any element in an array is true, and false otherwise, and `all`, which computes to true only if all the elements in an array are true.

Sometimes you want to design your function to check the inputs of a function to ensure that your function will be used properly. For example, the function `my_adder` in the previous chapter expects doubles as input. If the user inputs a `list` or a `string` as one of the input variables, then the function will throw an error or have unexpected results. To prevent this, you can put a check to tell the user the function has not been used properly. This and other techniques for controlling errors are explored further in Chapter 10. For the moment, you only need to know that we can use the `raise` statement with a `TypeError` exception to stop a function's execution and throw an error with a specific text.

EXAMPLE: Modify `my_adder` to throw out a warning if the user does not input numerical values. Try your function for nonnumerical inputs to show that the check works. When a statement is too long, we can use the "\" symbol to break a line into multiple lines.

```
In [11]: def my_adder(a, b, c):
             """
             Calculate the sum of three numbers
             author
             date
             """

             # Check for erroneous input
             if not (isinstance(a, (int, float)) \
                     or isinstance(b, (int, float)) \
                     or isinstance(c, (int, float))):
```

```
            raise TypeError("Inputs must be numbers.")
        # Return output
        return a + b + c

In [12]: x = my_adder(1,2,3)
         print(x)

6

In [13]: x = my_adder("1","2","3")
         print(x)


        ---------------------------------------------------------


        TypeError                 Traceback (most recent call last)

        <ipython-input-13-c3e353c636b0> in <module>
    ----> 1 x = my_adder("1","2","3")
          2 print(x)


        <ipython-input-11-0f3d29eecee0> in my_adder(a, b, c)
         10             or isinstance(b, (int, float)) \
         11             or isinstance(c, (int, float))):
    ---> 12         raise TypeError("Inputs must be numbers.")
         13     # Return output
         14     return a + b + c


        TypeError: Inputs must be numbers.
```

There is a large variety of erroneous inputs that your function may encounter from users, and it is unreasonable to expect that your function will catch them all. Therefore, unless otherwise stated, write your functions assuming the functions will be used properly.

The remainder of the section gives a few more examples of branching statements.

TRY IT! Write a function called `is_odd` that returns `"odd"` if the input is an odd number and `"even"` if it is even. You can assume that input will be a positive integer.

```
In [14]: def is_odd(number):
             """
             function returns "odd" if the input is odd,
```

```
            "even" otherwise
        author
        date
        :type number: Int
        :rtype: String
        """
        # use modulo to check if the input divisible by 2
        if number % 2 == 0:
            # if divisible by 2, then input is not odd
            return "even"
        else:
            return "odd"

In [15]: is_odd(11)

Out[15]: "odd"

In [16]: is_odd(2)

Out[16]: "even"
```

TRY IT! Write a function called my_circ_calc that takes a numerical number, r, and a string, calc, as input arguments. You may assume that r is positive, and that calc is either the string "area" or "circumference". The function my_circ_calc should compute the area of a circle with radius, r, if the string calc is the "area", and the circumference of a circle with radius, r, if calc is the "circumference".

```
In [17]: np.pi

Out[17]: 3.141592653589793

In [18]: def my_circ_calc(r, calc):
        """
        Calculate various circle measurements
        author
        date
        :type r: Int or Float
        :type calc: String
        :rtype: Int or Float
        """
        if calc == "area":
            return np.pi*r**2
```

```
        elif calc == "circumference":
            return 2*np.pi*r

In [19]: my_circ_calc(2.5, "area")

Out[19]: 19.634954084936208

In [20]: my_circ_calc(3, "circumference")

Out[20]: 18.84955592153876
```

Note! The function here is not limited to a single value input but can be executed using `NumPy` arrays as well (i.e., the same operation will apply on each item of the array). See the following example where we calculate the circumferences for radius as `[2, 3, 4]` using a `NumPy` array.

```
In [21]: my_circ_calc(np.array([2, 3, 4]), "circumference")

Out[21]: array([12.56637061, 18.84955592, 25.13274123])
```

4.2 TERNARY OPERATORS

Most programming languages have **ternary operators**, which are usually known as **conditional expressions**, that provide a mechanism using a one-line code to evaluate the first expression if the condition is true; otherwise, it evaluates the second expression. To implement the ternary operator in Python, use the construction presented below:

CONSTRUCTION: ternary operator in Python

```
expression_if_true if condition else expression_if_false
```

EXAMPLE: Ternary operator

```
In [1]: is_student = True
        person = "student" if is_student else "not student"
        print(person)
student
```

From the above example, we can see this one-line code is equivalent to the following block of codes.

```
In [2]: is_student = True
        if is_student:
            person = "student"
        else:
            person = "not student"
        print(person)
student
```

Ternary operators provide a simple way for branching and can make our codes concise. In the next chapter, we introduce its role in list comprehensions, and will prove it is useful.

4.3 SUMMARY AND PROBLEMS

4.3.1 SUMMARY

1. Branching (if-else) statements allow functions to take different actions under different circumstances.
2. Ternary operators allow single line branching statements.

4.3.2 PROBLEMS

1. Write a function `my_tip_calc(bill, party)` where `bill` is the total cost of a meal and `party` is the number of people in the group. The tip should be calculated as 15% for a party strictly less than six people, 18% for a party strictly less than eight, 20% for a party less than 11, and 25% for a party 11 or more. A couple of test cases are given below.

```
In [ ]: def my_tip_calc(bill, party):
            # write your function code here

            return tips
```

```
In [ ]: # t = 16.3935
        t = my_tip_calc(109.29,3)
        print(t)
```

```
In [ ]: # t = 19.6722
        t = my_tip_calc(109.29,7)
        print(t)
```

```
In [ ]: # t = 21.8580
        t = my_tip_calc(109.29,9)
        print(t)
```

```
In [ ]: # t = 27.3225
        t = my_tip_calc(109.29,12)
        print(t)
```

2. Write a function `my_mult_operation(a,b,operation)`. The input argument, `operation`, is a string that is either `"plus"`, `"minus"`, `"mult"`, `"div"`, or `"pow"`, and the function should compute: $a + b$, $a - b$, $a * b$, a/b, and a^b for the respective values for `operation`. A couple of test cases are given below.

```
In [ ]: def my_mult_operation(a,b,operation):
            # write your function code here

            return out

In [ ]: x = np.array([1,2,3,4])
        y = np.array([2,3,4,5])

In [ ]: # Output: [3,5,7,9]
        my_mult_operation(x,y,"plus")

In [ ]: # Output: [-1,-1,-1,-1]
        my_mult_operation(x,y,"minus")

In [ ]: # Output: [2,6,12,20]
        my_mult_operation(x,y,"mult")

In [ ]: # Output: [0.5,0.66666667,0.75,0.8]
        my_mult_operation(x,y,"div")

In [ ]: # Output: [1,8,81,1024]
        my_mult_operation(x,y,"pow")
```

3. Consider a triangle with vertices at (0, 0), (1, 0), and (0, 1). Write a function with the name `my_inside_triangle(x,y)` where the output is the string `"outside"` if the point (x, y) is outside of the triangle, `"border"` if the point is exactly on the border of the triangle, and `"inside"` if the point is on the inside of the triangle.

```
In [ ]: def my_inside_triangle(x,y):
            # write your function code here

            return position

In [ ]: # Output: "border"
        my_inside_triangle(.5,.5)

In [ ]: # Output: "inside"
        my_inside_triangle(.25,.25)

In [ ]: # Output: "outside"
        my_inside_triangle(5,5)
```

4. Write a function `my_make_size10(x)` where `x` is an array, and output is the first 10 elements of `x` if `x` has more than 10 elements, and output is the array `x` padded with enough zeros to make it length 10 if `x` has less than 10 elements.

```
In [ ]: def my_make_size10(x):
            # write your function code here

            return size10

In [ ]: # Output: [1,2,0,0,0,0,0,0,0,0]
        my_make_size10(range(1,2))

In [ ]: # Output: [1,2,3,4,5,6,7,8,9,10]
        my_make_size10(range(1,15))

In [ ]: # Output: [3,6,13,4,0,0,0,0,0,0]
        my_make_size10(5,5)
```

5. Can you write `my_make_size10` without using if-statements (i.e., using only logical and array operations)?

6. Write a function `my_letter_grader(percent)` where the grade is the string "A+" if `percent` is greater than 97, "A" if `percent` is greater than 93, "A-" if `percent` is greater than 90, "B+" if `percent` is greater than 87, "B" if `percent` is greater than 83, "B-" if `percent` is greater than 80, "C+" if `percent` is greater than 77, "C" if `percent` is greater than 73, "C-" if `percent` is greater than 70, "D+" if `percent` is greater than 67, "D" if `percent` is greater than 63, "D-" if `percent` is greater than 60, and "F" for any `percent` less than 60. Grades exactly on the division should be included in the higher grade category.

```
In [ ]: def my_letter_grader(percent):
            # write your function code here

            return grade

In [ ]: # Output: "A+"
        my_letter_grader(97)

In [ ]: # Output: "B"
        my_letter_grader(84)
```

7. Most engineering systems have a built-in redundancy. That is, an engineering system has fail-safes incorporated into the design to accomplish its purpose. Consider a nuclear reactor whose temperature is monitored by three sensors. An alarm should go off if any two of the sensor readings disagree. Write a function `my_nuke_alarm(s1,s2,s3)` where `s1`, `s2`, and `s3` are the temperature readings for sensor 1, sensor 2, and sensor 3, respectively. The output should be the string `"alarm!"` if any two of the temperature readings disagree by strictly more than 10 degrees and `"normal"` otherwise.

```
In [ ]: def my_nuke_alarm(s1,s2,s3):
            # write your function code here

            return response

In [ ]: #Output: "normal"
        my_nuke_alarm(94,96,90)
```

```
In [ ]: #Output: "alarm!"
        my_nuke_alarm(94,96,80)
```

```
In [ ]: #Output: "normal"
        my_nuke_alarm(100,96,90)
```

8. Let $Q(x)$ be the quadratic equation $Q(x) = ax^2 + bx + c$ for some scalar values a, b, and c. A root of $Q(x)$ is an r such that $Q(r) = 0$. The two roots of a quadratic equation can be described by the quadratic formula, which is

$$r = \frac{-b \pm \sqrt{b^2 - 4ac}}{2a}.$$

A quadratic equation has either two real roots (i.e., $b^2 > 4ac$), two imaginary roots (i.e., $b^2 < 4ac$), or one root $r = -\frac{b}{2a}$.
Write a function `my_n_roots(a,b,c)`, where a, b, and c are the coefficients of the quadratic $Q(x)$. The function should return two values: `n_roots` and `r`. Also `n_roots` is 2 if Q has two real roots, 1 if Q has one root, -2 if Q has two imaginary roots, and r is an array containing the roots of Q.

```
In [ ]: def my_n_roots(a,b,c):
            # write your function code here

            return n_roots, r
```

```
In [ ]: # Output: n_roots = 2, r = [3, -3]
        n_roots, r = my_n_roots(1,0,-9)
        print(n_roots, r)
```

```
In [ ]: # Output: n_roots = -2,
        # r = [-0.6667 + 1.1055i, -0.6667 - 1.1055i]
        my_n_roots(3,4,5)
```

```
In [ ]: # Output: n_roots = 1, r = [1]
        my_n_roots(2,4,2)
```

9. Write a function `my_split_function(f,g,a,b,x)`, where f and g are function objects `f(x)` and `g(x)`, respectively. The output should be `f(x)` if x ≤ a, `g(x)` if x ≥ b, and 0 otherwise. Assume that $b > a$.

```
In [ ]: def my_split_function(f,g,a,b,x):

            if x<=a:
                return f(x)
            elif x>=b:
                return g(x)
            else:
                return 0
```

```
In [ ]: # Output: 2.713
        my_split_function(np.exp,np.sin,2,4,1)
```

```
In [ ]: # Output: 0
        my_split_function(np.exp,np.sin,2,4,3)

In [ ]: # Output: -0.9589
        my_split_function(np.exp,np.sin,2,4,5)
```

CHAPTER

ITERATION

5

CONTENTS

5.1 FOR-LOOPS

A **for-loop** is a set of instructions that is repeated, or iterated, for every value in a sequence. Sometimes for-loops are referred to as **definite loops** because they have a predefined beginning and end as bounded by the sequence.

The general syntax of a for-loop block is as follows.

CONSTRUCTION: For-loop

```
for looping variable in sequence:
    code block
```

A `for-loop` assigns the **looping variable** to the first element of the sequence. It executes everything in the code block. Then it assigns the looping variable to the next element of the sequence and executes the code block again. It continues until there are no more elements in the sequence to assign.

TRY IT! What is the sum of every integer from 1 to 3?

```
In [1]: n = 0
        for i in range(1, 4):
            n = n + i

        print(n)
```


Python Programming and Numerical Methods. https://doi.org/10.1016/B978-0-12-819549-9.00014-2

```
6
```

WHAT IS HAPPENING?

0. First, the function `range(1, 4)` generates a sequence of numbers that begin at 1 and end at 3. Check the description of the function `range` and get familiar with how to use it. In a very simple form, it is `range(start, stop, step)`, and the `step` is optional with 1 as the default.
1. The variable `n` is assigned the value 0.
2. The variable `i` is assigned the value 1.
3. The variable `n` is assigned the value `n + i` $(0+1=1)$.
4. The variable `i` is assigned the value 2.
5. The variable `n` is assigned the value `n + i` $(1+2=3)$.
6. The variable `i` is assigned the value 3.
7. The variable `n` is assigned the value `n + i` $(3+3=6)$.
8. With no more values to assign in the list, the `for-loop` is terminated with `n = 6`.

Below are several more examples to give you a sense of how for-loops work. Other examples of sequences that we can iterate over include the elements of a tuple, the characters in a string, and other sequential data types.

EXAMPLE: Print all the characters in the string `"banana"`.

```
In [2]: for c in "banana":
            print(c)

b
a
n
a
n
a
```

Alternatively, you can use the index to get each character, but it is not as concise as the previous example. Recall that the length of a string can be determined by using the `len` function, and we can ignore the start by only giving one number as the stop.

```
In [3]: s = "banana"
        for i in range(len(s)):
            print(s[i])

b
a
n
a
```

```
n
a
```

EXAMPLE: Given a list of integers, `a`, add all the elements of `a`.

```
In [4]: s = 0
        a = [2, 3, 1, 3, 3]
        for i in a:
            s += i # note this is equivalent to s = s + i

        print(s)

12
```

The Python function `sum` has already been written to handle the previous example. What if you want to add the even indices numbers only? What change(s) would you make to the previous `for-loop` block to handle this restriction?

```
In [5]: s = 0
        for i in range(0, len(a), 2):
            s += a[i]

        print(s)

6
```

NOTE! We use `step` as 2 in the `range` function to get the even indexes for list `a`. A commonly used Python shortcut is the operator `+=`. In Python and many other programming languages, a statement like `i += 1` is equivalent to `i = i + 1` and is the same for other operators as `-=`, `*=`, `/=`.

EXAMPLE: Define a dictionary and loop through all the keys and values.

```
In [6]: dict_a = {"One":1, "Two":2, "Three":3}

        for key in dict_a.keys():
            print(key, dict_a[key])

One 1
Two 2
Three 3
```

In the above example, we used the method `keys` first to retrieve all keys. Next, we used the key to get access the value. Alternatively, we can use the `items` method in a dictionary, which returns an object containing a list of key and value pairs in tuple. We can assign them simultaneously to two variables (tuple assignment); see the example below.

```
In [7]: for key, value in dict_a.items():
            print(key, value)

One 1
Two 2
Three 3
```

Note that we can assign two different looping variables at the same time. There are other cases where we can assign tasks simultaneously. For example, if we have two lists with same length and we want to loop through them simultaneously, we use the `zip` function. See the example below. This function aggregates elements from two iterables and returns an iterator of tuples, where the `i`th tuple element contains the `i`th element of each of the iterables.

```
In [8]: a = ["One", "Two", "Three"]
        b = [1, 2, 3]

        for i, j in zip(a, b):
            print(i, j)

One 1
Two 2
Three 3
```

EXAMPLE: Let the function `have_digits` have a string as the input. The output `out` should take the value 1 if the string contains digits, and 0 otherwise. You can apply the `isdigit` method of the string to check if the character is a digit.

```
In [9]: def have_digits(s):

            out = 0

            # loop through the string
            for c in s:
                # check if the character is a digit
                if c.isdigit():
                    out = 1
                    break

            return out

In [10]: out = have_digits("only4you")
         print(out)
```

```
1

In [11]: out = have_digits("only for you")
         print(out)

0
```

The first step in the function have_digits assumes that there are no digits in the string s (i.e., the output is 0 or False).

Notice the new keyword break. If executed, the break keyword immediately stops the most immediate for-loop that contains it; that is, if it is contained in a nested for-loop, then it will only stop the innermost for-loop. In this particular case, the break command is executed if we ever find a digit in the string. The code will still function properly without this statement, but since the task is to find out if there are any digits in s, we do not have to keep looking if we find one. Similarly, if a human was given the same task for a long string of characters, that person would not continue looking for digits if he or she already found one. Break statements are used when anything happens in a for-loop that would make you want to stop the run early. A less intrusive command is the keyword continue, which skips the remaining code in the current iteration of the for-loop, and continues on to the next element of the looping array. See the following example where we use the keyword continue to skip the print function to print 2:

```
In [12]: for i in range(5):

             if i == 2:
                 continue

             print(i)

0
1
3
4
```

EXAMPLE: Let the function my_dist_2_points(xy_points, xy) where the input argument xy_points is a list of x–y coordinates of a point in Euclidean space, xy is a list that contain an x–y coordinate, and the output d is a list containing the distances from xy to the points contained in each row of xy_points.

```
In [13]: import math

         def my_dist_2_points(xy_points, xy):
             """
             Returns an array of distances between xy and the
```

```
        points contained in the rows of xy_points

        author
        date
        """
        d = []
        for xy_point in xy_points:
            dist = math.sqrt((xy_point[0]-xy[0])**2+(xy_point[1]-xy[1])**2)
            d.append(dist)

        return d

In [14]: xy_points = [[3,2], [2, 3], [2, 2]]
         xy = [1, 2]
         my_dist_2_points(xy_points, xy)

Out[14]: [2.0, 1.4142135623730951, 1.0]
```

Just like `if-statements`, `for-loops` can be nested.

EXAMPLE: Let `x` be a two-dimensional array, `[5 6;7 8]`. Use a nested `for-loop` to sum all the elements in `x`.

```
In [15]: x = np.array([[5, 6], [7, 8]])
         n, m = x.shape
         s = 0
         for i in range(n):
             for j in range(m):
                 s += x[i, j]

         print(s)

26
```

WHAT IS HAPPENING?

1. `s`, representing the running total sum, is set to 0.
2. The outer `for-loop` begins with looping variable, `i`, set to 0.
3. Inner `for-loop` begins with looping variable, `j`, set to 0.
4. `s` is incremented by `x[i,j] = x[0,0] = 5`. So `s = 5`.
5. Inner `for-loop` sets `j = 1`.
6. `s` is incremented by `x[i,j] = x[0,1] = 6`; therefore, `s = 11`.
7. Inner `for-loop` terminates.

8. Outer `for-loop` sets `i = 1`.
9. Inner `for-loop` begins with looping variable, `j`, set to 0.
10. `s` is incremented by `x[i,j] = x[1,0] = 7`; therefore, `s = 18`.
11. Inner `for-loop` sets `j = 1`.
12. `s` is incremented by `x[i,j] = x[1,1] = 8`. So `s = 26`.
13. Inner `for-loop` terminates.
14. Outer `for-loop` terminates with `s = 26`.

WARNING! Although it is possible to do so, do not try to change the looping variable inside of the `for-loop`. It will make your code very complicated and will likely result in errors.

5.2 WHILE LOOPS

A **while-loop** or **indefinite loop** is a set of instructions that is repeated as long as the associated logical expression is true. The following is the abstract syntax of a `while-loop` block.

CONSTRUCTION: While Loop

```
while <logical expression>:
    # Code block to be repeated until logical statement is false
    code block
```

When Python reaches a `while-loop` block, it first determines if the logical expression of the `while-loop` is true or false. If the expression is true, the code block will be executed. After it is executed, the program returns to the logical expression at the beginning of the `while statement`. If it is false, then the `while-loop` will terminate.

TRY IT! Determine the number of times 8 can be divided by 2 until the result is less than 1.

```
In [1]: i = 0
        n = 8

        while n >= 1:
            n /= 2
            i += 1

        print(f"n = {n}, i = {i}")

n = 0.5, i = 4
```

WHAT IS HAPPENING?

1. First the variable `i` (running count of divisions of `n` by 2) is set to 0.
2. `n` is set to 8 and represents the current value we are dividing by 2.
3. The while-loop begins.
4. Python evaluates expression $n \geq 1$ or $8 \geq 1$, which is true; therefore, the code block is executed.
5. `n` is assigned `n/2 = 8/2 = 4`.
6. `i` is incremented to 1.
7. Python evaluates expression $n \geq 1$ or $4 \geq 1$, which is true; therefore, the code block is executed.
8. `n` is assigned `n/2 = 4/2 = 2`.
9. `i` is incremented to 2.
10. Python evaluates expression $n \geq 1$ or $2 \geq 1$, which is true; therefore, the code block is executed.
11. `n` is assigned `n/2 = 2/2 = 1`.
12. `i` is incremented to 3.
13. Python evaluates expression $n \geq 1$ or $1 \geq 1$, which is true; therefore, the code block is executed.
14. `n` is assigned `n/2 = 1/2 = 0.5`.
15. `i` is incremented to 4.
16. Python evaluates expression $n \geq 1$ or $0.5 \geq 1$, which is false; therefore, the `while-loop` ends with `i = 4`.

You may ask, "What if the logical expression is true and never changes?"; an excellent question. If the logical expression is true and nothing in the `while-loop` code changes the expression, then the result is known as an **infinite loop**. Infinite loops run forever, or until your computer breaks, or runs out of memory.

EXAMPLE: Write a while-loop that causes an infinite loop.

```
In [ ]: n = 0
        while n > -1:
            n += 1
```

Since `n` will always be bigger than -1 no matter how many times the loop is run, this code will never end.

You can terminate the infinite while loop manually by pressing the `interrupt the kernel` – the black square button in the tool bar above shown in Fig. 5.1, or the drop down menu - `Kernel - Interrupt` in the notebook. Or if you are using the Python shell, press `cmd + c` on Mac or `Ctrl + c` on PC.

Can you change a single character so that the while-loop will run at least once but will not infinite loop?

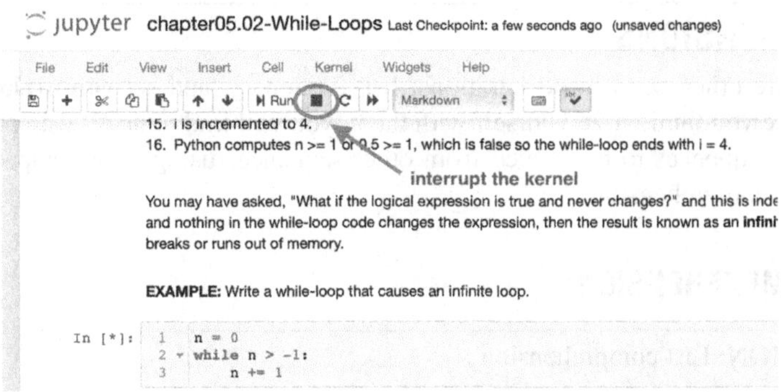

FIGURE 5.1

Interrupt the kernel by pressing the little square.

Infinite loops are not always easy to spot. Consider the next two examples: one performs infinite loops and one does not. Can you determine which is which? As your code becomes more complicated, it will become harder to detect.

EXAMPLE: Which `while-loop` causes an infinite loop?

```
In [ ]: # Example 1
        n = 1
        while n > 0:
            n /= 2
```

```
In [ ]: # Example 2
        n = 2
        while n > 0:
            if n % 2 == 0:
                n += 1
            else:
                n -= 1
```

Answer: The first example will not infinite loop because eventually `n` will be so small that Python cannot tell the difference between `n` and 0. This will be discussed in more detail in Chapter 9. The second example will infinite loop because `n` will oscillate between 2 and 3 indefinitely.

Now we know two types of loops: `for-loops` and `while-loops`. In some cases, either are appropriate, but sometimes one is better suited for the task than the other. In general, use `for-loops` when the number of iterations to be performed is well-defined; use `while-loops` statements when the number of iterations to be performed is indefinite or not well known.

5.3 COMPREHENSIONS

In Python, there are other ways to do iterations; list, dictionary, and set **comprehensions** are very popular ways. Once you familiarize yourself with them, you will find yourself using them a lot. Comprehensions allow sequences to be created from other sequences using very compact syntax. Let us first look at the list comprehension.

5.3.1 LIST COMPREHENSION

CONSTRUCTION: List comprehension

```
[Output Input_sequence Conditions]
```

EXAMPLE: If `x = range(5)`, square each number in `x`, and store it in a list `y`.

If we do not use list comprehension, the code will look something like this:

```
In [1]: x = range(5)
        y = []

        for i in x:
            y.append(i**2)
        print(y)

[0, 1, 4, 9, 16]
```

Using the list comprehension, we can write just one line.

```
In [2]: y = [i**2 for i in x]
        print(y)

[0, 1, 4, 9, 16]
```

In addition, we can also include conditions in the list comprehension. For example, if we just want to store the even numbers in the above example, we just add a condition in the list comprehension.

```
In [3]: y = [i**2 for i in x if i%2 == 0]
        print(y)

[0, 4, 16]
```

If we have two nested levels for loops, we can also use the list comprehensions. For example, we have the following two levels for loops that we can perform using the **list comprehension**.

```
In [4]: y = []
        for i in range(5):
            for j in range(2):
                y.append(i + j)
        print(y)

[0, 1, 1, 2, 2, 3, 3, 4, 4, 5]

In [5]: y = [i + j for i in range(5) for j in range(2)]
        print(y)

[0, 1, 1, 2, 2, 3, 3, 4, 4, 5]
```

5.3.2 DICTIONARY COMPREHENSION

Similarly, we can do the **dictionary comprehension** as well. See the following example.

```
In [6]: x = {"a": 1, "b": 2, "c": 3}
        {key:v**3 for (key, v) in x.items()}

Out[6]: {"a": 1, "b": 8, "c": 27}
```

Performing set comprehension is also possible in Python, but will not be explored here. This is a task that you should explore on your own.

5.4 SUMMARY AND PROBLEMS

5.4.1 SUMMARY

1. Loops provide a mechanism for a code to perform repetitive tasks, that is, iteration.
2. There are two kinds of loops: for-loops and while-loops.
3. Loops are important for constructing iterative solutions to problems.

5.4.2 PROBLEMS

1. What will the value of `y` be after the following code is executed?

```
In [ ]: y = 0
        for i in range(1000):
            for j in range(1000):
                if i == j:
                    y += 1
```

2. Write a function `my_max(x)` to return the maximum (largest) value in `x`. Do not use the built-in Python function `max`.

3. Write a function `my_n_max(x, n)` to return a list consisting of the `n` largest elements of `x`. You may use Python's `max` function. You may also assume that `x` is a one-dimensional list with no duplicate entries, and that `n` is a strictly positive integer smaller than the length of `x`

```
In [ ]: x = [7, 9, 10, 5, 8, 3, 4, 6, 2, 1]

        def my_n_max(x, n):
            # write your function code here

            return out

In [ ]: # Output = [10, 9, 8]
        out = my_n_max(x, n)
        print(out)
```

4. Let `m` be a matrix of positive integers. Write a function `my_trig_odd_even(m)` to return an array `q`, where `q[i, j] = sin(m[i, j])` if `m[i, j]` is even, and `q[i, j] = cos(m[i, j])` if `m[i, j]` is odd.

5. Let `P` be an $m \times p$ array and `Q` be a $p \times n$ array. As you will find later in this book, $M = P \times Q$ is defined as $M[i, j] = \sum_{k=1}^{p} P[i, k] \cdot Q[k, j]$. Write a function `my_mat_mult(P, Q)` that uses `for-loops` to compute `M`, the matrix product of `P * Q`. Hint: You may need up to three nested `for-loops`. Do not use the function `np.dot`.

```
In [ ]: import numpy as np

        def my_mat_mult(P, Q):
            # write your function code here

            return M

In [ ]: # Output:
        #  array([[3., 3., 3.],
        #         [3., 3., 3.],
        #         [3., 3., 3.]])

        P = np.ones((3, 3))
        my_mat_mult(P, P)

In [ ]: # Output:
        # array([[30, 30, 30],
        #        [70, 70, 70]])

        P = np.array([[1, 2, 3, 4], [5, 6, 7, 8]])
        Q = np.array([[1, 1, 1], [2, 2, 2], [3, 3, 3], [4, 4, 4]])
        my_mat_mult(P, Q)
```

6. The interest i on a principle, P_0, is a payment for allowing the bank to use your money. Compound interest is accumulated according to the formula $P_n = (1+i)P_{n-1}$, where n is the compounding period, usually in months or years. Write a function `my_saving_plan(P0, i, goal)` where the output is the number of years it will take `P0` to become `goal` at `i%` interest compounded annually.

```
In [ ]: def my_saving_plan(P0, i, goal):
            # write your function code here


            return years
```

```
In [ ]: # Output: 15
        my_saving_plan(1000, 0.05, 2000)
```

```
In [ ]: # Output: 11
        my_saving_plan(1000, 0.07, 2000)
```

```
In [ ]: # Output: 21
        my_saving_plan(500, 0.07, 2000)
```

7. Write a function with `my_find(M)` where the output is a list of indices `i` and where `M[i]` is 1. You may assume that `M` is a list of only ones and zeros. Do not use the built-in Python function `find`.

```
In [ ]: # Output: [0, 2, 3]

        M = [1, 0, 1, 1, 0]

        my_find(M)
```

8. Assume you are rolling two six-sided dice, with each side having an equal chance of occurring. Write a function `my_monopoly_dice()` where the output is the sum of the values of the two dice thrown but with the following extra rule: if the two dice rolls are the same, then another roll is made, and the new sum added to the running total. For example, if the two dice show 3 and 4, then the running total should be 7. If the two dice show 1 and 1, then the running total should be 2 plus the total of another throw. Rolls stop when the dice rolls are different.

9. A number is prime if it is divisible without remainder only by itself and 1. The number 1 is not a prime. Write a function `my_is_prime(n)` where the output is 1 if `n` is prime and 0 is otherwise. Assume that `n` is a strictly positive integer.

10. Write a function `my_n_primes(n)` where prime is a list of the first `n` primes. Assume that `n` is a strictly positive integer.

11. Write a function `my_n_fib_primes(n)` where the output `fib_primes` is a list of the first `n` numbers that are both a Fibonacci number and a prime. Note that 1 is not prime. Hint: Do not use the recursive implementation of Fibonacci numbers. A function to compute Fibonacci numbers is presented in Section 6.1. You may use the code freely.

```
In [ ]: def my_n_fib_primes(n):
            # write your function code here


            return fib_primes
```

```
In [ ]: # Output: [3, 5, 13, 89, 233, 1597, 28657, 514229]

        my_n_fib_primes(3)

In [ ]: # Output: [3, 5, 13]

        my_n_fib_primes(8)
```

12. Write a function `my_trig_odd_even(M)` where the output `Q[i,j] = sin(π/M[i,j])` if `M[i,j]` is odd, and `Q[i,j] = cos(π/M[i,j])` if `M[i,j]` is even. Assume that `M` is a two-dimensional array of strictly positive integers.

```
In [ ]: def my_trig_odd_even(M):
            # write your function code here

            return Q

In [ ]: # Output: [[0.8660, 0.7071], [0.8660, 0.4339]]
        M = [[3, 4], [6, 7]]
        my_trig_odd_even(M)
```

13. Let `C` be a square connectivity array containing zeros and ones. Point i has a connection to point j or i is connected to j if `C[i,j] = 1`. Note that connections in this context are one-directional, meaning `C[i,j]` is not necessarily the same as `C[j,i]`. For example, think of a one-way street from point A to point B. If A is connected to B, then B is not necessarily connected to A.
Write a function `my_connectivity_mat_2_dict(C, names)` where `C` is a connectivity array and `names` is a list of strings that denote the name of a point. That is, `names[i]` is the name of the name of the `i`th point.
The output variable `node` should be a dictionary with the key as the string in `names`, and value is a vector containing the indices `j`, such that `C[i,j] = 1`. In other words, it is a list of points that point `i` is connected to.

```
In [ ]: def my_connectivity_mat_2_dict(C, names):
            # write your function code here
            return node

In [ ]: C = [[0, 1, 0, 1], [1, 0, 0, 1], [0, 0, 0, 1], [1, 1, 1, 0]]
        names = ["Los Angeles", "New York", "Miami", "Dallas"]

In [ ]: # Output: node["Los Angeles"] = [2, 4]
        #         node["New York"] = [1, 4]
        #         node["Miami"] = [4]
        #         node["Dallas"] = [1, 2, 3]

        node = my_connectivity_mat_2_dict(C, names)
```

14. Turn the list `words` of lower case characters to upper case using the list comprehension.

```
In [ ]: words = ["test", "data", "analyze"]
```

CHAPTER

RECURSION

6

CONTENTS

6.1 RECURSIVE FUNCTIONS

A **recursive** function is a function that makes calls to itself. It works like the loops we described before. In some cases, however, it is preferable to use recursion than loops.

Every recursive function has two components: a **base case** and a **recursive step**. The **base case** is usually the smallest input and has an easily verifiable solution. This is also the mechanism that stops the function from calling itself forever. The **recursive step** is the set of all cases where a **recursive call**, or a function call to itself, is made.

As an example, we will demonstrate how recursion can be used to define and compute the factorial of an integer number. The factorial of an integer n is $1 \times 2 \times 3 \times \cdots \times (n-1) \times n$. The recursive definition can be written as follows:

$$f(n) = \begin{cases} 1 & \text{if } n = 1, \\ n \times f(n-1) & \text{otherwise.} \end{cases} \tag{6.1}$$

The base case is $n = 1$, which is trivial to compute: $f(1) = 1$. In the recursive step, n is multiplied by the result of a recursive call to the factorial of $n - 1$.

TRY IT! Write the factorial function using recursion. Use your function to compute the factorial of 3.

```
In [1]: def factorial(n):
            """Computes and returns the factorial of n,
            a positive integer.
            """
            if n == 1: # Base case!
```


Python Programming and Numerical Methods. https://doi.org/10.1016/B978-0-12-819549-9.00015-4

```
        return 1
    else: # Recursive step
        return n * factorial(n - 1) # Recursive call

In [2]: factorial(3)

Out[2]: 6
```

WHAT IS HAPPENING? First recall that when Python executes a function, it creates a workspace for the variables created in that function. Whenever a function calls another function, it will wait until that function returns an answer before continuing. In programming, this workspace is called a stack. Similar to a stack of plates in our kitchen cabinet, elements in a stack are added or removed from the top of the stack to the bottom, in a "last in, first out" order. For example, in the `np.sin(np.tan(x))`, `sin` must wait for `tan` to return an answer before it can be evaluated. Even though a recursive function makes calls to itself, the same rules apply.

1. A call is made to `factorial(3)`, whereby a new workspace is opened to compute `factorial(3)`.
2. Input argument value 3 is compared to 1. Since they are not equal, the "else" statement is executed.
3. `3*factorial(2)` must be computed. A new workspace is opened to compute `factorial(2)`.
4. Input argument value 2 is compared to 1. Since they are not equal, the "else" statement is executed.
5. `2*factorial(1)` must be computed. A new workspace is opened to compute `factorial(1)`.
6. Input argument value 1 is compared to 1. Since they are equal, the "if" statement is executed.
7. The return variable is assigned the value 1. `factorial(1)` terminates with output 1.
8. `2*factorial(1)` can be resolved to $2 \times 1 = 2$. The output is assigned the value 2. `factorial(2)` terminates with output 2.
9. `3*factorial(2)` can be resolved to $3 \times 2 = 6$. The output is assigned the value 6. Thus `factorial(3)` terminates with output 6.

The order of recursive calls can be depicted by a recursion tree shown in Fig. 6.1 for `factorial(3)`. A recursion tree is a diagram of the function calls connected by numbered arrows to depict the order in which the calls were made.

Fibonacci numbers were originally developed to model the idealized population growth of rabbits. Since then, they have been found to be significant in any naturally occurring phenomena. The Fibonacci numbers can be generated using the following recursive formula. Note that the recursive step contains two recursive calls and that there are also two base cases (i.e., two cases that cause the recursion to stop):

$$F(n) = \begin{cases} 1 & \text{if } n = 1, \\ 1 & \text{if } n = 2, \\ F(n-1) + F(n-2) & \text{otherwise.} \end{cases} \tag{6.2}$$

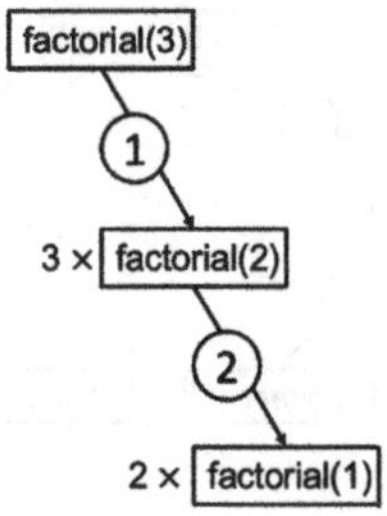

FIGURE 6.1

Recursion tree for `factorial(3)`.

TRY IT! Write a recursive function for computing the nth Fibonacci number. Use your function to compute the first five Fibonacci numbers. Draw the associated recursion tree.

```
In [3]: def fibonacci(n):
            """Computes and returns the Fibonacci of n,
            a positive integer.
            """
            if n == 1: # first base case
                return 1
            elif n == 2: # second base case
                return 1
            else: # Recursive step
                return fibonacci(n-1) + fibonacci(n-2)

In [4]: print(fibonacci(1))
        print(fibonacci(2))
        print(fibonacci(3))
        print(fibonacci(4))
        print(fibonacci(5))

1
1
2
3
5
```

As an exercise, consider the following modification to `fibonacci`, where the results of each recursive call are displayed to the screen (see Fig. 6.2).

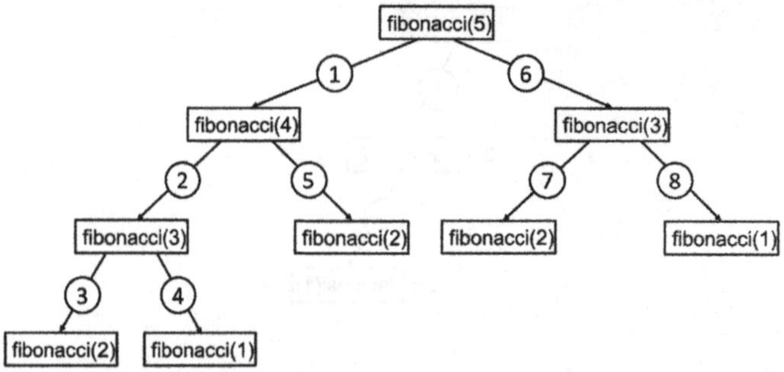

FIGURE 6.2

Recursion tree for `factorial(5)`.

EXAMPLE: Write a function `fib_disp(5)` based on modification of the `fibonacci` function. Can you determine the order in which the Fibonacci numbers will appear on the screen for `fib_disp(5)`?

```
In [5]: def fib_disp(n):
            """Computes and returns the Fibonacci of n,
            a positive integer.
            """
            if n == 1: # first base case
                out = 1
                print(out)
                return out
            elif n == 2: # second base case
                out = 1
                print(out)
                return out
            else: # Recursive step
                out = fib_disp(n-1) + fib_disp(n-2)
                print(out)
                return out # Recursive call
```

```
1
1
2
1
3
1
1
```

```
2
5

In [6]: fib_disp(5)

5
```

Note that the number of recursive calls becomes very large even for relatively small inputs for `n`. If you do not agree, try to draw the recursion tree for `fibonacci(10)`. If you try your unmodified function for inputs around 35, you will experience significant computation times.

```
In [7]: fibonacci(35)

Out[7]: 9227465
```

There is an iterative method for computing the `n`th Fibonacci number that requires only one workspace.

EXAMPLE: Iterative implementation for computing Fibonacci numbers.

```
In [8]: import numpy as np

        def iter_fib(n):
            fib = np.ones(n)

            for i in range(2, n):
                fib[i] = fib[i - 1] + fib[i - 2]

            return fib
```

TRY IT! Compute the `25`th Fibonacci number using `iter_fib` and `fibonacci`. Use the magic command `timeit` to measure the run time for each. Note the large difference in running times.

```
In [9]: %timeit iter_fib(25)
```

7.22 μs ± 171 ns per loop (mean ± std. dev. of 7 runs, 100000 loops each)

```
In [10]: %timeit fibonacci(25)
```

16.7 ms ± 910 μs per loop (mean ± std. dev. of 7 runs, 100 loops each)

You can see in the previous example that the iterative version runs much faster than the recursive counterpart. In general, iterative functions are faster than recursive functions performing the same task. So why use recursive functions at all? There are some solution methods that have a naturally recursive structure. In these cases, it is usually very hard to write a counterpart using loops. The primary value of writing recursive functions is that they can usually be written much more compactly than iterative functions. The cost of the improved compactness is added running time.

The relationship between the input arguments and the running time is discussed in more detail later in the chapter on complexity.

TIP! Try to write functions iteratively whenever it is convenient to do so. Your functions will run faster.

NOTE! When using a recursive call as shown above, we need to make sure that it can reach the base case, otherwise, it enters infinite recursion. In Python, executing a recursive function on a large output that cannot reach the base case will result in a "maximum recursion depth exceeded error". Try the following example to produce the `RecursionError`.

```
In []: factorial(5000)
```

We can work around the recursion limit using the `sys` module by setting a higher limit in Python. If you run the following code, the error message won't occur.

```
In []: import sys
        sys.setrecursionlimit(10**5)
        factorial(5000)
```

6.2 DIVIDE-AND-CONQUER

Divide-and-conquer is a useful strategy for solving difficult problems. Using divide-and-conquer, difficult problems are solved from solutions to many similar easy problems. In this way, difficult problems are broken up so they are more manageable. This section will present two classical examples of divide-and-conquer: the Tower of Hanoi Problem and the Quicksort algorithm.

6.2.1 TOWER OF HANOI

The Tower of Hanoi problem consists of three vertical rods (or towers) and N disks of different sizes, each with a hole in the center so that the rod can slide through it. The original configuration of the disks is that they are stacked on one of the towers in the order of descending size (i.e., the largest disc is on the bottom). The goal of the problem is to move all the disks to a different rod while complying with the following three rules:

1. Only one disk can be moved at a time.
2. Only the disk at the top of a stack may be moved.
3. A disk may not be placed on top of a smaller disk.

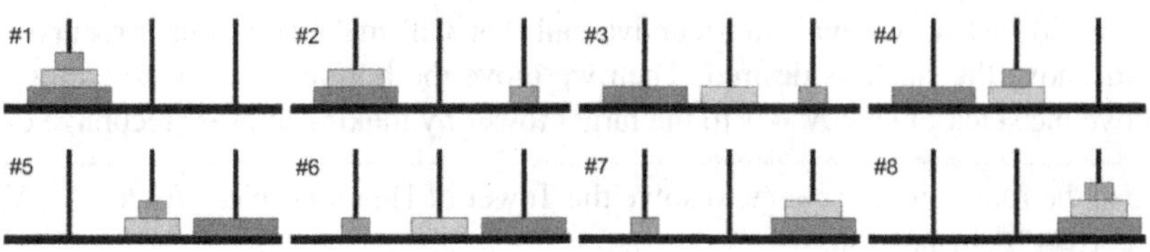

FIGURE 6.3

Illustration of the Tower of Hanoi: In eight steps, all disks are transported from pole 1 to pole 3, one at a time, by moving only the disk at the top of the current stack, and placing only smaller disks on top of larger disks.

Fig. 6.3 shows the steps of the solution to the Tower of Hanoi problem with three disks.

There is a legend saying that a group of Indian monks are in a monastery working to complete a Tower of Hanoi problem with 64 disks. When they complete the problem, the world will end. Fortunately, the number of moves required is $2^{64} - 1$, so even if they could move one disk per millisecond, it would take over 584 million years for them to finish.

The key to the Tower of Hanoi problem is breaking it down into smaller, easier-to-manage problems that we will refer to as **subproblems**. For this problem, it is relatively easy to see that moving a disk is easy (which has only three rules), but moving a tower is difficult, and the solution is not obvious. So we will assign moving a stack of size N to several subproblems of moving a stack of size $N - 1$.

Consider a stack of N disks that we wish to move from tower 1 to tower 3, and let `my_tower(N)` move a stack of size `N` to the desired tower (i.e., display the moves). How to write `my_tower` may not immediately be clear. If we think about the problem in terms of subproblems, we can see that we need to move the top $N - 1$ disks to the middle tower, then the bottom disk to the right tower, and then the $N - 1$ disks on the middle tower to the right tower. `my_tower` can display the instruction to move disk N and then make recursive calls to `my_tower(N-1)` to handle moving the smaller towers. The calls to `my_tower(N-1)` make recursive calls to `my_tower(N-2)`, and so on. A breakdown of the three steps is depicted in Fig. 6.4.

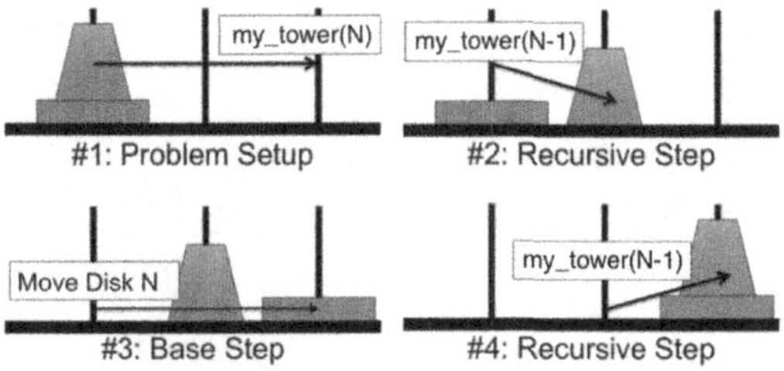

FIGURE 6.4

Breakdown of one iteration of the recursive solution of the Tower of Hanoi problem.

The code reproduced below is a recursive solution to the Tower of Hanoi problem. Note its compactness and simplicity. The code exactly reflects our intuition about the recursive nature of the solution: First we move a stack of size $N - 1$ from the original "from_tower" to the alternative "alt_tower". This is a difficult task, so instead we make a recursive call that will make subsequent recursive calls, but will, in the end, move the stack as desired. Then we move the bottom disk to the target "to_tower". Finally, we move the stack of size $N - 1$ to the target tower by making another recursive call.

TRY IT! Use the function `my_towers` to solve the Tower of Hanoi problem for `N = 3`. Verify the solution is correct by inspection.

```
In [1]: def my_towers(N, from_tower, to_tower, alt_tower):
            """
            Displays the moves required to move a tower of size
            N from the "from_tower" to the "to_tower".

            "from_tower", "to_tower" and "alt_tower" are
            either 1, 2, 3 referring to tower 1, 2, and 3.
            """

            if N != 0:
                # recursive call that moves N-1 stack from
                # starting tower to alternate tower
                my_towers(N-1, from_tower, alt_tower, to_tower)

                # display to screen movement of bottom disk from
                # starting tower to final tower
                print("Move disk %d from tower %d to tower %d."\
                          %(N, from_tower, to_tower))

                # recursive call that moves N-1 stack from
                # alternate tower to final tower
                my_towers(N-1, alt_tower, to_tower, from_tower)

In [2]: my_towers(3, 1, 3, 2)

Move disk 1 from tower 1 to tower 3.
Move disk 2 from tower 1 to tower 2.
Move disk 1 from tower 3 to tower 2.
Move disk 3 from tower 1 to tower 3.
Move disk 1 from tower 2 to tower 1.
Move disk 2 from tower 2 to tower 3.
Move disk 1 from tower 1 to tower 3.
```

By using divide-and-conquer, we have solved the Tower of Hanoi problem by making recursive calls to the slightly smaller Tower of Hanoi problems that, in turn, make recursive calls to yet smaller Tower of Hanoi problems. Together, the solutions form the solution to the whole problem. The actual work done by a single function call is actually quite small: two recursive calls and moving one disk. In other words, a function call does very little work (moving a disk), and then passes the rest of the work onto other calls, a skill you will find very useful throughout your engineering career.

6.2.2 QUICKSORT

A list of numbers, A, is **sorted** if the elements are arranged in ascending or descending order. Although there are many ways of sorting a list, **quicksort** is a divide-and-conquer approach, which is a very fast algorithm for sorting using a single processor (there are faster algorithms for multiple processors).

The *quicksort* algorithm starts with the observation that although sorting a list is hard, a comparison is relatively easy. So instead of sorting a list, we separate the list by comparing to a **pivot**. At each recursive call to *quicksort*, the input list is divided into three parts: elements that are smaller than the pivot, elements that are equal to the pivot, and elements that are larger than the pivot. Then a recursive call to *quicksort* is made on the two subproblems: the list of elements smaller than the pivot and the list of elements larger than the pivot. Eventually the subproblems are small enough (i.e., list size of length 1 or 0) so that sorting the list is now trivial.

Consider the following recursive implementation of *quicksort*.

```
In [3]: def my_quicksort(lst):

            if len(lst) <= 1:
                # list of length 1 is easiest to sort
                # because it is already sorted

                sorted_list = lst
            else:

                # select pivot as the first element of the list
                pivot = lst[0]

                # initialize lists for bigger and smaller
                # elements as well those equal to the pivot
                bigger = []
                smaller = []
                same = []

                # put elements into appropriate array

                for item in lst:
                    if item > pivot:
                        bigger.append(item)
                    elif item < pivot:
                        smaller.append(item)
                    else:
```

```
                same.append(item)

            sorted_list = my_quicksort(smaller) + same +
                          my_quicksort(bigger)

        return sorted_list

In [4]: my_quicksort([2, 1, 3, 5, 6, 3, 8, 10])

Out[4]: [1, 2, 3, 3, 5, 6, 8, 10]
```

As we did with the Tower of Hanoi problem, we have broken up the problem of sorting (hard) into many comparisons (easy).

6.3 SUMMARY AND PROBLEMS

6.3.1 SUMMARY

1. A recursive function is a function that calls itself.
2. Recursive functions are useful when problems have a hierarchical structure rather than an iterative structure.
3. Divide-and-conquer is a powerful problem-solving strategy that can be used to solve difficult problems.

6.3.2 PROBLEMS

1. Write a function `my_sum(lst)` where `lst` is a list, and the output is the sum of all the elements of `lst`. You can use a recursion or iteration function to solve the problem, but do not use Python's `sum` function.

```
In [ ]: def my_sum(lst):
            # Write your function code here

            return out

In [ ]: # Output: 6
        my_sum([1, 2, 3])

In [ ]: # Output: 5050
        my_sum(range(1,101))
```

2. Chebyshev polynomials are defined recursively and separated into two kinds: first and second. Chebyshev polynomials of the first kind, $T_n(x)$, and of the second kind, $U_n(x)$, are defined by the

following recurrence relations:

$$T_n(x) = \begin{cases} 1 & \text{if } n = 0, \\ x & \text{if } n = 1, \\ 2xT_{n-1}(x) - T_{n-2}(x) & \text{otherwise,} \end{cases} \tag{6.3}$$

$$U_n(x) = \begin{cases} 1 & \text{if } n = 0, \\ 2x & \text{if } n = 1, \\ 2xU_{n-1}(x) - U_{n-2}(x) & \text{otherwise.} \end{cases} \tag{6.4}$$

Write a function `my_chebyshev_poly1(n,x)` where the output `y` is the `n`th Chebyshev polynomial of the first kind evaluated at `x`. Be sure your function can take list inputs for `x`. You may assume that `x` is a list. The output variable, `y`, must be a list also.

```
In [ ]: def my_chebyshev_poly1(n,x):
            # Write your function code here

            return y

In [ ]: x = [1, 2, 3, 4, 5]

In [ ]: # Output: [1, 1, 1, 1, 1]
        my_chebyshev_poly1(0,x)

In [ ]: # Output: [1, 2, 3, 4, 5]
        my_chebyshev_poly1(1,x)

In [ ]: # Output: [1, 26, 99, 244, 485]
        my_chebyshev_poly1(3,x)
```

3. The Ackermann function, A, is a function quickly growing in popularity that is defined by the recursive relationship:

$$A(m, n) = \begin{cases} n+1 & \text{if } m = 0, \\ A(m-1, 1) & \text{if } m > 0 \text{ and } n = 1, \\ A(m-1, A(m, n-1)) & \text{if } m > 0 \text{ and } n > 0. \end{cases} \tag{6.5}$$

Write a function `my_ackermann(m,n)` where the output is the Ackermann function computed for `m` and `n`.

`my_ackermann(4,4)` is so large that it would be difficult to write down. Although the Ackermann function does not have many practical uses, the inverse Ackermann function has several uses in robotic motion planning.

```
In [ ]: def my_ackermann(m,n):
            # write your own function code here
            return out
```

```
In [ ]: # Output: 3
        my_ackermann(1,1)

In [ ]: # Output: 4
        my_ackermann(1,2)

In [ ]: # Output: 9
        my_ackermann(2,3)

In [ ]: # Output: 61
        my_ackermann(3,3)

In [ ]: # Output: 125
        my_ackermann(3,4)
```

4. The function, $C(n, k)$, computes the number of different ways of uniquely choosing k objects from n without repetition; it is commonly used in many statistics applications. For example, how many three-flavored ice cream sundaes are there if there are 10 ice-cream flavors? To solve this problem we would have to compute $C(10, 3)$, i.e., the number of ways of choosing three unique ice cream flavors from 10. The function C is commonly called "n choose k." You may assume that n and k are integers.
If $n = k$, then clearly $C(n, k) = 1$ because there is only way to choose n objects from n objects. If $k = 1$, then $C(n, k) = n$ because choosing each of the n objects is a way of choosing one object from n. For all other cases,

$$C(n, k) = C(n - 1, k) + C(n - 1, k - 1).$$

Can you see why?
Write a function `my_n_choose_k(n,k)` that computes the number of times k objects can be uniquely chosen from n objects without repetition.

```
In [ ]: def my_n_choose_k(n,k):
          # Write your own function code here
          return out

In [ ]: # Output: 10
        my_n_choose_k(10,1)

In [ ]: # Output: 1
        my_n_choose_k(10,10)

In [ ]: # Output: 120
        my_n_choose_k(10,3)
```

5. For any purchases paid in cash, the seller must return money that was overpaid. This is commonly referred to as "giving change." The bills and coins required to properly give change can be defined by a recursive relationship. If the amount paid is more than \$100 more than the cost, then a 100-dollar bill is returned, which is the result of a recursive call to the change function with \$100 subtracted from the amount paid. If the amount paid is more than \$50 over the cost of the item,

then a 50-dollar bill is returned, along with the result of a recursive call to the change function with \$50 subtracted. Similar clauses can be given for every denomination of US currency. The denominations of US currency, in dollars, are 100, 50, 20, 10, 5, 1, 0.25, 0.10, 0.05, and 0.01. For this problem, we will ignore the 2-dollar bill, which is not in common circulation. Assume that `cost` and `paid` are scalars and that `paid` ≥ `cost`. The output variable, `change`, must be a list, as shown in the test case.

Use recursion to program a function `my_change(cost, paid)` where `cost` is the cost of the item, `paid` is the amount paid, and the output `change` is a list of bills and coins that should be returned to the seller. Watch out for the base case!

```
In [ ]: def my_change(cost, paid):
            # Write your own function code here
            return change
```

```
In [ ]: #Output:[50.0,20.0,1.0,1.0,0.25,0.10,0.05,0.01,0.01,0.01]
        my_change(27.57, 100)
```

6. The golden ratio, ϕ, is the limit of $\frac{F(n+1)}{F(n)}$ as n goes to infinity and $F(n)$ is the nth Fibonacci number, which can be shown to be exactly $\frac{1+\sqrt{5}}{2}$ and is approximately 1.62. We say that $G(n) = \frac{F(n+1)}{F(n)}$ is the nth approximation of the golden ratio and $G(1) = 1$.
It can be shown that ϕ is also the limit of the continued fraction:

$$\varphi = 1 + \cfrac{1}{1 + \cfrac{1}{1 + \cfrac{1}{1 + \ddots}}}.$$

Write a recursive function with the header `my_golden_ratio(n)`, where the output is the nth approximation of the golden ratio according to the continued fraction recursive relationship. Use the continued fraction approximation for the golden ratio and not the $G(n) = F(n+1)/F(n)$ definition; however, for both definitions, $G(1) = 1$.
Studies have shown that rectangles with aspect ratio (i.e., length divided by width) close to the golden ratio are more pleasing to the eye than rectangles that are not. What is the aspect ratio of many wide-screen TVs and movie screens?

```
In [ ]: def my_golden_ratio(n):
            # Write your own function code here
            return out
```

```
In [ ]: # Output: 1.6181818181818182
        my_golden_ratio(10)
```

```
In [ ]: import numpy as np
        (1 + np.sqrt(5))/2
```

7. The greatest common divisor of two integers `a` and `b` is the largest integer that divides both numbers without remainder; the function to compute it is denoted by `gcd(a,b)`. The `gcd` function can be

written recursively. If b equals 0, then a is the greatest common divisor. Otherwise, gcd(a,b) = gcd(b,a%b) where a%b is the remainder of a divided by b. Assume that a and b are integers. Write a recursive function my_gcd(a,b) that computes the greatest common divisor of a and b. Assume that a and b are integers.

```
In [ ]: def my_gcd(a, b):
            # Write your own function code here
            return gcd

In [ ]: # Output: 2
        my_gcd(10, 4)

In [ ]: # Output: 11
        my_gcd(33, 121)

In [ ]: # Output: 1
        my_gcd(18, 1)
```

8. Pascal's triangle is an arrangement of numbers such that each row is equivalent to the coefficients of the binomial expansion of $(x+y)^{p-1}$, where p is some positive integer more than or equal to 1. For example, $(x+y)^2 = 1x^2 + 2xy + 1y^2$; therefore, the third row of Pascal's triangle is 1 2 1. Let R_m represent the mth row of Pascal's triangle and $R_m(n)$ be the nth element of the row. By definition, R_m has m elements, and $R_m(1) = R_m(n) = 1$. The remaining elements are computed using the following recursive relationship: $R_m(i) = R_{m-1}(i-1) + R_{m-1}(i)$ for $i = 2, \ldots, m-1$. The first few rows of Pascal's triangle are depicted in Fig. 6.5.

1
1 1
1 2 1
1 3 3 1
1 4 6 4 1
1 5 10 10 5 1

FIGURE 6.5

Pascal's triangle.

Write a function with my_pascal_row(m) where output variable row is the mth row of the Pascal triangle and must be a list. Assume that m is a strictly positive integer.

```
In [ ]: def my_pascal_row(m):
            # Write your own function code here
            return row

In [ ]: # Output: [1]
        my_pascal_row(1)

In [ ]: # Output: [1, 1]
        my_pascal_row(2)
```

```
In [ ]: # Output: [1, 2, 1]
        my_pascal_row(3)
```

```
In [ ]: # Output: [1, 3, 3, 1]
        my_pascal_row(4)
```

```
In [ ]: # Output: [1, 4, 6, 4, 1]
        my_pascal_row(5)
```

9. Consider an $n \times n$ matrix of the following form:

$$A = \begin{bmatrix} 1 & 1 & 1 & 1 & 1 \\ 1 & 0 & 0 & 0 & 0 \\ 1 & 0 & 1 & 1 & 0 \\ 1 & 0 & 0 & 1 & 0 \\ 1 & 1 & 1 & 1 & 0 \end{bmatrix}$$

where the ones form a right spiral. Write a function `my_spiral_ones(n)` that produces a $n \times n$ matrix of the given form. Make sure that the recursive steps are in the correct order (i.e., the ones go right, then down, then left, then up, then right, etc.).

```
In [ ]: def my_spiral_ones(n):
            # Write your own function code here
            return A
```

```
In [ ]: # Output: 1
        my_spiral_ones(1)
```

```
In [ ]: # Output:
        # array([[1, 1],
        #        [0, 1]])
        my_spiral_ones(2)
```

```
In [ ]: # Output:
        #array([[0, 1, 1],
        #       [0, 0, 1],
        #       [1, 1, 1]])
        my_spiral_ones(3)
```

```
In [ ]: # Output:
        #array([[1, 0, 0, 0],
        #       [1, 0, 1, 1],
        #       [1, 0, 0, 1],
        #       [1, 1, 1, 1]])
        my_spiral_ones(4)
```

```
In [ ]: # Output:
        #array([[1, 1, 1, 1, 1],
        #       [1, 0, 0, 0, 0],
```

```
#         [1, 0, 1, 1, 0],
#         [1, 0, 0, 1, 0],
#         [1, 1, 1, 1, 0]])
my_spiral_ones(5)
```

10. Rewrite `my_spiral_ones` without using recursion.

11. Draw the recursion tree for `my_towers(4)`.

12. Rewrite the Tower of Hanoi function in this chapter without using recursion.

13. Draw the recursion tree for `my_quicksort([5 4 6 2 9 1 7 3])`.

14. Rewrite the `quicksort` function in this chapter without using recursion.

CHAPTER

OBJECT-ORIENTED PROGRAMMING 7

CONTENTS

7.1 INTRODUCTION TO OOP

So far, all the codes we have written belong to the category of **procedure-oriented programming (POP)**, which consists of a list of instructions to tell the computer what to do; these instructions are then organized into functions. The program is divided into a collection of variables, data structures, and routines to accomplish different tasks. Python is a multiparadigm programming language, that is, it supports different programming approaches. One way to program in Python is to use **object-oriented programming (OOP)**. The learning curve is steeper, but it is extremely powerful and worth the time invested in mastering it. Note that you do not have to use OOP when programming in Python. You can still write very powerful programs using the POP. That said, the POP is good for simple and small programs, while the OOP is better suited for large programs. Let us take a closer look at object-oriented programming.

The object-oriented programming breaks the programming task into **objects**, which combine data (known as attributes) and **behaviors/functions** (known as methods). Thus, there are two main components of the OOP: **class** and **object**.

The class is a blueprint to define a logical grouping of data and functions. It provides a way to create data structures that model real-world entities. For example, we can create a `people` class that contains the data such as name, age, and some behavior functions to print out ages and genders of a group of people. While class is the blueprint, an object is an `instance` of the class with actual values.

Python Programming and Numerical Methods. https://doi.org/10.1016/B978-0-12-819549-9.00016-6

For example, a person named "Iron man" with age 35. Put it another way, a class is like a template to define the needed information, and an object is one specific copy that filled in the template. Also, objects instantiated from the same class are **independent** from each other. For example, if we have another person, say, "Batman" with age 33, it can be instantiated from the `people` class, but it remains an independent instance.

To implement the above example in Python, see the code below. Do not worry if you do not understand the syntax; the next section provides more helpful examples.

```
In [1]: class People():
            def __init__(self, name, age):
                self.name = name
                self.age = age

            def greet(self):
                print("Greetings, " + self.name)

In [2]: person1 = People(name = "Iron Man", age = 35)
        person1.greet()
        print(person1.name)
        print(person1.age)

Greetings, Iron Man
Iron Man
35


In [3]: person2 = People(name = "Batman", age = 33)
        person2.greet()
        print(person2.name)
        print(person2.age)

Greetings, Batman
Batman
33
```

In the above code example, we first defined a class, `People`, with `name` and `age` as the data, and a method `greet`. We then instantiated an object, `person1` with the specific name and age. We can clearly see that the `class` defines the whole structure, while the object is just an `instance` of the `class`.

The many benefits of using OOP are as follows: It provides a clear modular structure for programs that enhances code reusability. It provides a simple way to solve complex problems. It helps define more abstract data types to model real-world scenarios. It hides implementation details, leaving a clearly defined interface. It combines data and operations.

There are also other advantages of using OOP in a large project. We encourage you to search online to find out more. At this point, you may still not fully understand OOP's advantages until you are involved in complex large projects. We will continue to learn more about OOP during the course of this book, and its usefulness will become apparent.

7.2 CLASS AND OBJECT

The previous section introduced the two main components of OOP: *class*, which is a blueprint used to define a logical grouping of data and functions, and *object*, which is an instance of the defined class with actual values. In this section, we will get into greater detail of both of these components.

7.2.1 CLASS

A **class** is a definition of the structure that we want. Similar to a function, it is defined as a block of code, starting with the class statement. The syntax of defining a class is:

```
class ClassName(superclass):

    def __init__(self, arguments):
        # define or assign object attributes

    def other_methods(self, arguments):
        # body of the method
```

Note that the definition of a class is very similar to a function. It needs to be instantiated first before you can use it. For the class name, it is standard convention to use "CapWords." The **superclass** is used when you want create a new class to **inherit** the attributes and methods from another already defined class. We will talk more about **inheritance** in the next section. The `__init__` is one of the special methods in Python classes that is run as soon as an object of a class is instantiated (created). It assigns initial values to the object before it is ready to be used. Note the two underscores at the beginning and end of the `init`, indicating this is a special method reserved for special use in the language. In this `init` method, you can assign attributes directly when you create the object. The `other_methods` functions are used to define the instance methods that will be applied on the attributes, just like functions we discussed before. You may notice that there is an argument `self` for defining this method in the class. Why? A class instance method must have this extra argument as the first argument when you define it. This particular argument refers to the object itself; conventionally, we use `self` to name it. Instance methods can freely access attributes and other methods in the same object by using this `self` parameter. See the example below.

EXAMPLE: Define a class named `Student`, with the attributes `sid` (student id), `name`, `gender`, `type` in the `init` method, and a method called `say_name` to print out the student's name. All attributes will be passed in except `type`, which will have a value as "learning".

```
In [1]: class Student():

            def __init__(self, sid, name, gender):
                self.sid = sid
                self.name = name
                self.gender = gender
```

```
            self.type = "learning"

        def say_name(self):
            print("My name is " + self.name)
```

From the above example, we can see this simple class contains all the necessary parts mentioned previously. The `__init__` method will initialize the attributes when we create an object. We need to pass in the initial value for `sid`, `name`, and `gender`, while the attribute `type` is a fixed value as "learning". These attributes can be accessed by all the other methods defined in the class with `self.attribute`, for example, in the `say_name` method, we can use the `name` attribute with `self.name`. The methods defined in the class can be accessed and used in other different methods as well using `self.method`. Let us see the following example.

TRY IT! Add a method `report` that prints not only the student name, but also the student id. The method will have another parameter, `score`, that will pass in a number between 0 and 100 as part of the report.

```
In [2]: class Student():

            def __init__(self, sid, name, gender):
                self.sid = sid
                self.name = name
                self.gender = gender
                self.type = "learning"

            def say_name(self):
                print("My name is " + self.name)

            def report(self, score):
                self.say_name()
                print("My id is: " + self.sid)
                print("My score is: " + str(score))
```

7.2.2 OBJECT

As mentioned before, an **object** is an instance of the defined class with actual values. Many instances of different values associated with the class are possible, and each of these instances will be independent with each other (as seen previously). Also, after we create an object and call this instance method from the object, we do not need to give value to the `self` parameter because Python automatically provides it; see the following example.

EXAMPLE: Create two objects `("001", "Susan", "F")` and `("002", "Mike", "M")`, and call the method `say_name`.

```
In [3]: student1 = Student("001", "Susan", "F")
        student2 = Student("002", "Mike", "M")

        student1.say_name()
        student2.say_name()
        print(student1.type)
        print(student1.gender)

My name is Susan
My name is Mike
learning
F
```

In the above code, we created two objects, `student1` and `student2`, with two different sets of values. Each object is an instance of the `Student` class and has a different set of attributes. Type `student1.+TAB` to see the defined attributes and methods. To get access to one attribute, type `object.attribute`, e.g., `student1.type`. In contrast, to call a method, you need the parentheses because you are calling a function, such as `student1.say_name()`.

TRY IT! Call method `report` for student1 and student2 with scores of 95 and 90, respectively. Note that we do not need the "self" as an argument here.

```
In [4]: student1.report(95)
        student2.report(90)

My name is Susan
My id is: 001
My score is: 95
My name is Mike
My id is: 002
My score is: 90
```

We can see both methods calling to print out the data associated with the two objects. Note that the score value we passed in is only available to the method `report` (within the scope of this method). We can also see that the method `say_name` call in the `report` also works, as long as you call the method with the `self` in it.

7.2.3 CLASS VS INSTANCE ATTRIBUTES

The attributes we presented above are actually called instance attributes, which means that they are only belong to a specific instance; when you use them, you need to use the `self.attribute` within the class. There are other attributes called class attributes, which will be shared with all the instances created from this class. Let us see an example how to define and use a class attribute.

EXAMPLE: Modify the `Student` class to add a class attribute `n`, which will record how many object we are creating. Also, add a method `num_instances` to print out the number.

```
In [5]: class Student():

            n = 0

            def __init__(self, sid, name, gender):
                self.sid = sid
                self.name = name
                self.gender = gender
                self.type = "learning"
                Student.n += 1

            def say_name(self):
                print("My name is " + self.name)

            def report(self, score):
                self.say_name()
                print("My id is: " + self.sid)
                print("My score is: " + str(score))

            def num_instances(self):
                print(f"We have {Student.n}-instance in total")
```

In defining a class attribute, we must define it outside of all the other methods **without** using `self`. To use the class attributes, we use `ClassName.attribute`, which in this case is `Student.n`. This attribute will be shared with all the instances that are created from this class. Let us see the following code to show the idea.

```
In [6]: student1 = Student("001", "Susan", "F")
        student1.num_instances()
        student2 = Student("002", "Mike", "M")
        student1.num_instances()
        student2.num_instances()

We have 1-instance in total
We have 2-instance in total
We have 2-instance in total
```

As before, we created two objects, the instance attribute `sid`, `name`, but `gender` only belongs to the specific object. For example, `student1.name` is "Susan" and `student2.name` is "Mike". But when we print the class attribute `Student.n_instances` out after we created object `student2`, the one in the `student1` changes as well. This is the expectation we have for the class attribute because it is shared across all the created objects.

Now that we understand the difference between class and instance, we are in good shape to use basic OOP in Python. Before we can take full advantage of OOP, we still need to understand the concept of inheritance, encapsulation, and polymorphism. Let us start the next section!

7.3 INHERITANCE, ENCAPSULATION, AND POLYMORPHISM

We have already seen the modeling power of OOP using the class and object functions by combining data and methods. There are three more important concepts: (1) **inheritance**, which makes the OOP code more modular, easier to reuse, and capable of building a relationship between classes; (2) **encapsulation**, which can hide some of the private details of a class from other objects; and (3) **polymorphism**, which allows us to use a common operation in different ways. A brief discussion of how these concepts is discussed below.

7.3.1 INHERITANCE

Inheritance allows us to define a class that inherits all the methods and attributes from another class. Convention denotes the new class as **child class**, and the one that it inherits from is called the **parent class** or **superclass**. If we refer back to the definition of class structure, we can see the structure for basic inheritance is `class ClassName(superclass)`, which means the new class can access all the attributes and methods from the superclass. Inheritance builds a relationship between the child and parent classes. Usually, the parent class is a general type while the child class is a specific type. An example is presented below.

TRY IT! Define a class named `Sensor` with attributes `name`, `location`, and `record_date` that pass from the creation of an object and an attribute `data` as an empty dictionary to store data. Create one method `add_data` with `t` and `data` as input parameters to take in timestamp and data arrays. Within this method, assign `t` and `data` to the `data` attribute with "time" and "data" as the keys. In addition, create one `clear_data` method to delete the data.

```
In [1]: class Sensor():
            def __init__(self, name, location, record_date):
                self.name = name
                self.location = location
                self.record_date = record_date
                self.data = {}

            def add_data(self, t, data):
                self.data["time"] = t
```

```
            self.data["data"] = data
            print(f"We have {len(data)} points saved")

        def clear_data(self):
            self.data = {}
            print("Data cleared!")
```

Now we have a class to store general sensor information, we can create a sensor object to store data.

EXAMPLE: Create a sensor object.

```
In [2]: import numpy as np

        sensor1 = Sensor("sensor1", "Berkeley", "2019-01-01")
        data = np.random.randint(-10, 10, 10)
        sensor1.add_data(np.arange(10), data)
        sensor1.data

We have 10 points saved


Out[2]: {"time": array([0, 1, 2, 3, 4, 5, 6, 7, 8, 9]),
          "data": array([ 3, 2, 5, -1, 2, -2, 6, -1, 5, 4])}
```

7.3.1.1 *Inheriting and Extending New Method*

Suppose we have one different type of sensor, an accelerometer. It shares the same attributes and methods as `Sensor` class, but it also has different attributes or methods that need to be appended or modified from the original class. What should we do? Do we create a different class from scratch? This is where inheritance can be used to make life easier. This new class will inherit from the `Sensor` class with all the attributes and methods. We can think whether we want to extend the attributes or methods. Let us first create this new class, `Accelerometer`, and add a new method, `show_type`, to report what kind of sensor it is.

```
In [3]: class Accelerometer(Sensor):


            def show_type(self):
                print("I am an accelerometer!")


        acc = Accelerometer("acc1", "Oakland", "2019-02-01")
        acc.show_type()
        data = np.random.randint(-10, 10, 10)
        acc.add_data(np.arange(10), data)
        acc.data
```

```
I am an accelerometer!
We have 10 points saved


Out[3]: {"time": array([0, 1, 2, 3, 4, 5, 6, 7, 8, 9]),
         "data": array([ -1, 4, 7, -10, -2, -6, 2, -8, 9, 3])}
```

Creating this new **Accelerometer** class is very simple. It inherits from `Sensor` (denoted as a superclass), and the new class actually contains all the attributes and methods from the superclass. We then add a new method, `show_type`, which does not exist in the `Sensor` class, but we can successfully extend the child class by adding the new method. This shows the power of inheritance: we have reused most part of the `Sensor` class in a new class, and extended the functionality. Basically, the inheritance sets up a logical relationship for the modeling of the real-world entities: the `Sensor` class as the parent class is more general and passes all the characteristics to the child class `Accelerometer`.

7.3.1.2 Inheriting and Method Overriding

When we inherit from a parent class, we can change the implementation of a method provided by the parent class; this is called method overriding and is shown in the example below.

EXAMPLE: Create a class `UCBAcc` (a specific type of accelerometer that was created at UC Berkeley) that inherits from `Accelerometer` but replaces the `show_type` method that also prints out the name of the sensor.

```
In [4]: class UCBAcc(Accelerometer):

            def show_type(self):
                print(f"I am {self.name}, created at Berkeley!")

        acc2 = UCBAcc("UCBAcc", "Berkeley", "2019-03-01")
        acc2.show_type()

I am UCBAcc, created at Berkeley!
```

We see that our new `UCBAcc` class actually overrides the method `show_type` with new features. In this example, we are not only inheriting features from our parent class, but we are also modifying/improving some methods.

7.3.1.3 Inheriting and Updating Attributes With Super

Let us create a class **NewSensor** that inherits from `Sensor` class, but is updated the attributes by adding a new attribute `brand`. Of course, we can redefine the whole `__init__` method as shown below that is capable of overriding the parent function.

```
In [5]: class NewSensor(Sensor):
            def __init__(self,name,location,record_date,brand):
                self.name = name
                self.location = location
                self.record_date = record_date
                self.brand = brand
                self.data = {}

        new_sensor = NewSensor("OK", "SF", "2019-03-01", "XYZ")
        new_sensor.brand

Out[5]: "XYZ"
```

There is a better way to achieve the same result. If we use the `super` method, we can avoid referring to the parent class explicitly, as shown in the following example:

EXAMPLE: Redefine the attributes in inheritance.

```
In [6]: class NewSensor(Sensor):
            def __init__(self,name,location,record_date,brand):
                super().__init__(name, location, record_date)
                self.brand = brand

        new_sensor = NewSensor("OK", "SF", "2019-03-01", "XYZ")
        new_sensor.brand

Out[6]: "XYZ"
```

Now we can see with the `super` method, we have avoided listing all of the definition of the attributes; this helps keep your code maintainable for the foreseeable future. Because the child class does not implicitly call the `__init__` of the parent class, we must use `super().__init__`, as shown above.

7.3.2 ENCAPSULATION

Encapsulation is one of the fundamental concepts in OOP. It describes the idea of restricting access to methods and attributes in a class. Encapsulation hides complex details from the users and prevents data being modified by accident. In Python, this is achieved by using private methods or attributes using the underscore as prefix, i.e., single "_" or double "__", as shown the following example.

EXAMPLE:

```
In [7]: class Sensor():
    def __init__(self, name, location):
        self.name = name
```

```
        self._location = location
        self.__version = "1.0"

    # a getter function
    def get_version(self):
        print(f"The sensor version is {self.__version}")

    # a setter function
    def set_version(self, version):
        self.__version = version
```

```
In [8]: sensor1 = Sensor("Acc", "Berkeley")
print(sensor1.name)
print(sensor1._location)
print(sensor1.__version)

Acc
Berkeley


    ---------------------------------------------------------


    AttributeError          Traceback (most recent call last)

    <ipython-input-8-ca9b481690ba> in <module>
      2 print(sensor1.name)
      3 print(sensor1._location)
----> 4 print(sensor1.__version)


    AttributeError: 'Sensor' object has no attribute '__version'
```

The above example shows how the encapsulation works. With single underscore, we defined a private variable that should not be accessed directly. Note that this is convention and nothing stops you from actually accessing it. With double underscores, we can see that the attribute `__version` cannot be accessed or modified directly. To get access to the double underscore attributes, we need to use "getter" and "setter" functions to access it internally. A "getter" function is shown in the following example.

```
In [9]: sensor1.get_version()

The sensor version is 1.0
```

```
In [10]: sensor1.set_version("2.0")
sensor1.get_version()

The sensor version is 2.0
```

The single and double underscore(s) apply to private methods as well, which are not discussed because they are similar to the private attributes.

7.3.3 POLYMORPHISM

Polymorphism is another fundamental concept in OOP, which means multiple forms. Polymorphism allows the use of a single interface with different underlying forms, such as data types or classes. For example, we can have commonly named methods across classes or child classes. We have already seen one example above when we overrode the method `show_type` in the `UCBAcc`. Both the parent class `Accelerometer` and child class `UCBAcc` have a method named `show_type`, but they are implemented differently. This ability of using a single name with many forms acting differently in different situations greatly reduces our complexities. We will not expand this discussion on polymorphism, if you are interested, check more online to get a deeper understanding.

7.4 SUMMARY AND PROBLEMS

7.4.1 SUMMARY

1. OOP and POP are different. OOP has many benefits and is often more appropriate for use in large-scale projects.
2. Class is the blueprint of the structure that allows us to group data and methods, while object is an instance from the class.
3. The concept of "inheritance" is key to OOP, which allows us to refer attributes or methods from the superclass.
4. The concept of "encapsulation" allows us to hide some of the private details of a class from other objects.
5. The concept of "polymorphism" allows us to use a common operation in different ways for different data input.

7.4.2 PROBLEMS

1. Describe the differences between classes and objects.
2. Describe why we use "self" as the first argument in a method.
3. What is a constructor? And why do we use it?
4. Describe the differences between class and instance attributes.
5. The following is a definition of the class `Point` that takes in the coordinates `x`, `y`. Add a method `plot_point` that plots the position of a point.

```
import matplotlib.pyplot as plt

class Point():
    def __init__(self, x, y):
        self.x = x
        self.y = y
```

6. Use the class from Problem 5 and add a method `calculate_dist` which takes in `x` and `y` from another point, and returns the distance calculated between the two points.
7. What's inheritance?
8. How do we inherit from a superclass and add new methods?
9. When we inherit from a superclass, we need to replace a method with a new one; how do we do that?
10. What's the super method? Why do we need it?
11. Create a class to model some real world objects and create a new class to inherit from it. See the example below. You should use a different example and incorporate as many of the concepts we've learned so far as possible.

```
In [1]: class Car():
            def __init__(self, brand, color):
                self.brand = brand
                self.color = color

            def start_my_car(self):
                print("I am ready to drive!")

        class Truck(Car):
            def __init__(self, brand, color, size):
                super().__init__(brand, color)
                self.size = size

            def start_my_car(self, key):
                if key == "truck_key":
                    print("I am ready to drive!")
                else:
                    print("Key is not right")

            def stop_my_car(self, brake):
                if brake:
                    print("The engine is stopped!")
                else:
                    print("I am still running!")
```

```
truck1 = Truck("Toyota", "Silver", "Large")
truck1.start_my_car("truck_key")
truck1.stop_my_car(brake = False)
```

CHAPTER

COMPLEXITY

8

CONTENTS

8.1 COMPLEXITY AND BIG-O NOTATION

The **complexity** of a function is the relationship between the size of the input and the difficulty of running the function to completion. The size of the input is usually denoted by n. However, n usually describes something more tangible, such as the length of an array. The difficulty of a problem can be measured in several ways. One suitable way to describe the difficulty of the problem is to use **basic operations**: additions, subtractions, multiplications, divisions, assignments, and function calls. Although each basic operation takes different amount of time, the number of basic operations needed to complete a function is sufficiently related to the running time to be useful, and it is much easier to count.

TRY IT! Count the number of basic operations, in terms of n, required for the following function to terminate:

```
In [1]: def f(n):
            out = 0
            for i in range(n):
                for j in range(n):
                    out += i*j

            return out
```

Let us calculate the number of operations:
n^2 additions; 0 subtractions; n^2 multiplications; 0 divisions; $2n^2 + 1$ assignments; 0 function calls; $4n^2 + 1$ in total.

Python Programming and Numerical Methods. https://doi.org/10.1016/B978-0-12-819549-9.00017-8

The number of assignments is $2n^2 + n + 1$ because the line `out += i*j` is evaluated n^2 times, `j` is assigned n^2 times, `i` is assigned n times, and the line `out = 0` is assigned once. So, the complexity of the function `f` can be described as $4n^2 + n + 1$.

A common notation for complexity is called **Big-O notation**. Big-O notation establishes the relationship in the growth of the number of basic operations with respect to the size of the input as the input size becomes very large. Because hardware may be different on every machine, we cannot accurately calculate how long it will take to complete without also evaluating the hardware, which is only valid for that specific machine. How long it takes to calculate a specific set of input on a specific machine is not germane. What is germane is the "time to completion." Because this type of analysis is hardware independent, the basic operations grow in direct response to the increase in the size of the input. As n gets large, the highest power dominates; therefore, only the highest power term is included in Big-O notation. Additionally, coefficients are not required to characterize growth, and so coefficients are also dropped. In the previous example, we counted $4n^2 + n + 1$ basic operations to complete the function. In Big-O notation we would say that the function is $O(n^2)$ (pronounced "O of n-squared"). We say that any algorithm with complexity $O(n^c)$ where c is some constant with respect to n is **polynomial time**.

TRY IT! Determine the complexity of the iterative Fibonacci function in Big-O notation.

```
In [2]: def my_fib_iter(n):

            out = [1, 1]

            for i in range(2, n):
                out.append(out[i - 1] + out[i - 2])

            return out
```

Since the only lines of code that take more time as n grows are those in the for-loop, we can restrict our attention to the for-loop and the code block within it. The code within the for-loop does not grow with respect to n (i.e., it is constant). Therefore, the number of basic operations is Cn, where C is some constant representing the number of basic operations that occur in the for-loop, and these C operations run n times. This gives a complexity of $O(n)$ for `my_fib_iter`.

Assessing the exact complexity of a function can be difficult. In these cases, it might be sufficient to give an upper bound or even an approximation of the complexity.

TRY IT! Give an upper bound on the complexity of the recursive implementation of Fibonacci. Do you think it is a good approximation of the upper bound? Do you think that recursive Fibonacci can possibly be polynomial time?

```
In [3]: def my_fib_rec(n):

            if n < 2:
                out = 1
```

```
    else:
        out = my_fib_rec(n-1) + my_fib_rec(n-2)

    return out
```

As n gets large, we can say that the vast majority of function calls make two other function calls: one addition and one assignment to the output. The addition and assignment do not grow with n per function call, so we can ignore them in Big-O notation. However, the number of function calls grows approximately by 2^n, and so the complexity of `my_fib_rec` is upper bounded by $O(2^n)$.

There is an on-going debate whether or not $O(2^n)$ is a good approximation for the Fibonacci function.

Since the number of recursive calls grows exponentially with n, there is no way the recursive Fibonacci function can be polynomial. That is, for any c, there is an n such that `my_fib_rec` takes more than $O(n^c)$ basic operations to complete. Any function that is $O(c^n)$ for some constant c is said to be **exponential time.**

TRY IT! What is the complexity of the following function in Big-O notation?

```
In [4]: def my_divide_by_two(n):

            out = 0
            while n > 1:
                n /= 2
                out += 1

            return out
```

Again, only the while-loop runs longer for larger n, so we can restrict our attention there. Within the while-loop, there are two assignments: one division and one addition, both of which are constant time with respect to n. So the complexity depends only on how many times the while-loop runs.

The while-loop cuts n in half in every iteration until n is less than 1. So the number of iterations, I, is the solution to the equation $\frac{n}{2^I} = 1$. With some manipulation, the solution is $I = \log n$, so the complexity of `my_divide_by_two` is $O(\log n)$. If we recall log rules, it does not matter what the base of the log is because all logs are a scalar multiple of each other. Any function with complexity $O(\log n)$ is said to be **log-time**.

8.2 COMPLEXITY MATTERS

So why does complexity matter? Because differing complexities require different amounts of time to complete the task. Fig. 8.1 is a quick sketch showing you how the time changes with different input size for complexity $\log(n)$, n, n^2.

Let us look at another example. Assume you have an algorithm that runs in exponential time, say $O(2^n)$, and let N be the largest problem you can solve with this algorithm using the computational

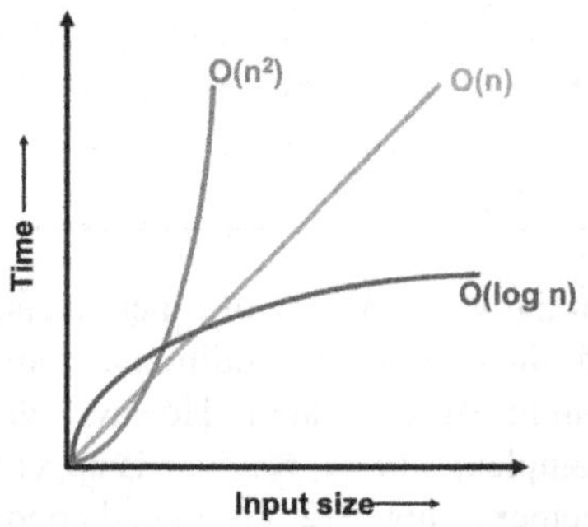

FIGURE 8.1

Illustration of running time for complexity $\log(n)$, n, and n^2.

resources you have, denoted by R. This R can be the amount of time you are willing to wait for the function to finish, or R can be the number of basic operations you watch the computer execute before you get sick of waiting. Using the same algorithm, how large of a problem can you solve given a new computer that is twice as fast?

If we establish $R = 2^N$, using our old computer, with our new computer we have $2R$ computational resources; therefore, we want to find N' such that $2R = 2^{N'}$. With some substitution, we can arrive at $2 \times 2^N = 2^{N'} \to 2^{N+1} = 2^{N'} \to N' = N + 1$. So with an exponential time algorithm, doubling your computational resources will allow you to solve a problem one unit larger than you can with your old computer. This is a very small difference. In fact, as N gets large, the relative improvement goes to 0.

With a polynomial time algorithm, you can do much better. This time let us assume that $R = N^c$, where c is some constant larger than one. Then $2R = N'^c$, which, if you use similar substitutions as before, will result in $N' = 2^{1/c}N$. So with a polynomial time algorithm with power c, you can solve a problem $\sqrt[c]{2}$ larger than you can with your old computer. When c is small, say less than 5, this is a much bigger difference than with the exponential algorithm.

Finally, let us consider a log-time algorithm. Let $R = \log N$. Then $2R = \log N'$; again with some substitution we obtain $N' = N^2$. So with the double resources, we can square the size of the problem we can solve!

The moral of the story is that exponential time algorithms do not scale well. That is, as you increase the size of the input, you will soon find that the function takes longer (much longer) than you are willing to wait. For one final example, `my_fib_rec(100)` would take on the order 2^{100} basic operations to complete the computation. If your computer can do 100 trillion basic operations per second (far faster than the fastest computer on earth), it would take your computer about 400 million years to complete; however, using `my_fib_iter(100)` would take less than 1 nanosecond to complete the same task.

There is both an exponential time algorithm (recursion) and a polynomial time algorithm (iteration) for computing Fibonacci numbers. Given a choice, we would clearly pick the polynomial time algorithm. However, there is a class of problems for which no one has ever discovered a polynomial time algorithm. In other words, there are only exponential time algorithms known for them. These problems are known as NP-complete; an ongoing investigation is trying to determine whether polynomial time algorithms exist for these problems. Examples of NP-complete problems include the Traveling Salesman, Set Cover, and Set Packing problems. Although theoretical in construction, solutions to these problems have numerous applications in logistics and operations research. In fact, some encryption algorithms that keep web and bank applications secure rely on the NP-completeness of breaking them. A further discussion of NP-complete problems and the theory of complexity is beyond the scope of this book, but these problems are very interesting and important to many engineering applications.

8.3 THE PROFILER

8.3.1 USING THE MAGIC COMMAND

Even if it does not change the Big-O complexity of a program, many programmers will spend long hours to make their code run twice as fast or to gain even smaller improvements.

There are ways to check the run time of the code in the Jupyter notebook. Introduced below are the `magic commands` to do that:

- `%time`: Get the run time of a single statement.
- `%timeit`: Get the repeated run time of a single statement.
- `%%time`: Get the run time of all the code in the cell.
- `%%timeit`: Get the repeated run time of a cell.

Note that the double percent magic command will measure the run time for all the code in a cell, while the single percent command only works for a single statement.

```
In [1]: %time sum(range(200))

    CPU times: user 6 µs, sys: 1 µs, total: 7 µs, wall time: 9.06 µs

Out[1]: 19900

In [2]: %timeit sum(range(200))

    1.24 µs ± 70.6 ns per loop (mean ± std. dev. of 7 runs, 1000000 loops each)

In [3]: %%time
        s = 0
        for i in range(200):
            s += i

    CPU times: user 15 µs, sys: 0 ns, total: 15 µs, wall time: 17.9 µs

In [4]: %%timeit
        s = 0
```

```
    for i in range(200):
        s += i
```

7.06 μs ± 414 ns per loop (mean ± std. dev. of 7 runs, 100000 loops each)

WARNING! Sometimes it may not be proper to use the `timeit`, since it will run many loops for the code, taking an inordinate amount of time to complete the task.

8.3.2 USE PYTHON PROFILER

You can also use the Python **profiler** (for additional discussion, read more in the Python documentation) to profile the code you write. In `Jupyter notebook`,

- `%prun`: Run a single statement through the Python code profiler.
- `%%prun`: Run a cell through the Python code profiler.

The following example sums random numbers over and over again:

```
In [6]: import numpy as np

In [7]: def slow_sum(n, m):

            for i in range(n):
                # we create a size m array of random numbers
                a = np.random.rand(m)

                s = 0
                # in this loop we iterate through the array
                # and add elements to the sum one by one
                for j in range(m):
                    s += a[j]

In [8]: %prun slow_sum(1000, 10000)
```

The results are shown in Fig. 8.2.

The table shows the following columns (from Python profiler):

- **ncalls** is the number of calls;
- **tottime** is the total time spent in performing the given function (and excluding time made in calls to subfunctions);
- **percall** is the quotient of dividing tottime by ncalls;
- **cumtime** is the total time spent in this and all subfunctions (from invocation till exit). This figure is accurate even for recursive functions;
- **percall** is the quotient of dividing cumtime by primitive calls.

```
         1004 function calls in 1.413 seconds

   Ordered by: internal time

   ncalls  tottime  percall  cumtime  percall filename:lineno(function)
        1    1.320    1.320    1.413    1.413 <ipython-input-20-cc5de53096ac>:1(slow_sum)
     1000    0.093    0.000    0.093    0.000 {method 'rand' of 'mtrand.RandomState' objects}
        1    0.000    0.000    1.413    1.413 {built-in method builtins.exec}
        1    0.000    0.000    1.413    1.413 <string>:1(<module>)
        1    0.000    0.000    0.000    0.000 {method 'disable' of '_lsprof.Profiler' objects}
```

FIGURE 8.2

The profiling result from prun.

8.3.3 USE LINE PROFILER

Many times we want to determine which line in a code script takes a long time so that we can rewrite this line to make it more efficient. This can be done using the `line_profiler`, which will profile the code line by line. This function is not shipped with Python; therefore, we need to install it, and then we can use the magic command:

- `%lprun`: Run the line by line profile on a single statement.

```
In [9]: # Note, you only need run this once.
        !conda install line_profiler
```

After you have installed this package, load the `line_profiler` extension:

```
In [10]: %load_ext line_profiler
```

The way we use the `line_profiler` to profile the code is shown as follows:

```
In [11]: %lprun -f slow_sum slow_sum(1000, 10000)
```

Running the above command will perform line by line profiling, as shown in Fig. 8.3.

The results include a summary for each line of the function. Note that lines 10 and 11 take the majority of the total running time.

Usually when code takes longer to run than you would like, there is a **bottleneck** where a majority of the time is being spent. That is, there is a line of code that is taking much longer to execute than the other lines in the program. Addressing the bottleneck in a program will usually lead to the biggest improvement in performance, even if there are other areas of your code that are more easily improved.

TIP! Start at the bottleneck when improving the performance of a code.

```
Timer unit: 1e-06 s

Total time: 6.1411 s
File: <ipython-input-20-cc5de53096ac>
Function: slow_sum at line 1

Line #      Hits         Time  Per Hit   % Time  Line Contents
==============================================================
     1                                           def slow_sum(n, m):
     2
     3      1001        301.0      0.3      0.0      for i in range(n):
     4                                                   # we create a size m array of random numbers
     5      1000      87876.0     87.9      1.4          a = np.random.rand(m)
     6
     7      1000        439.0      0.4      0.0          s = 0
     8                                                   # in this loop we iterate through the array
     9                                                   # and add elements to the sum one by one
    10  10001000    2463579.0      0.2     40.1          for j in range(m):
    11  10000000    3588901.0      0.4     58.4              s += a[j]
```

FIGURE 8.3

The line by line profiling result from `line_profiler`.

8.4 SUMMARY AND PROBLEMS

8.4.1 SUMMARY

1. The complexity of an algorithm is the relationship between the size of the input problem and the time it takes for the algorithm to terminate.
2. Big-O notation is a standard method of classifying algorithmic complexity in a way that is computer- and operating-system-independent.
3. Algorithms with log-complexity perform faster than algorithms with polynomial complexity, which are faster than algorithms with exponential complexity.
4. The Python profiler is a useful tool for determining where your code is running slowly so that you can improve its performance.

8.4.2 PROBLEMS

1. How would you define the size of the following tasks?

   - Solving a jigsaw puzzle.
   - Passing a handout to a class.
   - Walking to class.
   - Finding a name in a dictionary.

2. For the tasks given in the previous problem, what would you say is the Big-O complexity of the tasks in terms of the size definitions you gave?
3. You may be surprised to know that there is a log-time algorithm for finding a word in an n-word list. Instead of starting at the beginning of the list, you go to the middle. If this is the word you are looking for, then you are done. If the word comes after the word you are looking for, then look halfway between the current word and the end. If it is before the word you are looking for, then look

halfway between the first word and the current word. Keep repeating this process until you find the word. This algorithm is known as a binary search. It runs in log time because the search space is cut in half at each iteration; therefore, it requires, at most, $\log_2(n)$ iterations to find the word. Hence the increase in run time is only a log in the length of the list.
There is a way to look up a word in $O(1)$ or constant time. This means that no matter how long the list is, it takes the same amount of time! Can you think of how this is done? Hint: Research hash functions.

4. What is the complexity of the algorithms that compute the following recursive relationships? Classify the following algorithms as log time, polynomial time, or exponential time in terms of n given that the implementation is (a) recursive and (b) iterative.

Tribonacci, $T(n)$:

$$T(n) = T(n-1) + T(n-2) + T(n-3)$$
$$T(1) = T(2) = T(3) = 1.$$

Timmynacci, $t(n)$:

$$t(n) = t(n/2) + t(n/4)$$
$$t(n) = 1 \; if \; n < 1.$$

5. What is the Big-O complexity of the Towers of Hanoi problem given in Chapter 6? Is the complexity an upper bound or is it exact?

6. What is the Big-O complexity of the quicksort algorithm?

7. Run the following two iterative implementations of finding Fibonacci numbers in the `line_profiler` as well as using the magic command to get the repeated run time. The first implementation preallocates memory to an array that stores all the Fibonacci numbers. The second implementation expands the list at each iteration of the for-loop.

```
In [ ]: import numpy as np

        def my_fib_iter1(n):
            out = np.zeros(n)

            out[:2] = 1

            for i in range(2, n):
                out[i] = out[i-1] + out[i-2]

            return out

        def my_fib_iter2(n):

            out = [1, 1]
```

```
    for i in range(2, n):
        out.append(out[i-1]+out[i-2])

    return np.array(out)
```

CHAPTER

REPRESENTATION OF NUMBERS 9

CONTENTS

9.1 BASE-N AND BINARY

The **decimal system** is a way of representing numbers that you were introduced to in elementary school. In the decimal system, a number is represented by a list of digits from 0 to 9, where each digit represents the coefficient for a power of 10.

> **EXAMPLE:** Show the decimal expansion for 147.3.
> $147.3 = 1 \cdot 10^2 + 4 \cdot 10^1 + 7 \cdot 10^0 + 3 \cdot 10^{-1}$.

Since each digit is associated with a power of 10, the decimal system is also known as **base10** because it is based on 10 digits (0 to 9). There is nothing special about base10 numbers other than you are more accustomed to using them. For example, in base3 we have the digits 0, 1, and 2 and the number $121(\text{base3}) = 1 \cdot 3^2 + 2 \cdot 3^1 + 1 \cdot 3^0 = 9 + 6 + 1 = 16(\text{base10})$

For the purpose of this chapter, it is useful to denote a number's representation, i.e., every number will be followed by its representation in parentheses (e.g., 11(base10) means 11 in base10) unless the context is clear.

For computers, numbers are often represented in base2 or **binary** numbers. In binary, the only available digits are 0 and 1, and each digit is the coefficient of a power of 2. Digits in a binary number are also known as **bits**. Note that binary numbers are still numbers, and the processes of adding and multiplying them are exactly as you learned in grade school.

Python Programming and Numerical Methods. https://doi.org/10.1016/B978-0-12-819549-9.00018-X

TRY IT! Convert the number 11(base10) into binary. 11(base10) $= 8+2+1 = 1\cdot 2^3 + 0\cdot 2^2 + 1\cdot 2^1 + 1\cdot 2^0 =$ 1011(base2)

TRY IT! Convert 37(base10) and 17(base10) to binary. Add and multiply the resulting numbers in binary. Verify that the result is correct in base10.

Converting to binary:

37(base10) $= 32+4+1 = 1\cdot 2^5 + 0\cdot 2^4 + 0\cdot 2^3 + 1\cdot 2^2 + 0\cdot 2^1 + 1\cdot 2^0 =$ 100101(base2),

17(base10) $= 16+1 = 1\cdot 2^4 + 0\cdot 2^3 + 0\cdot 2^2 + 0\cdot 2^1 + 1\cdot 2^0 =$ 10001(base2).

Obtaining the results of addition and multiplication in decimal:

$37 + 17 = 54$,

$37 \times 17 = 629$.

Performing addition in binary (see Fig. 9.1).

Performing multiplication in binary (see Fig. 9.2).

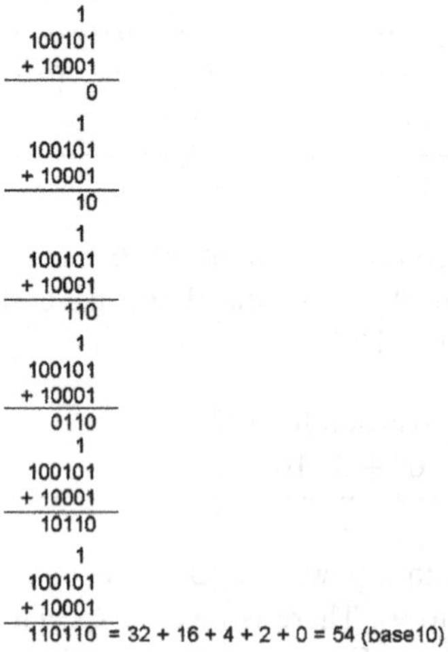

FIGURE 9.1

Binary addition.

Binary numbers are useful for computers because arithmetic operations on the digits 0 and 1 can be represented using AND, OR, and NOT, which can be computed quickly.

Unlike humans, who can abstract numbers to arbitrarily large values, computers have a fixed number of bits that they are capable of storing at one time. For example, a 32-bit computer can represent and process 32-digit binary numbers and no more. If all 32-bits are used to represent positive integer binary numbers, then this means that there are $\sum_{n=0}^{31} 2^n = 4,294,967,296$ numbers the computer can represent. This is not a lot of numbers at all and would be completely insufficient to perform anything more than basic calculations. For example, you can not compute the perfectly reasonable sum $0.5 + 1.25$ using this representation because all the bits are dedicated to only integers.

```
 100101
x 10001
-------

 100101
x 10001
-------
 100101

    100101
   x 10001
   -------
    100101
         0
        00
       000
+1001010000
-----------
1001110101 = 512 + 64 + 32 + 16 + 4 + 1 = 629 (base10)
```

FIGURE 9.2

Binary multiplication.

9.2 FLOATING POINT NUMBERS

The number of bits is usually fixed for any given computer. Using binary representation gives us an insufficient range and precision of numbers to do relevant engineering calculations. To achieve the range of values needed with the same number of bits, we use **floating point** numbers, or **floats** for short. Instead of utilizing each bit as the coefficient of a power of 2, floats allocate bits to three different parts: the **sign indicator**, s, which says whether a number is positive or negative; **characteristic** or **exponent**, e, which is the power of 2; and the **fraction**, f, which is the coefficient of the exponent. Almost all platforms map Python floats to the **IEEE754** double precision for a total of 64 bits. One bit is allocated to the sign indicator, 11 bits are allocated to the exponent, and 52 bits are allocated to the fraction. With 11 bits allocated to the exponent; this makes 2048 values that this number can use. Since we want to be able to make very precise numbers, we want some of these values to represent negative exponents (i.e., to allow numbers that are between 0 and 1 (base10)). To accomplish this, 1023 is subtracted from the exponent to normalize it. The value subtracted from the exponent is commonly referred to as the **bias**. The fraction is a number between 1 and 2. In binary, this means that the leading term will always be 1, and, therefore, it is a waste of bits to store it. To save space, the leading 1 is dropped. In Python, we can obtain the float information using the `sys` package as shown below:

```
In [1]: import sys
        sys.float_info

Out[1]: sys.float_info(max=1.7976931348623157e+308, max_exp=1024,
         max_10_exp=308, min=2.2250738585072014e-308,
         min_exp=-1021, min_10_exp=-307, dig=15,
         mant_dig=53, epsilon=2.220446049250313e-16,
         radix=2, rounds=1)
```

A float can then be represented as $n = (-1)^s 2^{e-1023}(1 + f)$, for 64-bit.

> **TRY IT!** What is the number
> 1 10000000010 1000000000000000000000000000000000000000000000000000
> (IEEE754) in base10?
>
> The exponent in decimal is $1 \cdot 2^{10} + 1 \cdot 2^1 - 1023 = 3$.
> The fraction is $1 \cdot \frac{1}{2^1} + 0 \cdot \frac{1}{2^2} + \cdots = 0.5$.
> Therefore, $n = (-1)^1 \cdot 2^3 \cdot (1 + 0.5) = -12.0$ (base10). See Fig. 9.3 for details.

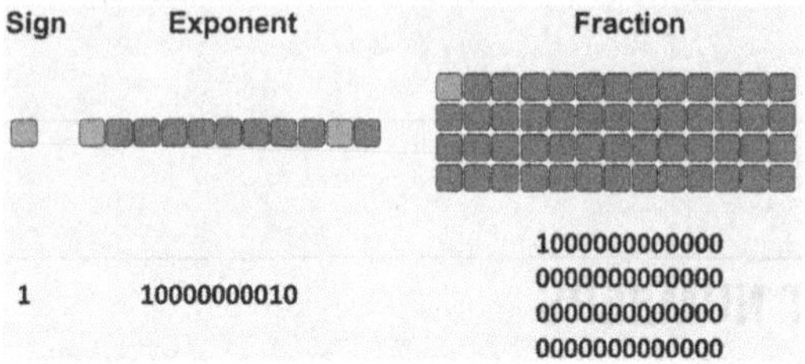

FIGURE 9.3

Illustration of -12.0 that is represented in computer with 64-bit. Each square is one bit, with the green square representing 1 and grey square as zero.

> **TRY IT!** What is 15.0(base10) in IEEE754? What is the largest number smaller than 15.0? What is the smallest number larger than 15.0?
>
> Since the number is positive, $s = 0$. The largest power of 2 that is smaller than 15.0 is 8, so the exponent is 3, therefore:
> $3 + 1023 = 1026$(base10) = 10000000010(base2).
> Then the fraction is:
> $15/8 - 1 = 0.875$(base10) $= 1 \cdot \frac{1}{2^1} + 1 \cdot \frac{1}{2^2} + 1 \cdot \frac{1}{2^3} =$
> 1110000000000000000000000000000000000000000000000000(base2).
> When combined, this produces the following conversion:
> 15.0(base10) =
> 0 10000000010 1110000000000000000000000000000000000000000000000000
> (IEEE754)
> The next smallest number is
> 0 10000000010 1101111111111111111111111111111111111111111111111111
> = 14.9999999999999982236431605997
> The next largest number is
> 0 10000000010 1110000000000000000000000000000000000000000000000001
> = 15.0000000000000017763568394003

Therefore, the IEEE754 number
0 10000000010 1110000000000000000000000000000000000000000000000000
not only represents the number 15.0, but also all real numbers halfway between its immediate neighbors. So any computation that has a result within this interval will be assigned as "15.0".

We call the distance from one number to the next the **gap**. Because the fraction is multiplied by 2^{e-1023}, the gap grows as the number represented grows. The gap at a given number can be computed using the function `spacing` in `NumPy`.

TRY IT! Use the `spacing` function to determine the gap at 1e9. Verify that adding a number to 1e9 that is less than half the gap at 1e9 results in the same number.

```
In [2]: import numpy as np

In [3]: np.spacing(1e9)

Out[3]: 1.1920928955078125e-07

In [4]: 1e9 == (1e9 + np.spacing(1e9)/3)

Out[4]: True
```

There are special cases for the value of a floating point number when $e = 0$ (i.e., $e =$ 00000000000(base2)) and when $e = 2047$, i.e., $e =$ 11111111111(base2), which are reserved. When the exponent is 0, the leading 1 in the fraction takes the value 0 instead. The result is a **subnormal number**, which is computed by $n = (-1)^s 2^{-1022}(0 + f)$ (note that it is -1022 instead of -1023). When the exponent is 2047 and f is nonzero, then the result is "Not a Number," which means that the number is undefined. When the exponent is 2047, then $f = 0$ and $s = 0$, and the result is positive infinity. When the exponent is 2047, then $f = 0$, and $s = 1$, and the result is minus infinity.

TRY IT! Compute the base10 value for
0 11111111110 1111111111111111111111111111111111111111111111111111
(IEEE754), the largest defined number for 64 bits, and for
0 00000000001 0000000000000000000000000000000000000000000000000000
(IEEE754), the smallest. Note that to comply with the previously stated rules, the exponent is $e = 2046$ and $e = 1$, respectively. Verify that Python agrees with these calculations using `sys.float_info.max` and `sys.float_info.min`.

```
In [5]: 1 = (2**(2046-1023))*((1 + sum(0.5**np.arange(1, 53))))
        1

Out[5]: 1.7976931348623157e+308
```

```
In [6]: sys.float_info.max

Out[6]: 1.7976931348623157e+308

In [7]: s = (2**(1-1023))*(1+0)
        s

Out[7]: 2.2250738585072014e-308

In [8]: sys.float_info.min

Out[8]: 2.2250738585072014e-308
```

Numbers that are larger than the largest floating point number capable of being represented result in **overflow**; Python handles this case by assigning the result to `inf`. Numbers that are smaller than the smallest subnormal number result in **underflow**; Python handles this case by assigning the result to zero.

TRY IT! Show that adding the maximum 64-bit float number with 2 results in the same number. Because the Python float does not have sufficient precision to store the + 2 for `sys.float_info.max`, the operation is essentially equivalent to adding zero. Also show that adding the maximum 64-bit float number with itself results in overflow, and that Python assigns this overflow number to `inf`.

```
In [9]: sys.float_info.max + 2 == sys.float_info.max

Out[9]: True

In [10]: sys.float_info.max + sys.float_info.max
Out[10]: inf
```

TRY IT! The smallest subnormal number in a 64-bit number has

$$s = 0, e = 00000000000$$

and

$$f = 0000000000000000000000000000000000000000000000000001.$$

Using the special rules for subnormal numbers, this results in the subnormal number $(-1)^0 2^{1-1023} 2^{-52} = 2^{-1074}$. Show that 2^{-1075} underflows to zero, and that the result cannot be distinguished from zero. Show that 2^{-1074} does not.

```
In [11]: 2**(-1075)
```

```
Out[11]: 0.0

In [12]: 2**(-1075) == 0

Out[12]: True

In [13]: 2**(-1074)

Out[13]: 5e-324
```

What have we gained by using IEEE754 versus binary? Using 64-bit binary gives us 2^{64} numbers. Since the number of bits does not change between binary and IEEE754, IEEE754 must also give us 2^{64} numbers. In binary, numbers have a constant spacing between them. As a result, you cannot have both range (i.e., large distance between minimum and maximum numbers capable of being represented) and precision (i.e., small spacing between numbers). Controlling these parameters would depend on where you put the decimal point in your number. IEEE754 overcomes this limitation by using very high precision at small numbers and very low precision at large numbers. This limitation is usually acceptable because the gap at large numbers is still small relative to the size of the number itself. Therefore, even if the gap is millions large, it is irrelevant to normal calculations if the number under consideration is in the trillions or higher.

This section introduced the representation of floating point numbers. This concept is described in detail *computer organization and design* by David Patterson and John Hennessy.

9.3 ROUND-OFF ERRORS

In the previous section, we talked about how the floating-point numbers are represented in computers as base2 fractions. This has a side effect that the floating-point numbers cannot be stored with perfect precision; instead the numbers are approximated by a finite number of bytes. Therefore, the difference between an approximation of a number used in computation and its correct (true) value is called a **round-off error**. It is one of the common errors found usually in numerical calculations. The other one is **truncation error**, which will be introduced in Chapter 18. The difference is that truncation error is the error made by truncating an infinite sum and approximating it by a finite sum.

9.3.1 REPRESENTATION ERROR

The most common form of a round-off error is the representation error in the floating-point numbers. A simple example will be how do we represent π? We know that π is an infinite number (i.e., has an infinite number of digits); typically, we only use a finite number of digits. For example, if you only use 3.14159265, there will be an error between this approximation and the true infinite number. Another example will be $1/3$, the true value will be 0.333333333... No matter how many decimal digits we choose, there is an round-off error as well.

When we round the numbers multiple times, the error will accumulate. For instance, if 4.845 is rounded to two decimal places, it is 4.85. Then if we round it again to one decimal place, it is 4.9, the total error will be 0.55. But if we only round one time to one decimal place, it is 4.8, and the error is 0.045.

9.3.2 ROUND-OFF ERROR BY FLOATING-POINT ARITHMETIC

From the above example, the error between 4.845 and 4.8 should be 0.055. But if you calculate it in Python, you will see the $4.9 - 4.845$ is not equal to 0.055.

```
In [1]: 4.9 - 4.845 == 0.055

Out[1]: False
```

Why does this happen? If we take a second look at $4.9 - 4.845$, we actually get 0.055000000000000604. This is because the floating point cannot be represented by the exact number because it is an approximation; when it is used in arithmetic, it causes a small error.

```
In [2]: 4.9 - 4.845

Out[2]: 0.055000000000000604

In [3]: 4.8 - 4.845

Out[3]: -0.04499999999999993
```

Another example shows below that $0.1 + 0.2 + 0.3$ is not equal to 0.6; the error is due to the same cause.

```
In [4]: 0.1 + 0.2 + 0.3 == 0.6

Out[4]: False
```

Though the numbers cannot be made closer to their intended exact values, the `round` function can be useful for post-rounding so that results with inexact values become comparable to one another:

```
In [5]: round(0.1 + 0.2 + 0.3, 5)  == round(0.6, 5)

Out[5]: True
```

9.3.3 ACCUMULATION OF ROUND-OFF ERRORS

When we are doing a sequence of calculations on an initial input with round-off errors due to inexact representation, the errors can be magnified or accumulated. As an example, if we have the number 1, add and then subtract $1/3$, it should give us 1. But if we add $1/3$ and subtract the same number of times $1/3$ over several iterations, do we still get the same number 1? No, we don't. In the example below, the more times you do this, the additional accumulates.

```
In [6]: # If we only do once
        1 + 1/3 - 1/3

Out[6]: 1.0

In [7]: def add_and_subtract(iterations):
            result = 1

            for i in range(iterations):
                result += 1/3

            for i in range(iterations):
                result -= 1/3
            return result

In [8]: # If we do this 100 times
        add_and_subtract(100)

Out[8]: 1.0000000000000002

In [9]: # If we do this 1000 times
        add_and_subtract(1000)

Out[9]: 1.0000000000000064

In [10]: # If we do this 10000 times
         add_and_subtract(10000)

Out[10]: 1.0000000000001166
```

9.4 SUMMARY AND PROBLEMS

9.4.1 SUMMARY

1. Numbers may be represented in several different ways, each representation scheme having advantages and disadvantages.
2. Computers must represent numbers using a finite number of digits (i.e., bits).
3. Binary and IEEE754 are finite representations of numbers used by computers.
4. The round-off error is an important error associated with numerical methods.

9.4.2 PROBLEMS

1. Write a function `my_bin_2_dec(b)` where `b` is a binary number represented by a list of ones and zeros. The last element of `b` represents the coefficient of 2^0, the second-to-last element of b repre-

sents the coefficient of 2^1, and so on. The output variable, d, should be the decimal representation of b. Test cases are provided below.

```
In [ ]: def my_bin_2_dec(b):
            # write your function code here
            return d

In [ ]: # Output: 7
        my_bin_2_dec([1, 1, 1])

In [ ]: # Output: 85
        my_bin_2_dec([1, 0, 1, 0, 1, 0, 1])

In [ ]: # Output: 33554431
        my_bin_2_dec([1]*25)
```

2. Write a function `my_dec_2_bin(d)` where d is a positive integer in decimal, and b is the binary representation of d. The output b must be a list of ones and zeros, and the leading term must be a 1 unless the decimal input value is 0. Test cases are provided below.

```
In [ ]: def my_dec_2_bin(d):
            # write your function code here
            return b

In [ ]: # Output: [0]
        my_dec_2_bin(0)

In [ ]: # Output: [1, 0, 1, 1, 1]
        my_dec_2_bin(23)

In [ ]: # Output: [1, 0, 0, 0, 0, 0, 1, 1, 0, 0, 0, 1]
        my_dec_2_bin(2097)
```

3. Use the two functions you wrote in Problems 1 and 2 to compute `d = my_bin_2_dec(my_dec_2_bin(12654))`. Do you get the same number?

4. Write a function `my_bin_adder(b1,b2)` where b1, b2 and the output variable b are binary numbers represented as in Problem 1. The output variable should be computed as `b = b1 + b2`. Do not use your functions from Problems 1 and 2 to write this function (i.e., do not convert b1 and b2 to decimals; add them and then convert the result back to binary). This function should be able to accept inputs b1 and b2 of any length (i.e., very long binary numbers), and b1 and b2 may not necessarily be of the same length.

```
In [ ]: def my_bin_adder(b1, b2):
            # write your function code here
            return b

In [ ]: # Output: [1, 0, 0, 0, 0, 0]
        my_bin_adder([1, 1, 1, 1, 1], [1])

In [ ]: # Output: [1, 1, 1, 0, 0, 1, 1]
        my_bin_adder([1, 1, 1, 1, 1], [1, 0, 1, 0, 1, 0, 0])
```

```
In [ ]: # Output: [1, 0, 1, 1]
        my_bin_adder([1, 1, 0], [1, 0, 1])
```

5. What is the effect of allocating more bits to the fraction versus the characteristic, and vice versa? What is the effect of allocating more bits to the sign?

6. Write a function `my_ieee_2_dec(ieee)` where `ieee` is a string that contains 64 characters of ones and zeros, representing a 64-bit IEEE754 number. The output should be `d`, which is the equivalent decimal representation of `ieee`. The input variable `ieee` will always be a 64-element string of ones and zeros defining a 64-bit float.

```
In [ ]: def my_ieee_2_dec(ieee):
            # Write your function here
            return d
```

```
In [ ]: # Output: -48
        ieee ="1100000001001000000000000000000000000000000000000000000000000000"
        my_ieee_2_dec(ieee)
```

```
In [ ]: # Output: 3.3999999999999999111821580299 9
        ieee ="0100000000001011001100110011001100110011001100110011001100110011"
        my_ieee_2_dec(ieee)
```

7. Write a function `my_dec_2_ieee(d)` where `d` is a number in decimal, and the output variable `ieee` is a string with 64 characters of ones and zeros, representing the 64-bit IEEE754 closest to `d`. Assume that `d` will not cause an overflow for 64-bit `ieee` numbers.

```
In [ ]: def my_dec_2_ieee(d):
            # write your function code here
            return ieee
```

```
In [ ]: #Output:"0100000000101110010111101010001110011100001100011010010001101000"

        d = 1.518484199625
        my_dec_2_ieee(d)
```

```
In [ ]: #Output:"1100000001110011010100100100010010010001001010011000100010010000"

        d = -309.141740
        my_dec_2_ieee(d)
```

```
In [ ]: #Output:"1100000011011000101010010000000000000000000000000000000000000000"

        d = -25252
        my_dec_2_ieee(d)
```

8. Define `ieee_baby` to be a representation of numbers using 6 bits, where the first bit is the sign bit, the second and third bits are allocated to the characteristic, and the fourth, fifth, and sixth bits are allocated to the fraction. The normalization for the characteristic is 1.
Write all the decimal numbers that can be represented by `ieee_baby`. What is the largest/smallest gap in `ieee_baby`?

9. Use the `np.spacing` function to determine the smallest number such that the gap is 1.
10. What are some of the advantages and disadvantages of using binary versus decimal?
11. Write the number 13(base10) in base1. How would you add and multiply in base1?
12. How high can you count on your fingers if you count in binary?
13. Let `b` be a binary number having `n` digits. Can you think of ways to multiply and divide `b` by 2 that does not involve any arithmetic? Hint: Think about how you multiply and divide a decimal number by 10.

CHAPTER

ERRORS, GOOD PROGRAMMING PRACTICES, AND DEBUGGING 10

CONTENTS

10.1 ERROR TYPES

We have noted errors before but have not yet talked about them in detail. There are three basic types of errors that programmers need to be concerned about: **syntax errors**, **runtime errors**, and **logical errors**. **Syntax** is the set of rules that govern a language. In written and spoken language, rules can be bent or even broken to accommodate the speaker or writer. However, in a programming language the rules are rigid. A syntax error occurs when the programmer writes an instruction using incorrect syntax; Python cannot understand what you are saying. For example, `1 = x` is not legal in the Python programming language because numbers cannot be assigned as variables. If the programmer tries to execute one of these instructions or any other syntactically incorrect statement, Python will return an error message to the programmer and point out the location where the error occurred. Note that although Python can identify the location where the syntax error is detected by the parser, it is possible that the syntax error causing the error may be far away from the specific line identified.

EXAMPLE: Syntax error examples.

```
In [1]: 1 = x
```


Python Programming and Numerical Methods. https://doi.org/10.1016/B978-0-12-819549-9.00019-1

```
      File "<ipython-input-1-7a7b257d8e3d>", line 1
    1 = x
         ^

SyntaxError: can't assign to literal

In [2]: (1]


      File "<ipython-input-2-800df0a5e99c>", line 1
    (1]
      ^

SyntaxError: invalid syntax

In [3]: if True
            print("Here")


      File "<ipython-input-3-025e9fce1ee3>", line 1
    if True
           ^

SyntaxError: invalid syntax
```

The last line of the error message shows what happened – `SyntaxError`, and the lines before indicate where the error happens in the context of the code. Overall, syntax errors are usually easily detectable, found, and fixed.

Even if all the syntax is correct in your code, it may still cause an error during execution of the code. Errors that occur during execution are called **exceptions** or **runtime errors**. Exceptions are more difficult to find and are only detectable when a program is run. Note that exceptions are not fatal. We will learn later how to handle them in Python. If we do not handle them, Python will terminate the program. Let us see some examples below.

```
In [4]: 1/0

    ---------------------------------------------------------------------------

    ZeroDivisionError      Traceback (most recent call last)

    <ipython-input-4-9e1622b385b6> in <module>
----> 1 1/0
```

```
        ZeroDivisionError: division by zero

In [5]: x = [2]
        x + 2


        ---------------------------------------------------


        TypeError           Traceback (most recent call last)

        <ipython-input-5-29a14b9fefb9> in <module>
          1 x = [2]
    ----> 2 x + 2


        TypeError: can only concatenate list (not "int") to list

In [6]: print(a)


        ---------------------------------------------------


        NameError            Traceback (most recent call last)

        <ipython-input-6-bca0e2660b9f> in <module>
    ----> 1 print(a)


        NameError: name "a" is not defined
```

As shown in the examples above, there are different built-in exceptions: `ZeroDivisionError`, `TypeError`, and `NameError`. You can find a complete list of built-in exceptions in the Python documentation.[1] In addition, you can define your own exception types, but we will not deal with this herein; if you are interested in how to define customized exceptions, check out the documentation.[2]

Most of the exceptions are easy to locate because Python will stop running and tell you where the problem is. After programming a function, seasoned programmers will usually run the function several times, allowing the function to "throw" any errors so that they can fix them. Note that no exception does not mean the function works correctly.

[1] https://docs.python.org/3/library/exceptions.html#bltin-exceptions.
[2] https://docs.python.org/3/tutorial/errors.html#user-defined-exceptions.

One of the most difficult errors to find are called **logic errors**. A logic error does not throw an error. Although the program will run smoothly, it is an error because the output obtained is not the solution you expect. For example, consider the following incorrect implementation of the factorial function.

```
In [7]: def my_bad_factorial(n):
            out = 0
            for i in range(1, n+1):
                out = out*i

            return out

In [8]: my_bad_factorial(4)

Out[8]: 0
```

This function will not produce a runtime error for any input that is valid for a correctly implemented factorial function; however, if you try using `my_bad_factorial`, you will find that the answer is always 0 because out is initialized to 0 instead of 1. Therefore, the line `out = 0` is a logic error. It does not produce a runtime error by Python, but it leads to an incorrect computation.

Although the logic errors seem unlikely to occur—or at least as easy to find as other kinds of errors—when programs become longer and more complicated, such errors are very easy to generate and notoriously difficult to find. When logic errors occur, you have no choice but to meticulously comb through each line of your code until you find the problem. For these cases, it is important to know exactly how Python will respond to every command you give and not make any assumptions. You can also use Python's debugger, which will be described in the last section of this chapter.

10.2 AVOIDING ERRORS

There are many techniques that can help prevent errors and make it easier for you to find them when they occur. Because becoming familiar with the types of mistakes common in programming is a "learning as you go" process; we cannot possibly list them all here. That said, we present a few of them in the following section to help you build good habits.

10.2.1 PLAN YOUR PROGRAM

When writing an essay, it is important to have a structure and a direction that you intend to follow. To help make your structure more tangible, writing an essay usually starts with an outline containing the main points you wish to address in your paper. This is even more important to do when programming because computers do not interpret what you write. When coding complicated programs, you should start with an outline of your program that addresses all the tasks you want your program to perform and in the order in which it should perform them.

Many novice programmers, eager to finish their assignments, will rush to the programming part without properly planning out the tasks needed to accomplish the given task. Haphazard planning results in equally haphazard code that is full of errors. Time spent planning out what you are trying to do is time well spent. Preplanning ensures that you will finish your program much faster than if you throw together a program without much thought.

So what does planning a program consist of? Recall in Chapter 3 that a function is defined as a sequence of instructions designed to perform a certain task. A **module** is a function or group of functions that perform a certain task. It is important to design your program in terms of modules, especially for tasks that need to be repeated over and over again. Each module should accomplish a small, well-defined task and know as little information about other functions as possible (i.e., have a very limited set of inputs and outputs).

A good rule of thumb is to plan from the top to bottom, and then program from the bottom to the top. That is, decide what the overall program is supposed to do, determine what code is necessary to complete the main tasks, and then break the main tasks into components until the module is small enough that you are confident you can write it without errors.

10.2.2 TEST EVERYTHING OFTEN

When coding in modules, you should test each module using test cases for which you know the answer, and code enough cases to be confident that the function is working properly (including corner cases). For example, if you are writing a function that tells you whether a number is prime or not, you should test the function for inputs of 0 (corner case), 1 (corner case), 2 (simple yes), 4 (simple no), and 97 (complicated no). If all of your test cases proceed without error, you can move on to other modules, confident that the current module works correctly. **This is especially important if subsequent modules depend on or call the module you are working on**. If you do not test your code and assume that your incorrect code is correct, when you begin to get error messages in later modules, you will not know whether the error is in the module you are working on or in a previous module; this will make finding the error exponentially more difficult.

You should test often, even within a single module or function. When you are working on a particular module that has several steps, you should perform intermediate tests to make sure it is correct up to that point. Then, if you ever get an error, it will probably be in the part of your code written since the last time you did test it. This tendency to rush one's coding is a mistake even seasoned programmers are guilty of; they will write pages and pages of code without testing and then will spend hours trying to find a small error somewhere.

10.2.3 KEEP YOUR CODE CLEAN

Just like good craftsmen keep their work area free of unnecessary clutter, so should you keep your code as clean as possible. There are many strategies you can implement to keep your code clean. First, you should write your code in the fewest instructions possible. For example,

```
y = x**2 + 2*x+1
```

is better than

```
y=x**2
y=y+2*x
y=y+1
```

Even if the outcome is the same, every character you type is a chance that you will make a mistake; therefore, reducing how much code you write will reduce your risk of introducing errors. Additionally, writing a complete expression will help you and other people understand what you are doing. In the previous example, in the first case it is clear that you are computing the value of a quadratic at x, while in the second case it is not clear. You can also keep your code "clean" by using variables rather than values.

EXAMPLE: Poor implementation of adding 10 random numbers.

```
In [1]: import numpy as np

        s = 0
        a = np.random.rand(10)
        for i in range(10):
            s = s + a[i]
```

EXAMPLE: Good implementation of adding 10 random numbers.

```
In [2]: n = 10
        s = 0
        a = np.random.rand(n)

        for i in range(n):
            s = s + a[i]
```

The second implementation is better for two reasons: first, it is easier for anyone reading your code that n represents the number of random numbers you want to add up, and it appears rationally where it is supposed to in the code (i.e., when creating the array of random numbers and when indexing the array in the for-loop); and second, if you ever wanted to change the number of random numbers you wish to add up, you would only have to change it in one place at the beginning. This reduces the chances of making mistakes while writing the code and when changing the value of n.

Again, this is not critical for such a small piece of code, but it will become very important when your code becomes more complicated and values must be reused many times.

EXAMPLE: An even better implementation of adding 10 random numbers.

```
In [3]: s = sum(np.random.rand(10))
```

When you become more familiar with Python, you will want to use as few lines as possible to do the same job; therefore, familiarity of the common functions and how to use them will make your code more concise and efficient.

You can also keep your code clean by assigning your variables short, descriptive names. For example, as noted earlier, `n` is a sufficient variable for such a simple task. The variable name `x` is probably a good name since `x` usually holds value of position rather than a number; but `theNumberOfRandomNumbersToBeAdded` is a poor variable name even though it is descriptive.

Finally, you can keep your code clean by commenting frequently. Although no commenting is certainly bad practice, over-commenting can be equally bad practice. Different programmers will disagree on exactly how much commenting is appropriate. It will be up to you to decide what level of commenting is appropriate.

10.3 TRY/EXCEPT

Often it is important to write programs that can handle certain types of errors or exceptions gracefully. More specifically, the error or exception must not cause a critical error that makes your program shut down. A **Try-Except statement** is a code block that allows your program to take alternative actions in case an error occurs.

CONSTRUCTION: Try-Exception Statement

```
try:
    code block 1
except ExceptionName:
    code block 2
```

Python will first attempt to execute the code in the `try` statement (code block 1). If no exception occurs, the `except` statement is skipped and the execution of the `try` statement is finished. If any exception occurs, the rest of the clause is skipped. Then if the exception type matches the exception named after the `except` keyword (`ExceptionName`), the code in the `except` statement will be executed (code block 2). If nothing in this block stops the program, it will continue to execute the rest of the code outside of the `try-except` code blocks. If the exception does not match the `ExceptionName`, it is passed on to outer `try` statements. If no other handler is found, then the execution stops with an error message.

EXAMPLE: Capture the exception.

```
In [1]: x = "6"
        try:
            if x > 3:
                print("X is larger than 3")
        except TypeError:
            print("Oops! x is not a valid number. Try again...")
```

```
Oops! x is not a valid number. Try again...
```

EXAMPLE: If your handler is trying to capture another exception type that the `except` does not capture, then an error occurs and the execution stops.

```
In [2]: x = "6"
        try:
            if x > 3:
                print("X is larger than 3")
        except ValueError:
            print("Oops! x is not a valid number. Try again...")

        ------------------------------------------------------


        TypeError            Traceback (most recent call last)

        <ipython-input-2-899d928e7a1f> in <module>
          1 x = "6"
          2 try:
    ----> 3     if x > 3:
          4         print("X is larger than 3")
          5 except ValueError:


        TypeError: ">" not supported between instances of "str"
                   and "int"
```

Of course, a `try` statement may have more than one except statement to handle different exceptions or you cannot specify the exception type so that the `except` will catch any exception.

```
In [3]: x = "s"

        try:
            if x > 3:
                print(x)
        except:
            print(f"Something is wrong with x = {x}")

Something is wrong with x = s
```

EXAMPLE: Handling multiple exceptions.

```
In [4]: def test_exceptions(x):
            try:
                x = int(x)
                if x > 3:
                    print(x)
            except TypeError:
                print("x was not a valid number. Try again...")
            except ValueError:
                print("Cannot convert x to int. Try again...")
            except:
                print("Unexpected error")

In [5]: x = [1, 2]
        test_exceptions(x)

x was not a valid number. Try again...

In [6]: x = "s"
        test_exceptions(x)

Cannot convert x to int. Try again...
```

Another useful thing in Python is that we can raise some exceptions in certain cases using `raise`. For example, if we need x to be less than or equal to 5, we can use the following code to raise an exception if x is larger than 5. The program will display our exception and stop the execution.

```
In [7]: x = 10

        if x > 5:
            raise(Exception("x should be <= 5"))


        ---------------------------------------------------------


        Exception                 Traceback (most recent call last)

        <ipython-input-7-99b32b52c4f8> in <module>
          2
          3 if x > 5:
    ----> 4     raise(Exception("x should be <= 5"))
```

```
Exception: x should be <= 5
```

WARNING! `Try-except` statements should never be used in place of good programming practice. For example, you should not code sloppily and then encase your program in a `try-except` statement until you have taken every measure you can think of to ensure that your function is working properly.

10.4 TYPE CHECKING

Python is both a strongly and dynamically typed programming language. This means that any variable can take on any data type at any time (this is dynamically typed part); however, once a variable is assigned a type, it cannot change. For example, you can write `x = 1` immediately followed by `x = "s"` because Python is a dynamically typed language. You cannot run `"3" + 5`, because it is a strongly typed language, i.e., the string "3" cannot convert in runtime to an integer. In statically typed programming languages, you must declare what kind of data type your variable is to execute before you use it, and the data type usage of your variable cannot change within the scope of a function.

In the case of Python, there is no way to ensure that the user of your function is inputting variables of the data type you expect. For example, the function `my_adder` in Chapter 3 is designed to add three numbers together. However, the user can input strings, lists, dictionaries, or functions, each of which will cause different levels of problems. You can have your function type check the input variables before continuing and force an error using the error function.

TRY IT! Modify `my_adder` to type check that the input variables are floats. If any of the input variables are not floats, the function should return an appropriate error to the user using the `raise` function. Try your function for erroneous input arguments to verify that they are checked.

```
In [1]: def my_adder(a, b, c):
            # type check
            if isinstance(a, float) and isinstance(b, float) and
               isinstance(c, float):
                pass
            else:
                raise(TypeError("Inputs must be floats"))

            out = a + b + c
            return out

In [2]: my_adder(1.0, 2.0, 3.0)
```

```
Out[2]: 6.0

In [3]: my_adder(1.0, 2.0, "3.0")


        ---------------------------------------------------


        TypeError         Traceback (most recent call last)

        <ipython-input-3-14e4b71b8c1d> in <module>
    ----> 1 my_adder(1.0, 2.0, "3.0")


        <ipython-input-1-c2a54d39e3d9> in my_adder(a, b, c)
    ----> 6         raise(TypeError("Inputs must be floats"))
          7
          8     out = a + b + c
          9     return out


        TypeError: Input arguments must be floats

In [4]: my_adder(1, 2, 3)


        -----------------------------------------------------


        Exception           Traceback (most recent call last)

        <ipython-input-4-fc54adcab3d7> in <module>
    ----> 1 my_adder(1, 2, 3)


        <ipython-input-1-c2a54d39e3d9> in my_adder(a, b, c)
          4         pass
          5     else:
    ----> 6         raise(TypeError("Inputs must be floats"))
          7
          8     out = a + b + c


        TypeError: Inputs must be floats
```

Note that 1, 2, 3 are integers instead of floats, therefore, it raised an error message, and we need to change the function to make sure that any numbers will be added.

```
In [5]: def my_adder(a, b, c):
            # type check
            if isinstance(a, (float, int, complex)) and
               isinstance(b, (float, int, complex)) and
               isinstance(c, (float, int, complex)):
                pass
            else:
                raise(TypeError("Inputs must be numbers"))

            out = a + b + c
            return out

In [6]: my_adder(1, 2, 3)

Out[6]: 6

In [7]: my_adder(1.0, 2, 3)

Out[7]: 6.0

In [8]: my_adder(1j, 2+2j, 3+2j)

Out[8]: (5+5j)
```

10.5 DEBUGGING

Debugging is the process of systematically removing errors, or bugs, from your code. Python has functionalities that can assist you when debugging. The standard debugging tool in Python is `pdb` (Python DeBugger) for interactive debugging. It lets you step through the code line by line to find out what might be causing a difficult error. The IPython version of this is `ipdb` (IPython DeBugger). We will not cover too much about it herein; check out the documentation[3] for details. In this section, basic debug steps in `Jupyter notebook` will be introduced. We will show you how to use two really useful magic commands `%debug` and `%pdb` to find the code causing trouble.

There are two ways you can debug your code: (1) activate the debugger when you run into an exception; and (2) activate debugger before running the code.

10.5.1 ACTIVATING DEBUGGER AFTER RUNNING INTO AN EXCEPTION

If we run the code which stops at an exception, we can call `%debug`. For example, we have a function that squares the input number and then adds to itself, as shown below:

[3] https://docs.python.org/3/library/pdb.html.

```
In [1]: def square_number(x):

            sq = x**2
            sq += x

            return sq

In [2]: square_number("10")


        ---------------------------------------------------


        TypeError             Traceback (most recent call last)


        <ipython-input-2-e0b77a2957d5> in <module>
    ----> 1 square_number("10")



        <ipython-input-1-3fc6a3900214> in square_number(x)
          1 def square_number(x):
          2
    ----> 3     sq = x**2
          4     sq += x
          5



        TypeError: unsupported operand type(s) for ** or pow():
                   "str" and "int"
```

After we locate this exception, we can activate the debugger by using the magic command `%debug`, which opens an interactive debugger, at which point you can type in commands in the debugger to get useful information.

```
In [3]: %debug

> <ipython-input-1-3fc6a3900214>(3)square_number()
      1 def square_number(x):
      2
----> 3     sq = x**2
      4     sq += x
      5
```

```
ipdb> h

Documented commands (type help <topic>):
========================================
EOF    cl         disable  interact  next    psource  rv          unt
a      clear      display  j         p       q        s           until
alias  commands   down     jump      pdef    quit     source      up
args   condition  enable   l         pdoc    r        step        w
b      cont       exit     list      pfile   restart  tbreak      whatis
break  continue   h        ll        pinfo   return   u           where
bt     d          help     longlist  pinfo2  retval   unalias
c      debug      ignore   n         pp      run      undisplay

Miscellaneous help topics:
==========================
exec  pdb

ipdb> p x
'10'
ipdb> type(x)
<class 'str'>
ipdb> p locals()
{'x': '10'}
ipdb> q
```

You can see that after we activate the `ipdb`, we can type commands to get the information of the code. In the example above, we typed the following commands:

- `h` to get a list of help functions
- `p x` to print the value of x
- `type(x)` to get the type of x
- `p locals()` to print out all the local variables

There are some most frequent commands you can type in the pdb, like:

- `n(ext)` line and run this one
- `c(ontinue)` running until next breakpoint
- `p(rint)` print variables
- `l(ist)` where you are
- `Enter` repeat the previous command
- `s(tep)` step into a subroutine
- `r(eturn)` return out of a subroutine
- `h(elp)` get help
- `q(uit)` the debugger

10.5.2 ACTIVATING DEBUGGER BEFORE RUNNING THE CODE

We can also turn on the debugger before we run the code and then turn it off once we are finished running the code.

```
In [4]: %pdb on

Automatic pdb calling has been turned ON


In [5]: square_number("10")


        ---------------------------------------------------------


        TypeError                 Traceback (most recent call last)

        <ipython-input-5-e0b77a2957d5> in <module>
    ----> 1 square_number("10")


        <ipython-input-1-3fc6a3900214> in square_number(x)
          1 def square_number(x):
          2
    ----> 3     sq = x**2
          4     sq += x
          5


        TypeError: unsupported operand type(s) for ** or pow():
                   "str" and "int"


> <ipython-input-1-3fc6a3900214>(3)square_number()
      1 def square_number(x):
      2
----> 3     sq = x**2
      4     sq += x
      5

ipdb> p x
"10"
ipdb> c
```

```
In [6]: # let's turn off the debugger
        %pdb off

Automatic pdb calling has been turned OFF
```

10.5.3 ADD A BREAKPOINT

It is often very useful to insert a breakpoint into your code. A breakpoint is a line in your code at which Python will stop when the function is run.

```
In [7]: import pdb

In [8]: def square_number(x):

            sq = x**2

            # we add a breakpoint here
            pdb.set_trace()

            sq += x

            return sq

In [9]: square_number(3)

> <ipython-input-8-e48ec2675aea>(8)square_number()
-> sq += x
(Pdb) l
  3         sq = x**2
  4
  5         # we add a breakpoint here
  6         pdb.set_trace()
  7
  8  ->     sq += x
  9
 10         return sq
[EOF]
(Pdb) p x
3
(Pdb) p sq
9
```

```
(Pdb) c

Out[9]: 12
```

After we added `pdb.set_trace()`, the program stopped at this line and activated the `pdb` debugger. We can now check all the variable values that were assigned before this line and use the command `c` to continue the execution.

Using the Python's debugger can be extremely helpful in finding and fixing errors in your code. We encourage you to use the debugger for large programs.

10.6 SUMMARY AND PROBLEMS

10.6.1 SUMMARY

1. Errors are inevitable when coding. Errors are important because they tell you that something is not working the way you intended.
2. There are three types of errors: syntax errors, exceptions, and logical errors.
3. You can reduce the numbers of errors in your coding with good coding practice.
4. Try-except statements can be used to handle exceptions without stopping your code.
5. The Debugger is a Python tool for helping you find errors.

10.6.2 PROBLEMS

None, have fun with what we learned in this chapter.

CHAPTER

READING AND WRITING DATA 11

CONTENTS

11.1 TXT FILES

So far, we have used the `print` function to display the data to the screen. But there are many ways to store data onto your disk and share it with other programs or colleagues. For example, if we have some strings in this notebook but we want to use them in another notebook, the easiest way is to store the strings in a text file first, and then open it in another notebook. A **text file**, which often appears with an extension **.txt**, is a file containing only plain text, which is defined as a pure sequence of character codes. Note that programs that you write and programs that read your text files will usually expect the text file to be in a certain format, that is, organized in a specific way.

To work with text files, we need to use the `open` function, which returns a `file object`. It is commonly used with two arguments:

Python Programming and Numerical Methods. https://doi.org/10.1016/B978-0-12-819549-9.00020-8

```
f = open(filename, mode)
```

The `f` above is the returned file object. The `filename` takes a string that tells the computer the location of the file you want to open, and `mode` is another string containing a few characters, which describes the way in which the file will be used. The common modes are:

- `"r"`, this is the default mode, which opens a file for reading.
- `"w"`, this mode opens a file for writing. If the file does not exist, it creates a new file.
- `"a"`, opens a file in append mode, so that you can append data to end of file. If the file does not exist, it creates a new file.
- `"b"`, opens a file in binary mode.
- `"r+"`, opens a file (does not create) for reading and writing.
- `"w+"`, opens or creates a file for writing and reading, discards existing contents.
- `"a+"`, opens or creates a file for reading and writing, and appends data to the end of file.

11.1.1 WRITING TO A FILE

TRY IT! Create a text file called `test.txt` and write a couple lines in it.

```
In [1]: f = open("test.txt", "w")
        for i in range(5):
            f.write(f"This is line {i}\n")

        f.close()
```

In the above code, we first opened a file object `f` with the file name `"test.txt"`. We used `"w"` for the mode, which indicates we want to write code. We wrote five lines (note the newline `\n` at the end of the string), and then we closed the file object. The content of the file is shown in Fig. 11.1.

NOTE! It is good practice to close the file using `f.close()` at the end of the file. If you do not close them yourself, Python will eventually close them for you. Note that sometimes when writing to a file, the data may not write to disk until you close the file. The longer you keep the file open, the greater the chance you will lose your data.

11.1.2 APPENDING A FILE

Next, we will append a string to the `test.txt` file. It is very similar to how we write the file, with only one difference: we change the mode to "a" instead. See the results in Fig. 11.2.

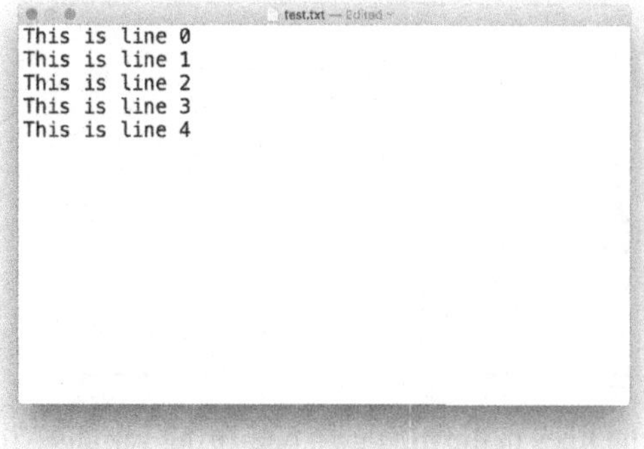

FIGURE 11.1

The content in the text file we write.

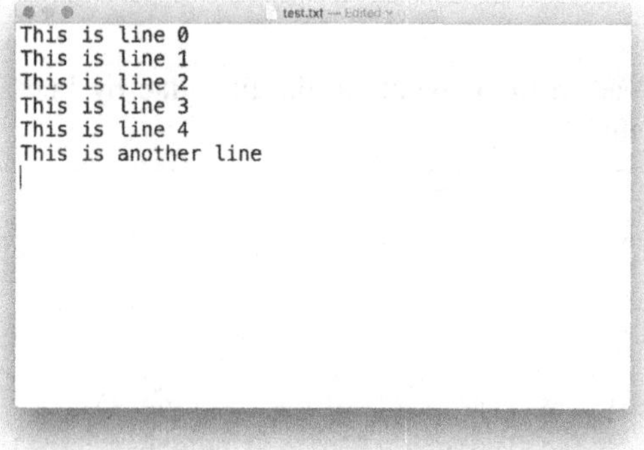

FIGURE 11.2

Append a line to the end of an existing file.

```
In [2]: f = open("test.txt", "a")
        f.write(f"This is another line\n")
        f.close()
```

11.1.3 READING A FILE

We can read a file from disk and store all the contents to a variable. Let us read in the `test.txt` file we created above and store all the contents in the file to a variable `content`.

```
In [3]: f = open("./test.txt", "r")
        content = f.read()
```

```
        f.close()
        print(content)

This is line 0
This is line 1
This is line 2
This is line 3
This is line 4
This is another line
```

In this way, we can store all the lines in the file into a one-string variable. To prove that this is a string, we can verify that variable `content`.

```
In [4]: type(content)

Out[4]: str
```

Sometimes we want to read in the contents in the files line by line and store it in a list: using `f.readlines()` will achieve this.

```
In [5]: f = open("./test.txt", "r")
        contents = f.readlines()
        f.close()
        print(contents)

["This is line 0\n","This is line 1\n","This is line 2\n",
"This is line 3\n","This is line 4\n","This is another line\n"]


In [6]: type(contents)

Out[6]: list
```

11.1.4 DEALING WITH NUMBERS AND ARRAYS

Since we will be working with numerical methods later and will often work with numbers or arrays, we can use the above methods to save the numbers or arrays to a file and read it back to the memory. Although viable, it is a bit clumsy. Instead, programmers commonly use the `NumPy` package to directly save/read an array. An example is presented below.

TRY IT! Store an array `[[1.20, 2.20, 3.00], [4.14, 5.65, 6.42]]` to a file named `my_array.txt` and read it back to a variable called `my_arr`.

```
In [7]: import numpy as np
```

```
In [8]: arr = np.array([[1.20, 2.20, 3.00], [4.14, 5.65, 6.42]])
        arr

Out[8]: array([[1.2 , 2.2 , 3.  ],
               [4.14, 5.65, 6.42]])

In [9]: np.savetxt("my_arr.txt", arr, fmt="%.2f",
                   header = "Col1 Col2 Col3")
```

The above example demonstrates how to save a 2D array into a text file using `np.savetxt`. The first argument is the file name, the second argument is the object we wish to save, and the third argument is to define the format for the output (we use `"%.2f"` to indicate we want the output numbers with two decimals). The fourth argument is the header we wish to write into the file. See the results in Fig. 11.3.

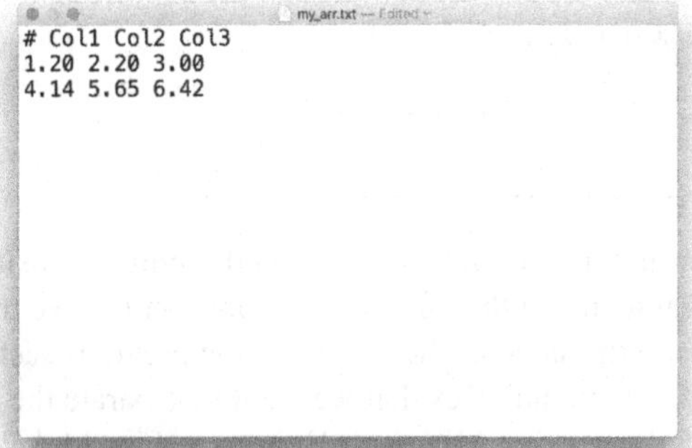

FIGURE 11.3

The `NumPy` array we saved in the file.

```
In [10]: my_arr = np.loadtxt("my_arr.txt")
         my_arr

Out[10]: array([[1.2 , 2.2 , 3.  ],
                [4.14, 5.65, 6.42]])
```

Reading the file directly into an array is very simple by using the `np.loadtxt` function, which skips the first header. There are many different arguments that can control how a file is read. Additional discussion on this topic will be explored in the next section; however, we are not going into much detail herein beyond what is presented above. Check the documentation or use the question mark if you need help.

11.2 CSV FILES

Often, scientific data are stored in the **comma-separated values** (CSV) file format, which is a delimited text file that uses a comma to separate values. It is a very useful format that can store large tables of data (numbers and text) in plain text. Each line (row) in the data is one data record, and each record consists of one or more fields separated by commas. It also can be opened using Microsoft Excel to visualize the rows and columns.

Python has its own CSV module to read and write a CSV file; we are not going to discuss this further. Details on this module is available in the documentation.[1] Instead, we will use the NumPy package to deal with CSV files, because it is preferred to read a CSV file directly into a NumPy array.

11.2.1 WRITING AND OPENING A CSV FILE

Presented below is a simple example of generating 100 rows and 5 columns of data.

```
In [1]: import numpy as np

In [2]: data = np.random.random((100,5))

In [3]: np.savetxt("test.csv", data, fmt = "%.2f",
            delimiter=",", header = "c1, c2, c3, c4, c5")
```

First, we generated some random data for 100 rows and 5 columns using the `np.random` function and assigned it to a `data` variable, using the `np.savetxt` function to save the data to a CSV file. Note that the first three arguments are the same as those used in the previous section, but in this case we set the delimiter argument to ",", which indicates that we want to separate the data using a comma.

Now, we can open the CSV file using Microsoft Excel; see Fig. 11.4. We can also open the CSV file using a text editor; note that the values are separated by commas (Fig. 11.5).

11.2.2 READING A CSV FILE

As before, we can read in the CSV file using the `np.loadtxt` function. If we read in the CSV file we just saved to the disk to a variable `my_csv` and output the first 5 rows, we need to use the `delimiter` again to specify that the data in the file is separated by commas.

```
In [4]: my_csv = np.loadtxt("./test.csv", delimiter=",")
        my_csv[:5, :]

Out[4]: array([[0.84, 0.99, 0.56, 0.24, 0.71],
               [0.33, 0.8 , 0.32, 0.28, 0.83],
               [0.89, 0.19, 0.25, 0.63, 0.84],
               [0.08, 0.49, 0.76, 0.34, 0.69],
               [0.66, 0.65, 0.73, 0.48, 0.12]])
```

[1] https://docs.python.org/3/library/csv.html.

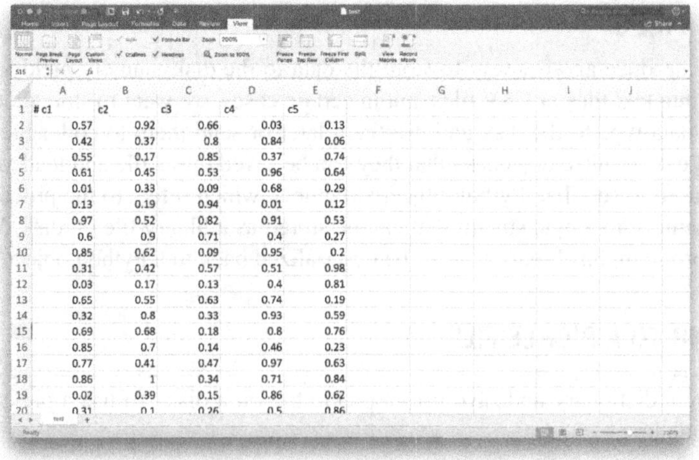

FIGURE 11.4

Open the csv file using Microsoft Excel.

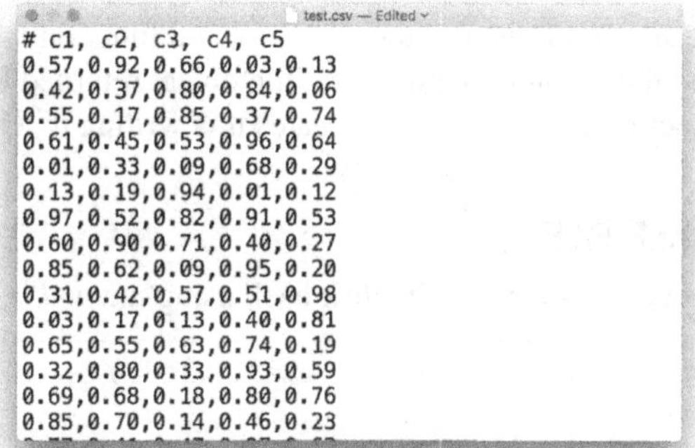

FIGURE 11.5

Open the csv file using a text editor.

11.2.3 BEYOND `NUMPY`

Although `NumPy` is very convenient when dealing with CSV files, there are numerous packages capable of handling CSV files. The `Pandas` package is popular and can easily deal with the tabular data in the `Dataframe`. We encourage you to explore on your own these different ways of handling CSV files.

11.3 PICKLE FILES

This section will introduce another way to store the data to the disk, namely **pickle**. We have talked about saving data into text files or CSV files, but in certain cases, we want to store dictionaries, tuples, lists, or any other data type to the disk and use them later or send them to colleagues. This is where pickle comes in; it can serialize objects so that they can be saved into a file and loaded again later.

Pickle can be used to serialize Python object structures, which refers to the process of converting an object in the memory to a byte stream that can be stored as a binary file on disk. When we load it back to a Python program, this binary file can be deserialized back to a Python object.

11.3.1 WRITING TO A PICKLE FILE

TRY IT! Create a dictionary and save it to a pickle file on a disk. To use a pickle, we need to import the module first.

```
In [1]: import pickle

In [2]: dict_a = {"A":0, "B":1, "C":2}
        pickle.dump(dict_a, open("test.pkl", "wb"))
```

To use pickle to serialize an object, we use the `pickle.dump` function which takes two arguments: the first one is the object, and the second argument is a file object returned by the `open` function. Note that the mode of the `open` function is `"wb"` which indicates it is writing to a binary file.

11.3.2 READING A PICKLE FILE

Next, we load the pickle file we just saved on the disk back using the *pickle.load* function.

```
In [3]: my_dict = pickle.load(open("./test.pkl", "rb"))
        my_dict

Out[3]: {"A": 0, "B": 1, "C": 2}
```

We can see the loading of a pickle file is very similar to the saving process, but here the mode of the `open` function is `"rb"`, which indicates it is reading the binary file. This function deserializes the binary file back to the original object, which in this case is a dictionary. This is also one of the reason that the "pickle" format is popular; it is very easy to store and load a Python data structure without adding extra code to change it.

11.3.3 READING IN PYTHON 2 PICKLE FILE

Sometimes, you may need to open a pickle file from some colleague who generated it using Python 2 instead of Python 3. You can either unpickle it using Python 2, or use Python 3 with the `encoding="latin1"` in the `pickle.load` function.

```
infile = open(filename,"rb")
new_dict = pickle.load(infile, encoding="latin1")
```

WARNING! One drawback of pickle files is that they are not in a universal file format, which means that it is not easy for other programming languages to use it. TXT and CSV files can be easily shared with other colleagues who are not using Python, and they can open them using R, Matlab®, Java, and so on. But pickle files are specially designed for Python and the data is not designed to work with other programming languages.

11.4 JSON FILES

JSON is another format we are introducing. It stands for **JavaScript Object Notation**. A JSON file usually ends with the extension ".json". Unlike pickle, which is Python dependent, JSON is a language-independent data format, making it attractive to use. In addition, it usually takes up less space on the disk, and manipulating a JSON file is faster compared to pickle; search online for more detail regarding the advantages and disadvantages of using JSON versus pickle. Thus, it is a good practice to store your data using JSON. This section briefly explores how to handle JSON files in Python.

11.4.1 JSON FORMAT

Text in JSON is represented using quoted strings that contain value in key-value pairs within {}; the structure should appear very similar as it is nearly identical to the dictionary using in Python. For example,

```
{
  "school": "UC Berkeley",
  "address": {
    "city": "Berkeley",
    "state": "California",
    "postal": "94720"
  },

  "list":[
      "student 1",
      "student 2",
      "student 3"
      ]
}
```

11.4.2 WRITING A JSON FILE

The easiest way to handle JSON in Python is to use the `json` library. There are several other libraries that are available, such as `simplejson`, `jyson`, etc. This section will be limited to using `json`, which is natively supported by Python to enable you to write and load JSON files.

TRY IT! Create a dictionary and save it to a JSON file on the disk. We need to import the pickle module first.

```
In [1]: import json
```

```
In [2]: school = {
  "school": "UC Berkeley",
  "address": {
    "city": "Berkeley",
    "state": "California",
    "postal": "94720"
  },

  "list":[
      "student 1",
      "student 2",
      "student 3"
      ],

  "array":[1, 2, 3]
}
json.dump(school, open("school.json", "w"))
```

To serialize an object using JSON, we use the `json.dump` function which takes two arguments: the first one is the object, and the second argument is a file object returned by the `open` function. Note that the mode of the `open` function is `"w"`, indicating that it is a "write" file.

11.4.3 READING A JSON FILE

Now we load the JSON file just saved on the disk after using the `json.load` function.

```
In [3]: my_school = json.load(open("./school.json", "r"))
my_school
```

We can see the use of `json` is actually very similar to `pickle` in the last section. JSON supports strings and numbers, as well as nested lists, tuples, and objects. We suggest exploring these different options on your own.

11.5 HDF5 FILES

Scientific computing often needs to store large amounts of data with quick access; the file formats we introduced earlier are not applicable. To store large amounts of data, **HDF5** (Hierarchical Data Format) is the solution. It is a powerful binary data format with no limit on the file size. It provides parallel IO (input/output) and carries out a bunch of low-level optimizations "under the hood" to speed up the queries and minimize the storage requirements.

An HDF5 file saves two types of objects: `datasets`, which are array-like collections of data (like `NumPy` arrays), and `groups`, which are folder-like containers that hold datasets and other groups. There are also attributes that can be associated with the datasets and groups to describe some properties. The so called *hierarchical* in HDF5 refers to the fact that the data can be saved like a file system, with folder-like structures, such as folder, subfolder (in HDF5, it is called group, subgroup). Groups operate like dictionaries with the `keys` and `values`: `keys` are names of the groups, and `values` are the subgroups or datasets.

In order to use read/write HDF5 in Python, there are some packages or wrappers that serve this purpose. The most common two packages are `PyTables`[2] and `h5py`.[3] We will only introduce the `h5py` here. You can install `h5py` by using `conda` (refer back to Chapter 1 if you need a refresher).

Once `h5py` is installed, follow the *Quick Start Guide* in `h5py` documentation.[4] For your information, presented below is one example to demonstrate how to create and read an HDF5 file. We will import the `NumPy` and `h5py` first.

```
In [1]: import numpy as np
        import h5py
```

EXAMPLE: Assume we have deployed instruments to monitor the accelerations and GPS location in San Francisco Bay Area, California. We have deployed two accelerometers at Berkeley and Oakland, as well as one GPS station in San Francisco. They record data at different sampling rates, with the accelerometer at Berkeley sampling the data every 0.04 s, and the sensor in Oakland at every 0.01 s. The GPS samples the location every 60 s in San Francisco. We want to store the two types of data into an HDF5, as well as some attributes to indicate where the data has been recorded, the start time of the recording, station name, and the sampling interval.

```
In [2]: # Generate random data for recording
        acc_1 = np.random.random(1000)
        station_number_1 = "1"
        # unix timestamp
        start_time_1 = 1542000276
        # time interval for recording
        dt_1 = 0.04
```

2 https://www.pytables.org.
3 https://www.h5py.org.
4 http://docs.h5py.org/en/latest/quick.html.

```
        location_1 = "Berkeley"

        acc_2 = np.random.random(500)
        station_number_2 = "2"
        start_time_2 = 1542000576
        dt_2 = 0.01
        location_2 = "Oakland"

In [3]: hf = h5py.File("station.hdf5", "w")

In [4]: hf["/acc/1/data"] = acc_1
        hf["/acc/1/data"].attrs["dt"] = dt_1
        hf["/acc/1/data"].attrs["start_time"] = start_time_1
        hf["/acc/1/data"].attrs["location"] = location_1

        hf["/acc/2/data"] = acc_2
        hf["/acc/2/data"].attrs["dt"] = dt_2
        hf["/acc/2/data"].attrs["start_time"] = start_time_2
        hf["/acc/2/data"].attrs["location"] = location_2

        hf["/gps/1/data"] = np.random.random(100)
        hf["/gps/1/data"].attrs["dt"] = 60
        hf["/gps/1/data"].attrs["start_time"] = 1542000000
        hf["/gps/1/data"].attrs["location"] = "San Francisco"

In [5]: hf.close()
```

The above code shows the core concepts in HDF5: the groups, datasets, and attributes. First, we created an HDF5 object for writing: `station.hdf5`. Then we stored the data into two top-level groups: `acc` and `gps`. Both of these top-levels groups contain subgroups labeled 1 or 2 to indicate the station names. Each station will contain the next level subgroup, i.e., the data, which is used to store the array data we collected. Next, we add attributes to the groups or the data. In this case, we have added the `dt`, `start_time`, and `location` as the attributes to the datasets stored here. You can see that it is quite similar to folder-like structure, with data `acc_1` saved at `/acc/1/data`. Last, we close the file object.

Saving data in HDF5 is easy. We can also use the function `create_dataset` and `create_group` as shown in the quick start.[5]

11.5.1 READING AN HDF5 FILE

Now suppose you send the `station.hdf5` to a colleague, who wants to get access to the data. Here is how he/she will do it.

[5] http://docs.h5py.org/en/latest/quick.html.

```
In [6]: hf_in = h5py.File("station.hdf5", "r")

In [7]: list(hf_in.keys())

Out[7]: ["acc", "gps"]

In [8]: acc = hf_in["acc"]

In [9]: list(acc.keys())

Out[9]: ["1", "2"]

In [10]: data_1 = hf_in["acc/1/data"]

In [11]: data_1.value[:10]

Out[11]: array([0.41820889, 0.89832446, 0.40229251, 0.41287538,
                0.16173359, 0.75855904, 0.89288185, 0.82944522,
                0.84228139, 0.50365515])

In [12]: list(data_1.attrs)

Out[12]: ["dt", "start_time", "location"]

In [13]: data_1.attrs["dt"]

Out[13]: 0.04

In [14]: data_1.attrs["location"]

Out[14]: "Berkeley"
```

Reading an HDF5 turns out to be easy as well using `h5py`. After we read in the HDF5 to `hf_in`, we can see what groups are in the HDF5 using the `keys` function. Then we can get access to the group members and see what is contained in the subgroups as the `hf_in["acc"]`, or directly specify the path to the datasets as `hf_in["acc/1/data"]` and get the array data. Remember that the attributes associated with the data can also be accessed as a dictionary.

11.6 SUMMARY AND PROBLEMS

11.6.1 SUMMARY

1. Data must often be stored to a disk for a later Python session or for reading by other programs.
2. Data created by other programs may have to be read by Python.
3. Python has functions to read and write data in several standard forms: TXT, CSV, pickle, and HDF5.

11.6.2 PROBLEMS

1. Create a list and save it in a text file so that each of the items in the list will take one line.
2. Save the same list in Problem 1 to a CSV file.
3. Create a 2D `NumPy` array, then save it to a CSV file and read it back to a 2D array.
4. Save the same array in Problem 2 to a pickle file and load it back.
5. Create a dictionary and save it to a JSON file.
6. Create a 1D `NumPy` array, and save it to a JSON file with the key named "data." Then load it back.

CHAPTER

VISUALIZATION AND PLOTTING 12

CONTENTS

12.1 2D PLOTTING

In Python, the **matplotlib** is the most important package when plotting data. View the matplotlib gallery[1] to get a sense of what can be done. Usually, the first thing you should do is to import the matplotlib package. In `Jupyter notebook`, we can show the figure directly within the notebook and also have the interactive operations like pan, zoom in/out, and so on using the magic command – `%matplotlib notebook`. Below are some examples.

```
In [1]: import numpy as np
        import matplotlib.pyplot as plt
        %matplotlib notebook
```

The basic plotting function is `plot(x,y)`. The `plot` function takes in two lists/arrays, `x` and `y`, and produces a visual display of the respective points in `x` and `y`.

TRY IT! Given the lists `x = [0, 1, 2, 3]` and `y = [0, 1, 4, 9]`, use the plot function to produce a plot of `x` versus `y`.

```
In [2]: x = [0, 1, 2, 3]
        y = [0, 1, 4, 9]
        plt.plot(x, y)
        plt.show()
```

[1] https://matplotlib.org/gallery/index.html#gallery.

Python Programming and Numerical Methods. https://doi.org/10.1016/B978-0-12-819549-9.00021-X

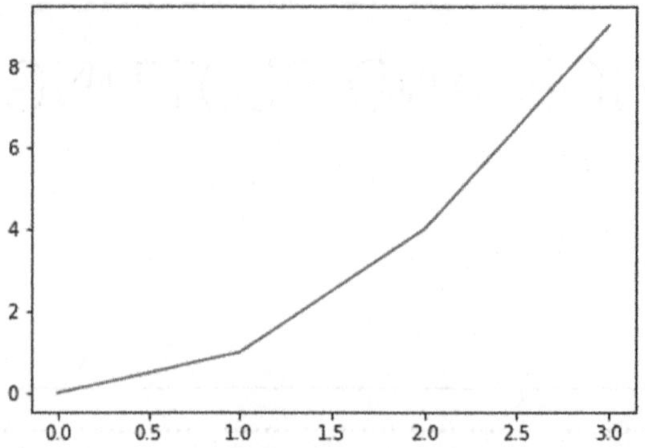

FIGURE 12.1

Plot of x versus y.

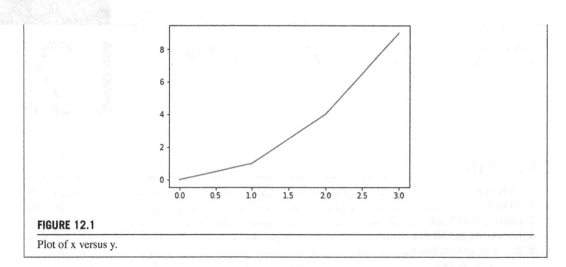

Note in Fig. 12.1 that by default, the plot function connects each point with a blue line. To make the function look smooth, you can use finer discretization points. The `plt.plot` function did the main job to plot the figure, and `plt.show()` is telling Python that we are done plotting and asks to show the figure. The buttons beneath the plot allow you to move the line, zoom in or out, or save the figure. Note that before you plot Fig. 12.2, you need to turn off the interactive plot by pressing the `stop interaction` button on the top right of the figure, otherwise, the next figure will be plotted in the same frame, or use the magic function `%matplotlib inline` to turn off the interactive features.

TRY IT! Make a plot of the function $f(x) = x^2$ for $-5 \le x \le 5$.

```
In [3]: %matplotlib inline

In [4]: x = np.linspace(-5,5, 100)
        plt.plot(x, x**2)
        plt.show()
```

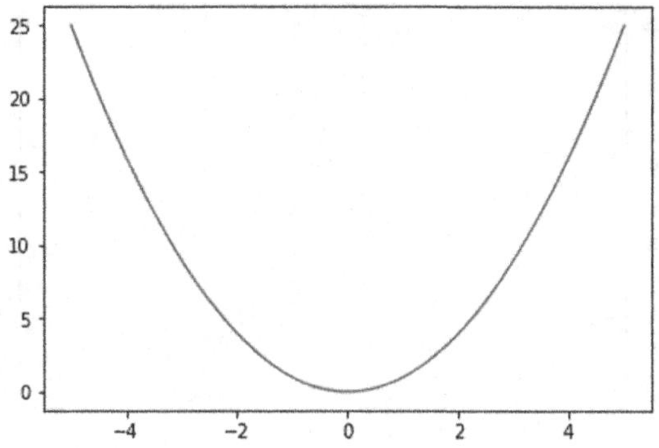

FIGURE 12.2

Plot of the function $f(x) = x^2$ for $-5 \le x \le 5$.

To change the marker or line, add a third input argument into plot, which is a string that specifies the color and line style to be used in the plot. For example, `plot(x,y,"ro")` will plot the elements of `x` against the elements of `y` using red, `"r"`, circles, `"o"`. The possible specifications are shown below in the table (see also Fig. 12.3).

Symbol	Description	Symbol	Description
`b`	blue	`T`	T
`g`	green	`s`	square
`r`	red	`d`	diamond
`c`	cyan	`v`	triangle (down)
`m`	magenta	`^`	triangle (up)
`y`	yellow	`<`	triangle (left)
`k`	black	`>`	triangle (right)
`w`	white	`p`	pentagram
`.`	point	`h`	hexagram
`o`	circle	`-`	solid
`x`	x-mark	`:`	dotted
`+`	plus	`-.`	dashed–dotted
`*`	star	`-`	dashed

TRY IT! Make a plot of the function $f(x) = x^2$ for $-5 \le x \le 5$ using a dashed green line.

```
In [5]: x = np.linspace(-5,5, 100)
        plt.plot(x, x**2, "g-")
        plt.show()
```

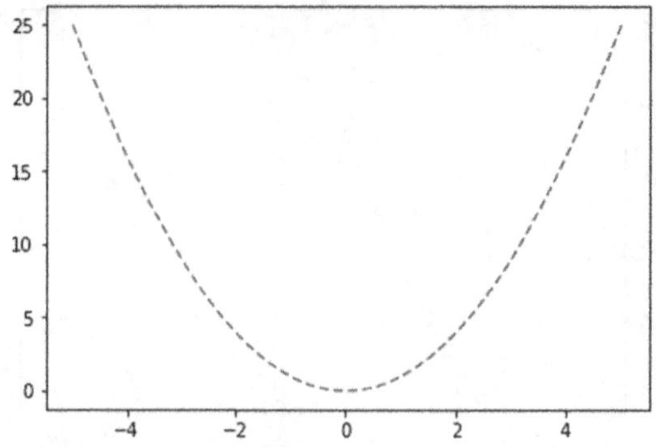

FIGURE 12.3

Plot of the function $f(x) = x^2$ for $-5 \le x \le 5$ using dashed line.

Before the `plt.show()` statement, you can add in and plot more datasets within one figure (see also Fig. 12.4).

TRY IT! Make a plot of the function $f(x) = x^2$ and $g(x) = x^3$ for $-5 \le x \le 5$. Use different colors and markers for each function.

```
In [6]: x = np.linspace(-5,5,20)
        plt.plot(x, x**2, "ko")
        plt.plot(x, x**3, "r*")
        plt.show()
```

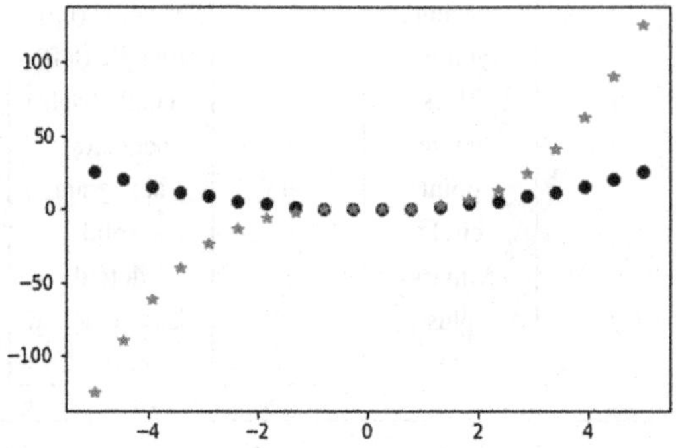

FIGURE 12.4

Using different colors and markers.

It is customary in engineering and science to always provide your plot with both title and axis labels so that people know what your plot is about. Sometimes you want to change the size of the figure as well. You can add a title to your plot using the `title` function, which takes as input a string and puts that string as the title of the plot. The functions `xlabel` and `ylabel` work in the same way to name your axis labels. To change the size of the figure, create a figure object and resize it. Note that every time we call `plt.figure` function, we create a new figure object and draw on it.

TRY IT! Add a title and axis labels to the previous plot. Make the figure larger so that it is 10 inches wide and 6 inches high.

```
In [7]: plt.figure(figsize = (10,6))

        x = np.linspace(-5,5,20)
        plt.plot(x, x**2, "ko")
        plt.plot(x, x**3, "r*")
        plt.title(f"Plot of Various Polynomials from {x[0]} to
                     {x[-1]}")
        plt.xlabel("X axis", fontsize = 18)
        plt.ylabel("Y axis", fontsize = 18)
        plt.show()
```

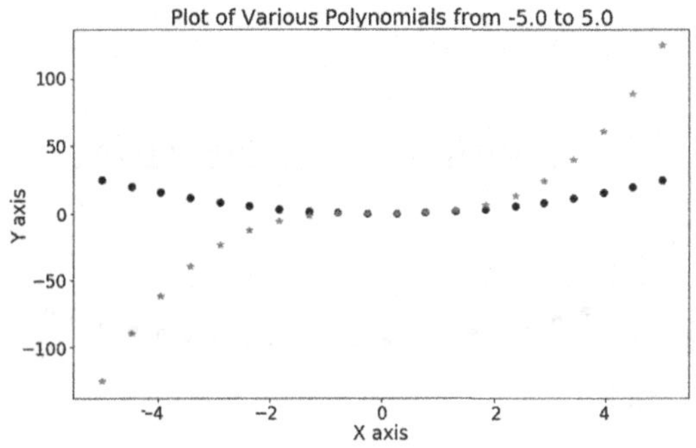

FIGURE 12.5

Adding title and axis labels to a figure.

As demonstrated in Fig. 12.5, we can change any part of the figure, such as the x and y axis label size by specifying a `fontsize` argument in the `plt.xlabel` function. There are some predefined styles that automatically change the style. Here is the list of the styles.

```
In [8]: print(plt.style.available)

["seaborn-dark", "seaborn-darkgrid", "seaborn-ticks",
"fivethirtyeight", "seaborn-whitegrid", "classic",
"_classic_test", "fast", "seaborn-talk",
"seaborn-dark-palette", "seaborn-bright",
"seaborn-pastel", "grayscale", "seaborn-notebook",
```

```
"ggplot", "seaborn-colorblind", "seaborn-muted",
"seaborn’, "Solarize_Light2", "seaborn-paper",
"bmh", "tableau-colorblind10", "seaborn-white",
"dark_background", "seaborn-poster", "seaborn-deep"]
```

One of my favorite predefined styles is the `seaborn` style; we can change it using the `plt.style.use` function. If we change it to `"seaborn-poster"`, it will make everything bigger (see also Fig. 12.6).

```
In [9]: plt.style.use("seaborn-poster")

In [10]: plt.figure(figsize = (10,6))

         x = np.linspace(-5,5,20)
         plt.plot(x, x**2, "ko")
         plt.plot(x, x**3, "r*")
         plt.title(f"Plot of Various Polynomials from {x[0]} to
                   {x[-1]}")
         plt.xlabel("X axis")
         plt.ylabel("Y axis")
         plt.show()
```

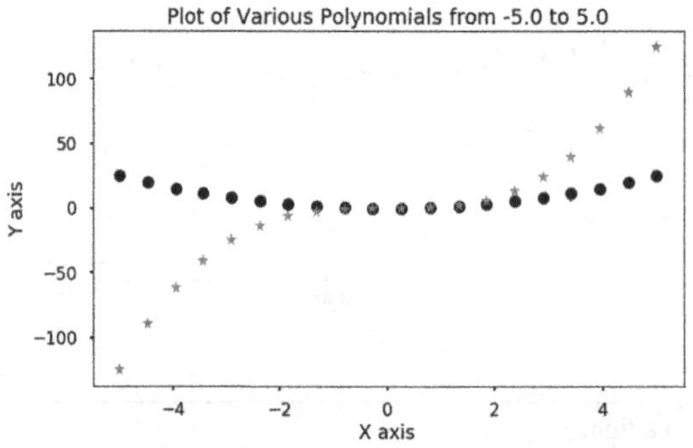

FIGURE 12.6

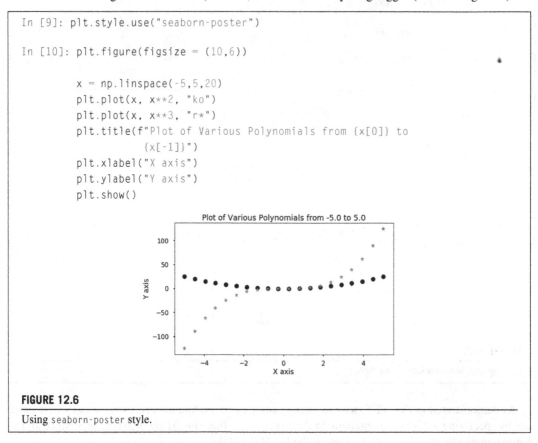

Using `seaborn-poster` style.

You can create a legend for your plot by using the `legend` function, by adding the `label` argument in the `plot` function. The legend function also takes argument of `loc` to indicate where to put the legend; change it from 0 to 10 and see what happens (see also Fig. 12.7).

```
In [11]: plt.figure(figsize = (10,6))

         x = np.linspace(-5,5,20)
         plt.plot(x, x**2, "ko", label = "quadratic")
```

```
plt.plot(x, x**3, "r*", label = "cubic")
plt.title(f"Plot of Various Polynomials from {x[0]} to
          {x[-1]}")
plt.xlabel("X axis")
plt.ylabel("Y axis")
plt.legend(loc = 2)
plt.show()
```

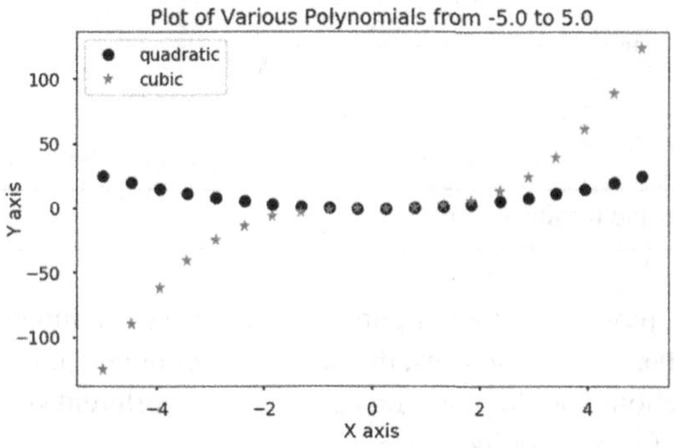

FIGURE 12.7

Using a legend in the figure.

Finally, you can further customize the appearance of your plots by changing the limits of each axis using the `xlim` or `ylim` function. Using the `grid` function will turn on the grid of Fig. 12.8.

TRY IT! Change the limits of the plot so that *x* is visible from −6 to 6, and *y* is visible from −10 to 10. Turn on the grid.

```
In [12]: plt.figure(figsize = (10,6))

         x = np.linspace(-5,5,100)
         plt.plot(x, x**2, "ko", label = "quadratic")
         plt.plot(x, x**3, "r*", label = "cubic")
         plt.title(f"Plot of Various Polynomials from {x[0]} to
                   {x[-1]}")
         plt.xlabel("X axis")
         plt.ylabel("Y axis")
         plt.legend(loc = 2)
         plt.xlim(-6.6)
         plt.ylim(-10,10)
         plt.grid()
         plt.show()
```

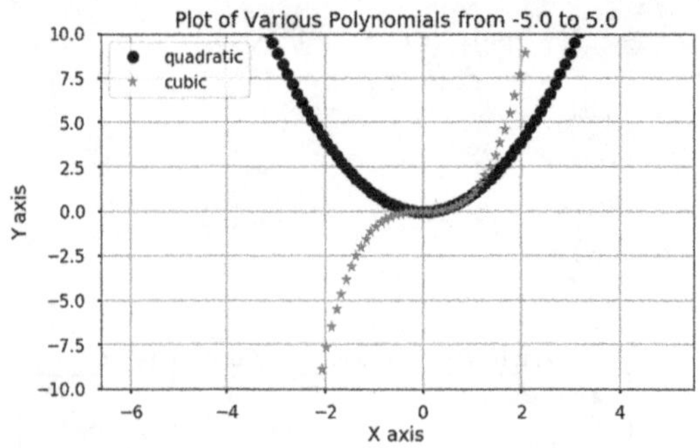

FIGURE 12.8

Changing limits of the figure and turning on grid.

We can create a table of plots on a single figure using the `subplot` function. The `subplot` function takes three inputs: the number of rows of plots, the number of columns of plots, and designating which plot all calls to plotting functions should plot. You can move to a different subplot by calling the subplot again with a different entry for the plot location.

There are several other plotting functions that plot x versus y data. Some of them are `scatter`, `bar`, `loglog`, `semilogx`, and `semilogy`. Function `scatter` works exactly the same as `plot` except it defaults to red circles (i.e., `plot(x,y,"ro")` is equivalent to `scatter(x,y)`). The `bar` function plots bars centered at `x` with height `y`. The `loglog`, `semilogx`, and `semilogy` functions plot the data in `x` and `y`, with the `x` and `y` axis on a log-scale, the `x` axis on a log-scale and the `y` axis on a linear scale, and the `y` axis on a log-scale and the `x` axis on a linear scale, respectively (see also Fig. 12.9).

TRY IT! Given the lists x = np.arange(11) and $y = x^2$, create a 2 × 3 subplot where each subplot plots x versus y using `plot`, `scatter`, `bar`, `loglog`, `semilogx`, and `semilogy`. Title and label each plot appropriately. Use a grid, but a legend is not necessary here.

```
In [13]: x = np.arange(11)
         y = x**2

         plt.figure(figsize = (14, 8))

         plt.subplot(2, 3, 1)
         plt.plot(x,y)
         plt.title("Plot")
         plt.xlabel("X")
         plt.ylabel("Y")
         plt.grid()

         plt.subplot(2, 3, 2)
```

```
plt.scatter(x,y)
plt.title("Scatter")
plt.xlabel("X")
plt.ylabel("Y")
plt.grid()

plt.subplot(2, 3, 3)
plt.bar(x,y)
plt.title("Bar")
plt.xlabel("X")
plt.ylabel("Y")
plt.grid()

plt.subplot(2, 3, 4)
plt.loglog(x,y)
plt.title("Loglog")
plt.xlabel("X")
plt.ylabel("Y")
plt.grid(which="both")

plt.subplot(2, 3, 5)
plt.semilogx(x,y)
plt.title("Semilogx")
plt.xlabel("X")
plt.ylabel("Y")
plt.grid(which="both")

plt.subplot(2, 3, 6)
plt.semilogy(x,y)
plt.title("Semilogy")
plt.xlabel("X")
plt.ylabel("Y")
plt.grid()

plt.tight_layout()

plt.show()
```

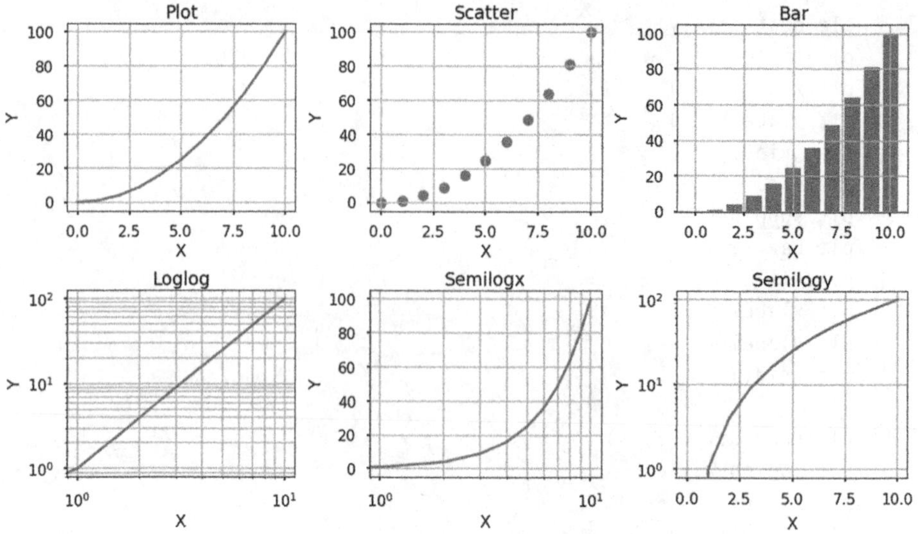

FIGURE 12.9

Different type of plots.

We can see that at the end of our plot, we used `plt.tight_layout` to ensure that the subfigures did not overlap with each other. Rerun this plot and see the effect without this statement.

Sometimes, you want to save the figures in a specific format, such as pdf, jpeg, png, and so on. You can do this with the function `plt.savefig` (see also Fig. 12.10).

```
In [14]: plt.figure(figsize = (8,6))
         plt.plot(x,y)
         plt.xlabel("X")
         plt.ylabel("Y")
         plt.savefig("image.pdf")
```

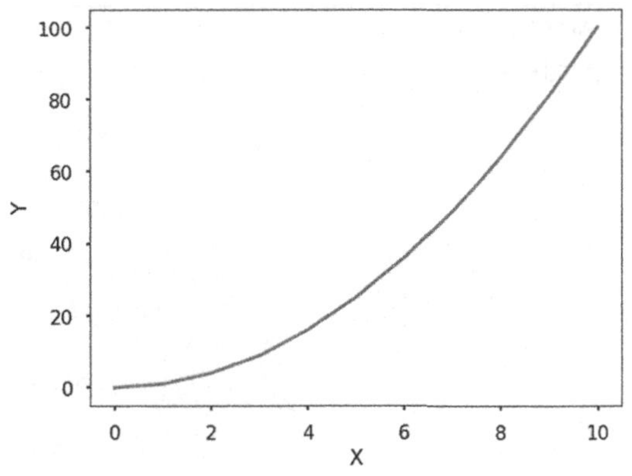

FIGURE 12.10

Saving a figure.

Finally, there are other functions for plotting data in 2D. The `errorbar` function plots `x` versus `y` data but with error bars for each element. The `polar` function plots θ versus `r` rather than `x` versus `y`. The `stem` function plots stems at `x` with height at `y`. The `hist` function makes a histogram of a dataset; `boxplot` gives a statistical summary of a dataset; and `pie` makes a pie chart. The usage of these functions are left for you to explore on your own. Remember to check the examples on the `matplotlib` gallery.[2]

12.2 3D PLOTTING

To plot three-dimensional (3D) figures, first we need to import the `mplot3d` toolkit, which adds the simple 3D plotting capabilities to `matplotlib`.

```
In [1]: import numpy as np
        from mpl_toolkits import mplot3d
        import matplotlib.pyplot as plt
        plt.style.use("seaborn-poster")
```

Once we have imported the `mplot3d` toolkit, we can create 3D axes and add data to the axes. Let us first create 3D axes (see also Fig. 12.11).

2 https://matplotlib.org/gallery/index.html#gallery.

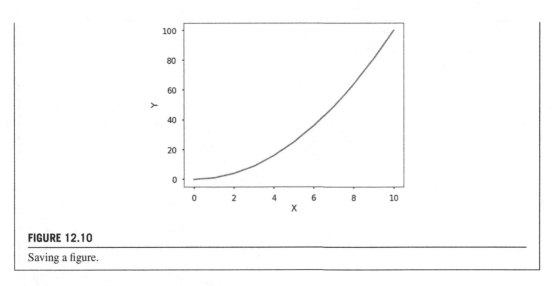

```
In [2]: fig = plt.figure(figsize = (10,10))
        ax = plt.axes(projection="3d")
        plt.show()
```

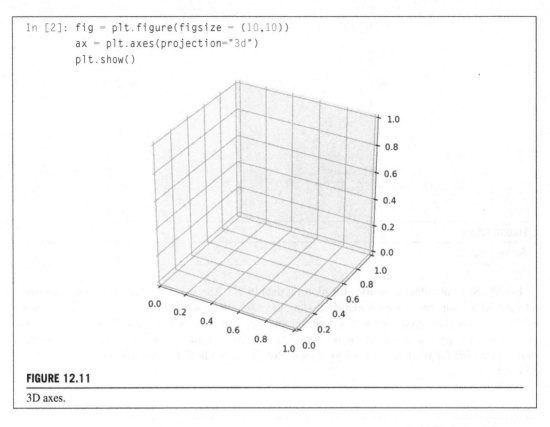

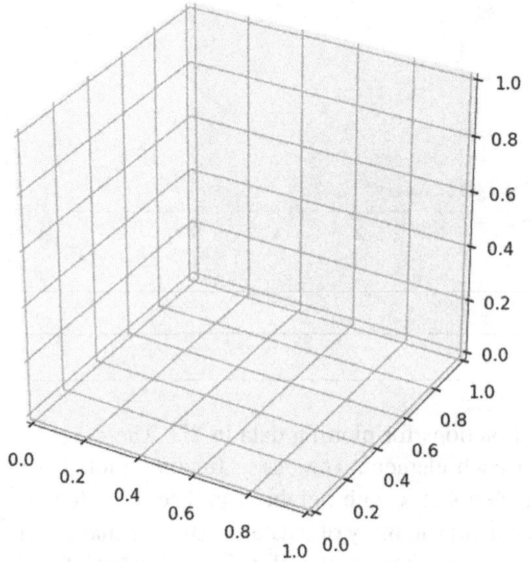

FIGURE 12.11

3D axes.

The `ax = plt.axes(projection="3d")` created a 3D-axes object. To add data to it, use `plot3D` function. It is possible to change the title, set the x, y, z labels for the plot as well.

TRY IT! Consider the parameterized dataset: t is a vector from 0 to 10π with a step $\pi/50$, $x = \sin(t)$, and $y = \cos(t)$. Make a 3D plot of the (x, y, t) dataset using `plot3`. Turn on the grid, make the axis equal, and add axis labels and a title. Activate the interactive plot using `%matplotlib notebook` so that you can move and rotate the figure as well.

```
In [3]: %matplotlib notebook

In [4]: fig = plt.figure(figsize = (8,8))
        ax = plt.axes(projection="3d")
        ax.grid()
        t = np.arange(0, 10*np.pi, np.pi/50)
        x = np.sin(t)
        y = np.cos(t)
```

```
ax.plot3D(x, y, t)
ax.set_title("3D Parametric Plot")

# Set axes label
ax.set_xlabel("x", labelpad=20)
ax.set_ylabel("y", labelpad=20)
ax.set_zlabel("t", labelpad=20)

plt.show()
```

See the interactive examples in the notebook.

Try to rotate the above figure to get a 3D view of the plot. You may notice that we also set the `labelpad=20` to the *3-axis* labels, which will make it so that the label does not overlap with the tick texts.

We can also plot a 3D scatter plot using the `scatter` function (see also Fig. 12.12).

TRY IT! Make a 3D scatter plot with randomly generated 50 data points for `x`, `y`, and `z`. Set the point color as red and size of the point as 50.

```
In [5]: # Turn off the interactive plot
        %matplotlib inline

In [6]: x = np.random.random(50)
        y = np.random.random(50)
        z = np.random.random(50)

        fig = plt.figure(figsize = (10,10))
        ax = plt.axes(projection="3d")
        ax.grid()

        ax.scatter(x, y, z, c = "r", s = 50)
        ax.set_title("3D Scatter Plot")

        # Set axes label
        ax.set_xlabel("x", labelpad=20)
        ax.set_ylabel("y", labelpad=20)
        ax.set_zlabel("z", labelpad=20)

        plt.show()
```

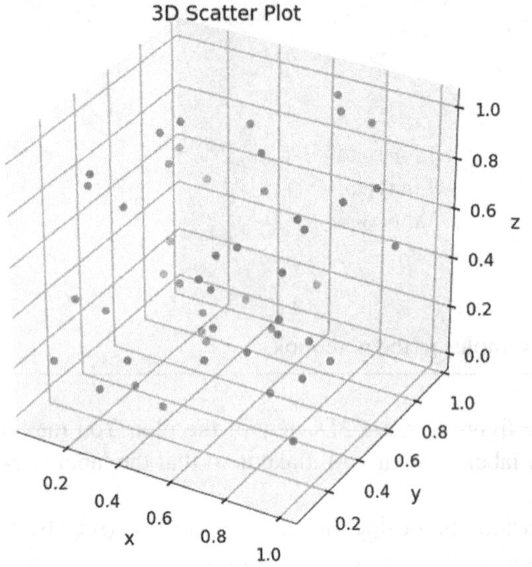

FIGURE 12.12

3D scatter plot.

When plotting in three dimensions, sometimes it is desirable to use a surface plot rather than a line plot. In 3D surface plotting, we wish to make a graph of some relationship $f(x, y)$. In surface plotting all (x, y) pairs must be given. This is not straightforward to do using vectors; therefore, in surface plotting, the first data structure you must create is called a mesh. Given lists/arrays of `x` and `y` values, a mesh is a listing of all the possible combinations of `x` and `y`. In Python, the mesh is given as two arrays `X` and `Y`, where `X[i,j]` and `Y[i,j]` that define the possible (x, y) pairs. A third array, `Z`, can then be created such that `Z[i,j]=f(X[i,j],Y[i,j])`. A mesh can be created using the `np.meshgrid` function in Python. The `meshgrid` function has the inputs `x` and `y`, which are lists containing the independent dataset. The output variables `X` and `Y` were described earlier.

TRY IT! Create a mesh for `x = [1, 2, 3, 4]` and `y = [3, 4, 5]` using the `meshgrid` function.

```
In [7]: x = [1, 2, 3, 4]
        y = [3, 4, 5]

        X, Y = np.meshgrid(x, y)
        print(X)

[[1 2 3 4]
 [1 2 3 4]
```

```
 [1 2 3 4]]


In [8]: print(Y)

[[3 3 3 3]
 [4 4 4 4]
 [5 5 5 5]]
```

In addition, we can plot 3D surfaces in Python: the function to plot 3D surfaces is `plot_surface(X,Y,Z)`, where `X` and `Y` are the output arrays from `meshgrid`, and `Z=f(X,Y)` or `Z[i,j]=f(X[i,j],Y[i,j])` (see Fig. 12.13).

TRY IT! Make a plot of the surface $f(x, y) = \sin(x) \cdot \cos(y)$ for $-5 \le x \le 5, -5 \le y \le 5$ using the `plot_surface` function. Take care to use a sufficiently fine discretization in `x` and `y` so that the plot looks smooth.

```
In [9]: fig = plt.figure(figsize = (12,10))
        ax = plt.axes(projection="3d")

        x = np.arange(-5, 5.1, 0.2)
        y = np.arange(-5, 5.1, 0.2)

        X, Y = np.meshgrid(x, y)
        Z = np.sin(X)*np.cos(Y)

        surf = ax.plot_surface(X, Y, Z, cmap = plt.cm.cividis)

        # Set axes label
        ax.set_xlabel("x", labelpad=20)
        ax.set_ylabel("y", labelpad=20)
        ax.set_zlabel("z", labelpad=20)

        fig.colorbar(surf, shrink=0.5, aspect=8)

        plt.show()
```

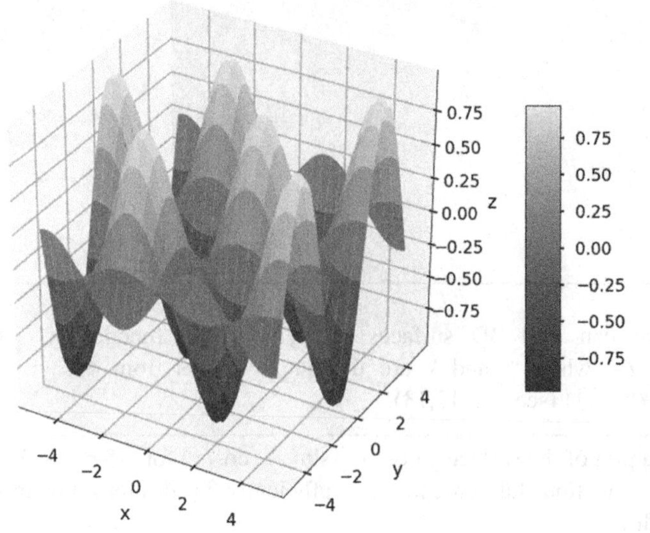

FIGURE 12.13

3D surface plot.

Note that the surface plot shows different colors for different elevations, yellow for higher and blue for lower (light grey for higher and dark grey for lower in print version); this is because we used the colormap `plt.cm.cividis` in the surface plot. You can change to different color schemes for the surface plot. These are left for you to complete as additional exercises. We also plotted a colorbar to show the corresponding colors to different values.

We can have subplots of different 3D plots as well (see Fig. 12.14). To do that, use the `add_subplot` function from the figure object we created to generate the subplots for 3D cases.

TRY IT! Make a 1 × 2 subplot to plot the above `X`, `Y`, `Z` data in a wireframe plot and a surface plot.

```
In [10]: fig = plt.figure(figsize=(12,6))


         ax = fig.add_subplot(1, 2, 1, projection="3d")
         ax.plot_wireframe(X,Y,Z)
         ax.set_title("Wireframe plot")

         ax = fig.add_subplot(1, 2, 2, projection="3d")
         ax.plot_surface(X,Y,Z)
         ax.set_title("Surface plot")
```

```
    plt.tight_layout()

    plt.show()
```

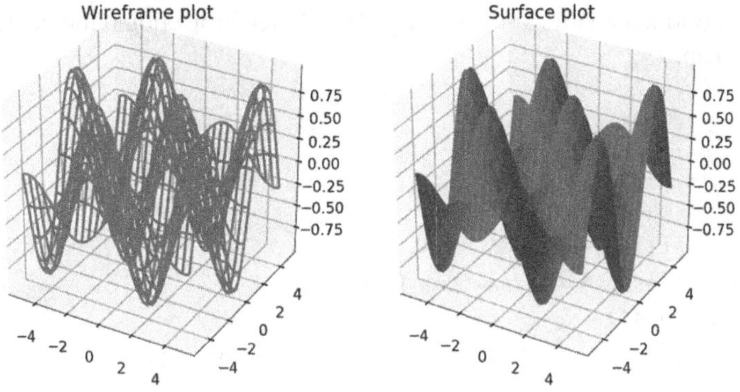

FIGURE 12.14

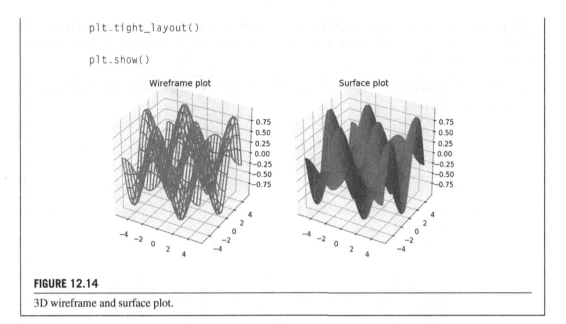

3D wireframe and surface plot.

Given there are many more functions related to plotting in Python, this is in no way an exhaustive list; however, it should be sufficient to get you started so that you can discover which plotting functions in Python suit you best and provide you with enough background to learn how to use them properly. You can find more examples of different types 3D plots on the mplot3d tutorial website.[3]

12.3 WORKING WITH MAPS

Often in engineering and science, we have to interact with maps or use geographical representation of data. There are many different Python packages that can draw maps, such as basemap,[4] cartopy,[5] folium,[6] etc. The `folium` package allows you to plot interactive maps for webpages. But most of the time, we only need to plot a static map to show spatial features, and `basemap` and `cartopy` will do the job. This section is a brief introduction into how to draw maps with data using `cartopy`. First, install the `cartopy` with `conda install cartopy`.

The basics of a map are simple: it is a 2D plot with specific projections.[7] The x-axis is the longitude ranging from -180 to 180, which specifies the east–west position of a point on the Earth's surface. The y-axis is the latitude ranging from -90 to 90, which describes a point's south–north position. If you specify a latitude and longitude pair, you can uniquely determine the location of the point on Earth.

[3] https://matplotlib.org/mpl_toolkits/mplot3d/tutorial.html.
[4] https://matplotlib.org/basemap/.
[5] https://scitools.org.uk/cartopy/docs/latest/.
[6] https://github.com/python-visualization/folium.
[7] https://en.wikipedia.org/wiki/Map_projection.

The package cartopy has very nice API to interact with matplotlib to plot a map; we only need to tell the matplotlib to use a specific map projection, and then we can add other map features to the plot (see also Fig. 12.15).

TRY IT! Plot a world map with cartopy using Plate Carree projection (Google it), and draw the coastline on the map.

```
In [1]: import cartopy.crs as ccrs
        import matplotlib.pyplot as plt
        %matplotlib inline

In [2]: plt.figure(figsize = (12, 8))
        ax = plt.axes(projection=ccrs.PlateCarree())
        ax.coastlines()
        ax.gridlines(draw_labels=True)
        plt.show()
```

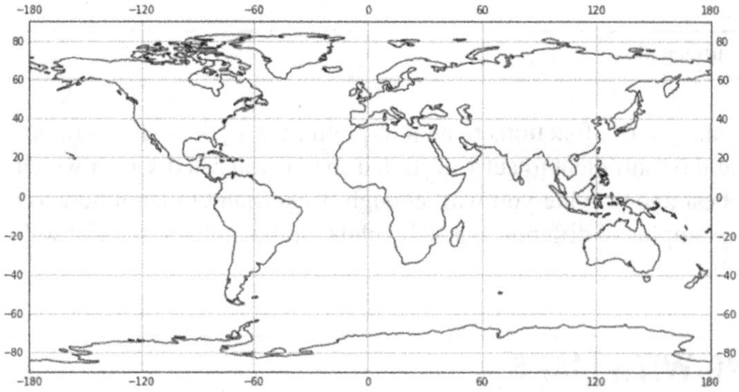

FIGURE 12.15

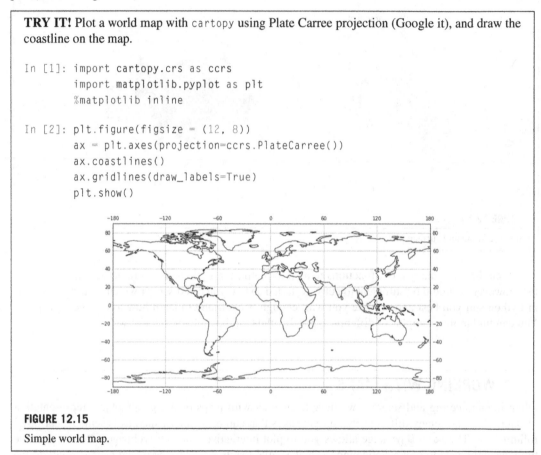

Simple world map.

The above example plotted the map with the Plate Carree projection. In addition, we turned on the grid lines and drew the labels on the maps. We suggest you check out other cartopy supported projections.[8]

The map background we drew above is blank; however, we can easily add a nice map background in cartopy using stock_img (see also Fig. 12.16).

[8] https://scitools.org.uk/cartopy/docs/v0.16/crs/projections.html#cartopy-projections.

```
In [3]: plt.figure(figsize = (12, 8))
        ax = plt.axes(projection=ccrs.PlateCarree())
        ax.coastlines()
        ax.stock_img()
        ax.gridlines(draw_labels=True)
        plt.show()
```

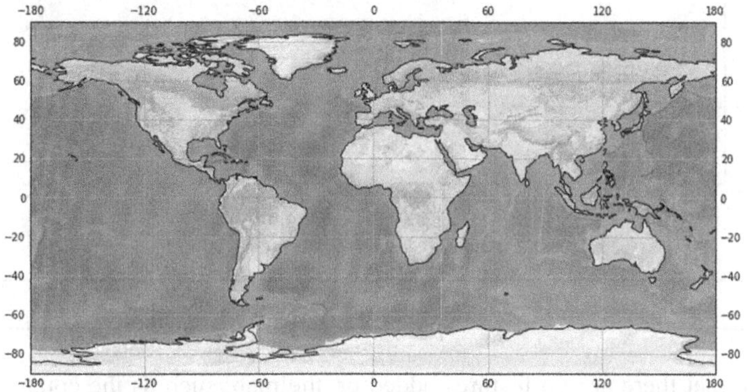

FIGURE 12.16

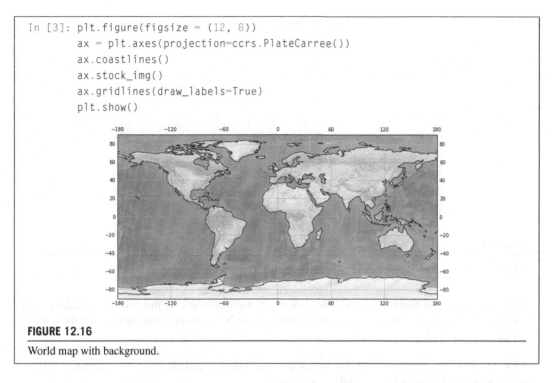

World map with background.

We can zoom in the map to any places on the Earth using the `ax.set_extent` function, which takes a list with the first two numbers that are the x-axis limits and the last two numbers are the y-axis limits (see also Fig. 12.17).

TRY IT! Zoom in the map to the United States.

```
In [4]: plt.figure(figsize = (10, 5))
        ax = plt.axes(projection=ccrs.PlateCarree())
        ax.coastlines()
        ax.set_extent([-125, -75, 25, 50])
        ax.gridlines(draw_labels=True)
        plt.show()
```

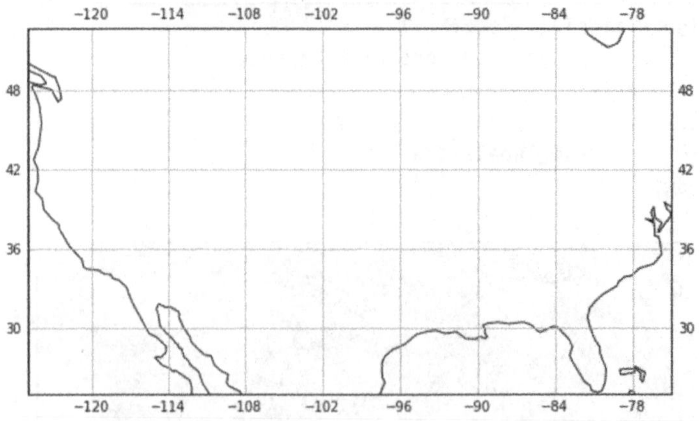

FIGURE 12.17

Simple US map.

Did you notice that there are no features added on the map, such as the country boundary, state boundary, lakes/water, etc.? In `cartopy`, these features must be specified if we want to add them to our map (see also Fig. 12.18).

TRY IT! For the map of the United States we made above, add the following features: land, ocean, states and country borders, lakes, and rivers.

```
In [5]: import cartopy.feature as cfeature

In [6]: plt.figure(figsize = (10, 5))
        ax = plt.axes(projection=ccrs.PlateCarree())
        ax.coastlines()
        ax.set_extent([-125, -75, 25, 50])

        ax.add_feature(cfeature.LAND)
        ax.add_feature(cfeature.OCEAN)
        ax.add_feature(cfeature.STATES, linestyle=":")
        ax.add_feature(cfeature.BORDERS)
        ax.add_feature(cfeature.LAKES, alpha=0.5)
        ax.add_feature(cfeature.RIVERS)

        plt.show()
```

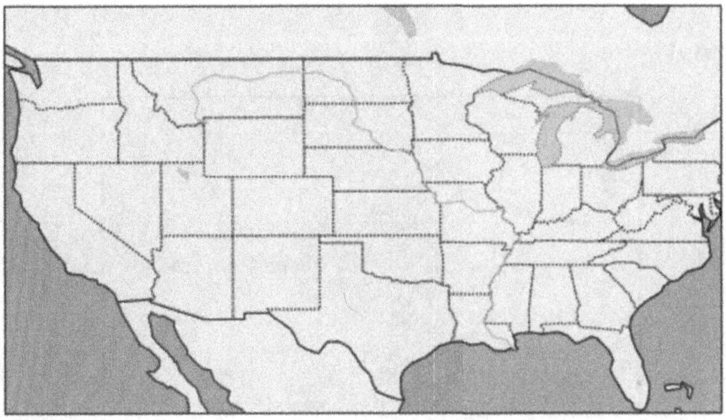

FIGURE 12.18

US map with various features.

It is possible to zoom in further to a smaller area, but we would need to download and use the high-resolution coastlines and land to have a decent-looking map (see also Fig. 12.19).

TRY IT! Plot the San Francisco Bay Area with the 10 m-resolution coastlines and land. Try to change one of them to 50 m and see what happens.

```
In [7]: plt.figure(figsize = (10, 8))
        ax = plt.axes(projection=ccrs.PlateCarree())
        ax.coastlines(resolution="10m")
        ax.set_extent([-122.8, -122, 37.3, 38.3])

        # we can add high-resolution land and water
        LAND =
          cfeature.NaturalEarthFeature("physical", "land", "10m",
                       edgecolor="face",
                       facecolor=cfeature.COLORS["land"],
                       linewidth=.1)

        OCEAN =
          cfeature.NaturalEarthFeature("physical","ocean","10m",
                       edgecolor="face",
                       facecolor=cfeature.COLORS["water"],
                       linewidth=.1)

        ax.add_feature(LAND, zorder=0)
        ax.add_feature(OCEAN, zorder=0)
```

```
plt.show()
```

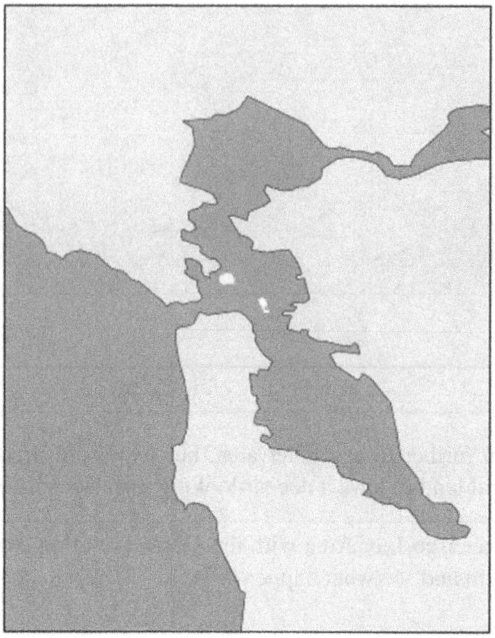

FIGURE 12.19

Bay area map with higher resolution.

In many cases, we want to plot our data onto the map and show the spatial location of different entities. Data can be added to it in exactly the same way as with normal `matplotlib` axes. By default, the coordinate system of any additional data is the same as the coordinate system of the axes we defined at the beginning of the plot. Let us first try to add some data to the map above (see also Fig. 12.20).

TRY IT! Add the location of UC Berkeley and Stanford University on the Bay Area map above.

```
In [8]: plt.figure(figsize = (10, 8))

        # plot the map related stuff
        ax = plt.axes(projection=ccrs.PlateCarree())
        ax.coastlines(resolution="10m")
        ax.set_extent([-122.8, -122, 37.3, 38.3])

        ax.add_feature(LAND, zorder=0)
        ax.add_feature(OCEAN, zorder=0)
```

```
# plot the data related stuff
berkeley_lon, berkeley_lat = -122.2585, 37.8719
stanford_lon, stanford_lat = -122.1661, 37.4241

# plot the two universities as blue dots
ax.plot([berkeley_lon, stanford_lon],
         [berkeley_lat, stanford_lat],
        color="blue", linewidth=2, marker="o")

# add labels for the two universities
ax.text(berkeley_lon + 0.16, berkeley_lat - 0.02,
        "UC Berkeley", horizontalalignment="right")

ax.text(stanford_lon + 0.02, stanford_lat - 0.02,
        "Stanford", horizontalalignment="left")

plt.show()
```

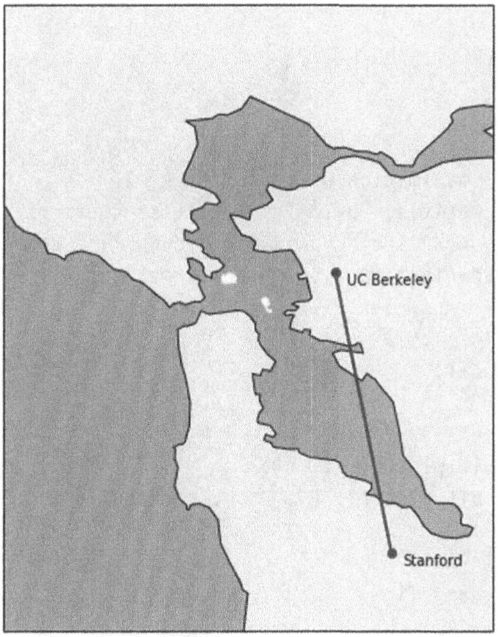

FIGURE 12.20

Add more entities on the map.

The `cartopy` package is extremely versatile. The official example is available at the gallery online.[9] We suggest exploring this on your own to augment your map-making abilities.

12.4 ANIMATIONS AND MOVIES

An **animation** is a sequence of still frames, or plots, that are displayed in fast enough succession to create the illusion of continuous motion. Animations and movies often convey information better than individual plots. You can create animations in Python by calling a plot function inside of a loop (usually a for-loop). The main tools for making animations in Python is the `matplotlib.animation.Animation` base class, which provides a framework around which the animation functionality is built. See the example below.

TRY IT! Create an animation of a red circle following a blue sine wave.

```
In [1]: import numpy as np
        import matplotlib.pyplot as plt
        import matplotlib.animation as manimation

In [2]: n = 1000
        x = np.linspace(0, 6*np.pi, n)
        y = np.sin(x)

        # Define the meta data for the movie
        FFMpegWriter = manimation.writers["ffmpeg"]
        metadata = dict(title="Movie Test", artist="Matplotlib",
            comment="a red circle following a blue sine wave")
        writer = FFMpegWriter(fps=15, metadata=metadata)

        # Initialize the movie
        fig = plt.figure()

        # plot the sine wave line
        sine_line, = plt.plot(x, y, "b")
        red_circle, = plt.plot([], [], "ro", markersize = 10)
        plt.xlabel("x")
        plt.ylabel("sin(x)")

        # Update the frames for the movie
        with writer.saving(fig, "writer_test.mp4", 100):
```

9 https://scitools.org.uk/cartopy/docs/latest/gallery/index.html.

```
    for i in range(n):
        x0 = x[i]
        y0 = y[i]
        red_circle.set_data(x0, y0)
        writer.grab_frame()
```

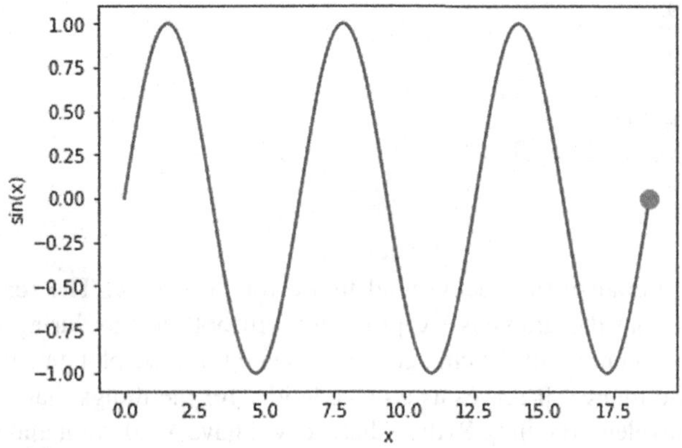

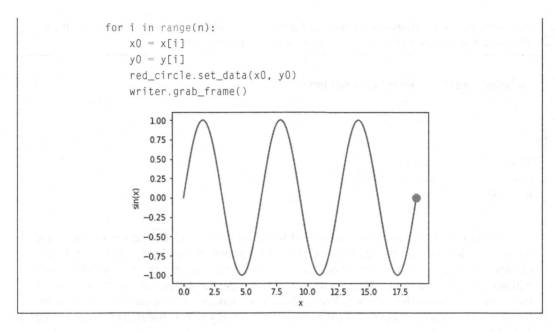

Before we make a movie, it is advised to first follow the next three steps:

- Define the meta data for the movie.
- Decide what in the background does not need changing.
- Decide which objects needs changing in each movie frame.

Once these parameters are determined, it is relatively easy to make the movie using Python following the next three steps:

- Define the meta data for the movie.
- Initialize the movie background figure.
- Update the frames for the movie.

Using the example above, we can clearly see how the code relates to these three steps.

1. Defining the meta data for the movie

```
FFMpegWriter = manimation.writers["ffmpeg"]
metadata = dict(title="Movie Test", artist="Matplotlib",
         comment="a red circle following a blue sine wave")
writer = FFMpegWriter(fps=15, metadata=metadata)
```

This block of code is telling Python that we want to create a movie using a movie writer and need to provide the title, artist, and any necessary commentary. In addition, we need to tell

Python the rate of the frames in the movie, that is, `fps=15`, which means that we want to display 15 consecutive frames within 1 s (`fps` stands for frames per second).

2. Initializing the movie background figure

```
fig = plt.figure()

# plot the sine wave line
sine_line, = plt.plot(x, y, "b")
red_circle, = plt.plot([], [], "ro", markersize = 10)
plt.xlabel("x")
plt.ylabel("sin(x)")
```

First we need to initialize the background figure for the movie. The reason we call it the background figure is that the graphics we plot here will not change during the movie. In this example, the sine wave curve will not change. At the same time, we plot an empty red dot (which will not appear on the figure). It serves as a place holder for the things that will change later in the movie; this is equivalent to telling Python that we will have a red point and we will update the location of the point later. In this case, the `x` and `y` axes labels will not change, and, therefore, are plotted them here.

3. Updating the frames for the movie

```
with writer.saving(fig, "writer_test.mp4", 100):
    for i in range(n):
        x0 = x[i]
        y0 = y[i]
        red_circle.set_data(x0, y0)
        writer.grab_frame()
```

This block of code specifies the name of the output file, format, and resolution of the figure (`dpi` – dots per inch). In this case, we want the output file to have a name `"writer_test"` with the format `"mp4"`, and we want a dpi of 100 for this figure. Next we code the core part of the movie: repeated updating of the figure, i.e., to create the "motion." We use a for-loop to update the figure, and in each loop, we change the location (`x`, `y` location) of the red circle. The `writer.grab_frame` function will capture this change in each frame and display it based on the fps we set.

This is how to make a simple movie.

Numerous examples on how to make a movie are available in `matplotlib` movie tutorial.[10] Run through some of the examples to get a better understanding of how to make movies in Python.

[10] https://matplotlib.org/api/animation_api.html.

12.5 SUMMARY AND PROBLEMS

12.5.1 SUMMARY

1. Visualizing data is an essential tool in engineering and science applications.
2. Python has many different packages of plotting tools that can be used to visualize data.
3. Two- and three-dimensional plots and maps are usually used in engineering and science to communicate research results.
4. Videos are a sequence of static images that are displayed at a certain speed.

12.5.2 PROBLEMS

1. A cycloid is the curve traced by a point located on the edge of a wheel rolling along a flat surface. The (x, y) coordinates of a cycloid generated from a wheel with radius, r, can be described by the parametric equations:
   $x = r(\phi - \sin\phi)$,
   $y = r(1 - \cos\phi)$,
   where ϕ is the number of radians that the wheel has rolled through.
   Generate a plot of the cycloid for $0 \le \phi \le 2\pi$ using 1000 increments and $r = 3$. Give your plot a title and labels. Turn on the grid and modify the axis limits to make the plot neat and attractive.

2. Consider the following function:

   $$y(x) = \sqrt{\frac{100(1-0.01x^2)^2 + 0.02x^2}{(1-x^2)^2 + 0.1x^2}}.$$

   Generate a 2×2 subplot of $y(x)$ for $0 \le x \le 100$ using `plot`, `semilogx`, `semilogy`, and `loglog`. Use a fine enough discretization in `x` to make the plot appear smooth. Give each plot axis labels and a title. Turn on the grid. Which plot seems to convey the most information?.

3. Plot the functions $y_1(x) = 3 + \exp(-x)\sin(6x)$ and $y_2(x) = 4 + \exp(-x)\cos(6x)$ for $0 \le x \le 5$ on a single axis. Give the plot axis labels, a title, and a legend.

4. Generate 1000 normally distributed random numbers using the `np.random.randn` function. Look up the help for the `plt.hist` function. Use the `plt.hist` function to plot a histogram of the randomly generated numbers. Use the `plt.hist` function to distribute the randomly generated numbers into 10 bins. Create a bar graph of the output using the `plt.bar` function. It should look very similar to the plot produced by `plt.hist`.
   Do you think that the `np.random.randn` function is a good approximation of a normally distributed number?

5. Let the number of students with A's, B's, C's, D's, and F's be contained in the list `grade_dist = [42, 85, 67, 20, 5]`. Use the `plt.pie` function to generate a pie chart of `grade_dist`. Put a title and legend on the pie chart.

6. Let $-4 \le x \le 4$, $-3 \le y \le 3$, and $z(x, y) = \frac{xy(x^2-y^2)}{x^2+y^2}$. Create arrays `x` and `y` with 100 evenly spaced points over the interval. Create meshgrids `X` and `Y` for `x` and `y` using the `meshgrid` func-

tion. Compute the matrix Z from X and Y . Create a 1×2 subplot where the first figure is the 3D surface Z plotted using `plt.plot_surface` and the second figure is the 3D wireframe plot using `plt.plot_wireframe`, respectively. Give each axis a title and label the axes.

7. Write a function `my_polygon(n)` that plots a regular polygon with n sides and radius 1. Recall that the radius of a regular polygon is the distance from its centroid to the vertices. Use `plt.axis("equal")` to make the polygon look regular. Remember to give the axes a label and a title. Use `plt.title` to title the plot according to the number of sides. Hint: This problem is significantly easier if you think in polar coordinates. Recall that a complete revolution around the unit circle is 2π radians. Note that the first and last point on the polygon should be the point associated with the polar coordinate angles, 0 and 2π, respectively.
Test case:

```
my_polygon(5)
```

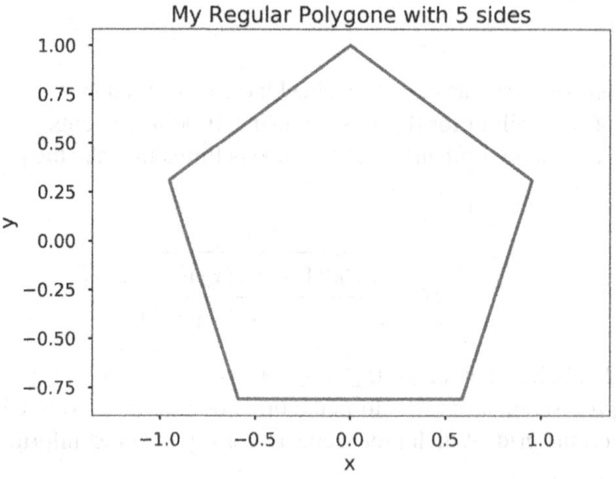

8. Write a function `my_fun_plotter(f, x)` where f is a lambda function and x is an array. The function should plot f evaluated at x. Remember to label the x- and y-axes.
Test case:

```
my_fun_plotter(lambda x: np.sqrt(x) + np.exp(np.sin(x)),
               np.linspace(0, 2*np.pi, 100))
```

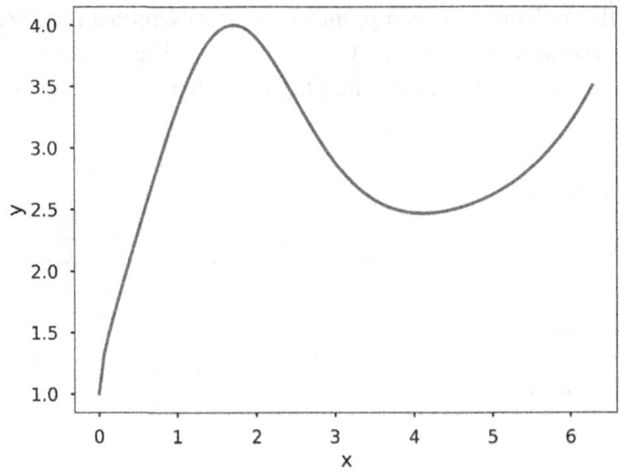

9. Write a function with `my_poly_plotter(n,x)` that plots the polynomials $p_k(x) = x^k$ for $k = 1, \ldots, n$. Make sure your plot has axis labels and a title.
Test case:

```
my_poly_plotter(5, np.linspace(-1, 1, 200))
```

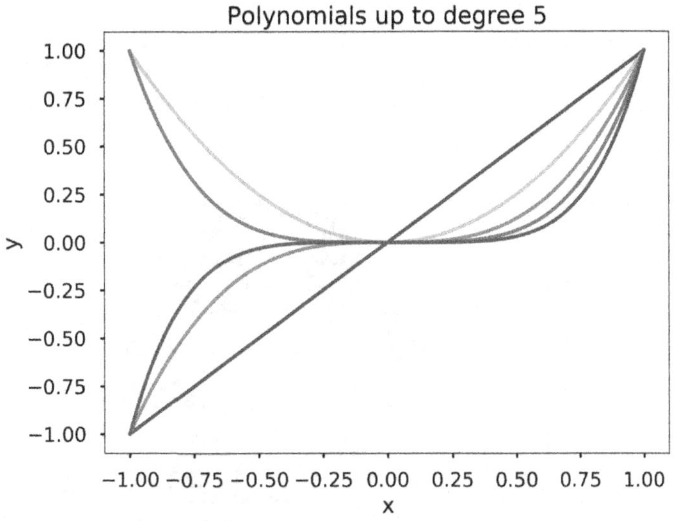

10. Assume you have three points at the corner of an equilateral triangle, $P_1 = (0, 0)$, $P_2 = (0.5, \sqrt{2}/2)$, and $P_3 = (1, 0)$. Generate another set of points $p_i = (x_i, y_i)$ such that $p_1 = (0, 0)$ and p_{i+1} is the midpoint between p_i and P_1 with 33% probability, the midpoint between p_i and P_2 with 33%

probability, and the midpoint between p_i and P_3 with 33% probability. Write a function `my_sierpinski(n)` that generates the points p_i for $i = 1, \ldots, n$. The function should make a plot of the points using blue dots (i.e., `"b."` as the third argument to the `plt.plot` function).
Test cases:

```
my_sierpinski(100)
```

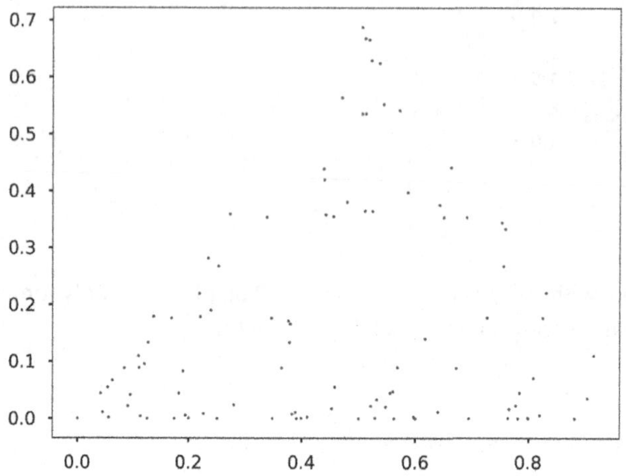

```
my_sierpinski(10000)
```

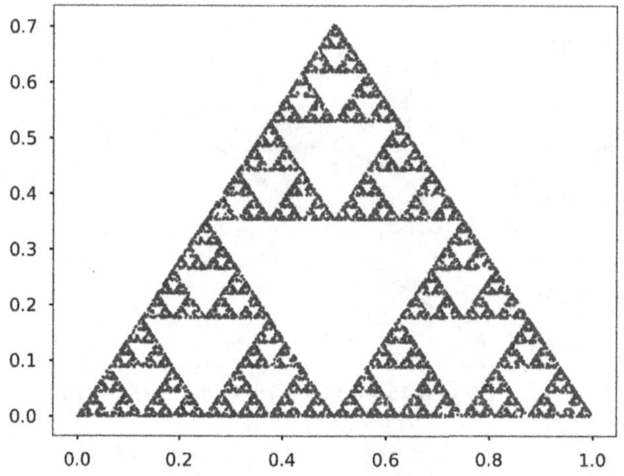

11. Assume you are generating a set of points (x_i, y_i) where $x_1 = 0$ and $y_1 = 0$. The points (x_i, y_i) for $i = 2, \ldots, n$ are generated according to the following probabilistic relationship:

With 1% probability:
$x_i = 0,$
$y_i = 0.16y_{i-1};$
With 7% probability:
$x_i = 0.2x_{i-1} - 0.26y_{i-1},$
$y_i = 0.23x_{i-1} + 0.22y_{i-1} + 1.6;$
With 7% probability:
$x_i = -0.15x_{i-1} + 0.28y_{i-1},$
$y_i = 0.26x_{i-1} + 0.24y_{i-1} + 0.44;$
With 85% probability:
$x_i = 0.85x_{i-1} + 0.04y_{i-1},$
$y_i = -0.04x_{i-1} + 0.85y_{i-1} + 1.6.$
Write a function `my_fern(n)` that generates the points (x_i, y_i) for $i = 1, \ldots, n$ and plots them using blue dots. Also use `plt.axis("equal")` and `plt.axis("off")` to make the plot more attractive.
Test cases:

```
my_fern(100)
```

Try your function for `n = 10000`. The image generated is called a stochastic fractal. Many times it is cheaper (i.e., requires less space) to store the fractal generating code rather than the image. This makes stochastic fractals useful for compressing images.

```
my_fern(10000)
```

12. Write a function `my_parametric_plotter (x,y,t)` where `x` and `y` are function objects `x(t)` and `y(t)`, respectively, and `t` is a one-dimensional array. The function `my_parametric_plotter` should produce the curve $(x(t), y(t), t)$ in a 3D plot. Be sure to give your plot a title and label your axes. Test case:

```
from mpl_toolkits import mplot3d
f = lambda t: np.sin(t)
g = lambda t: t**2
my_parametric_plotter(f, g, np.linspace(0, 6*np.pi, 100))
```

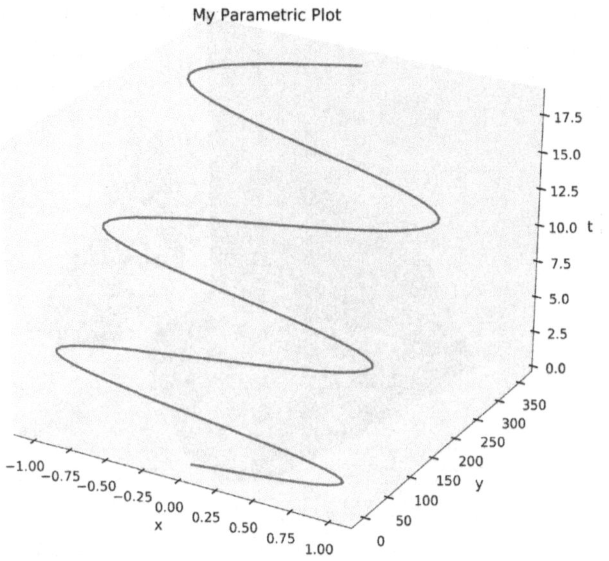

13. Write a function `my_surface_plotter(f, x, y, option)` where `f` is a function object `f(x,y)`. The function `my_surface_plotter` should produce a 3D surface plot of `f(x,y)` using `plot_surface` if the option used is the string `"surface"`. It should produce a contour plot of `f(x,y)` if the option used is the string `"contour"`. Assume that `x` and `y` are one-dimensional arrays or lists. Remember to give the plot a title and label your axes.
Test cases:

```
from mpl_toolkits import mplot3d
f = lambda x, y: np.cos(y)*np.sin(np.exp(x))
my_surface_plotter(f, np.linspace(-1, 1, 20),
                   np.linspace(-2, 2, 40), "surface")
my_surface_plotter(f, np.linspace(-1, 1, 20),
                   np.linspace(-2, 2, 40), "contour")
```

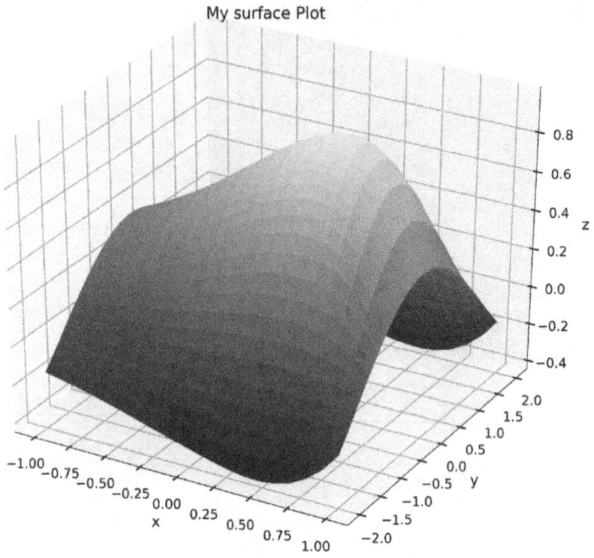

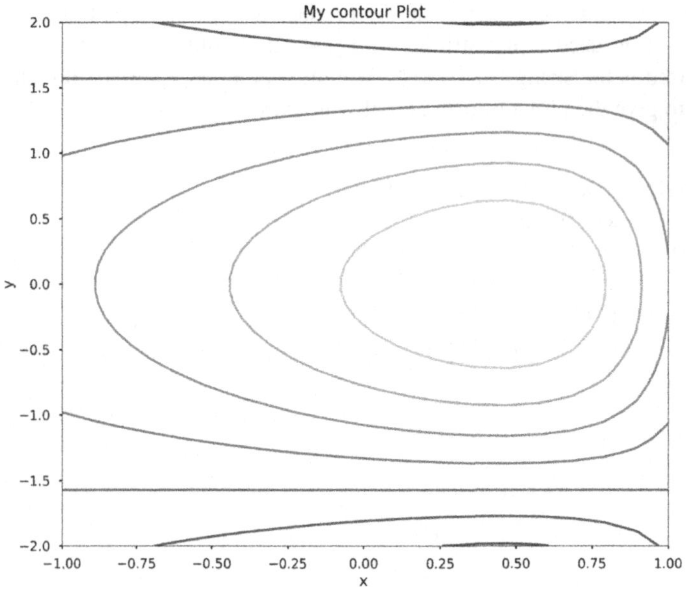

14. Write a line of code that generates the following error:

```
ValueError: x and y must have the same first dimension, ...
```

15. We can make maps in `cartopy` using the online program *Web Map Tile Service* (WMTS) online. Plot an Earth-night map shown below for the main part of North America, with the extent latitude from 19.50139 to 64.85694 and longitude from −128.75583 to −68.01197. Hint: Check out the gallery on `cartopy` website.

CHAPTER

PARALLELIZE YOUR PYTHON 13

CONTENTS

13.1 PARALLEL COMPUTING BASICS

We now have a working knowledge of Python, and soon we will start to use it to analyze data and numerical analysis. Before we go deeper, we need to cover **parallel computing** in Python. This means you will be able to run your code simultaneously on multiple cores on your CPU processor (or multiple CPU processors) or increase the speed by taking advantage of the wasted CPU cycles while your program is waiting for external resources (i.e., downloading files, API calls, etc.). The fundamental idea of parallel computing is rooted in doing multiple tasks at the same time to reduce the running time of your program. Fig. 13.1 illustrates the simple idea of doing parallel computing versus serial computing, which is what we have discussed so far. For example, if you have 1 million data files and need to apply the same operations to each one of them, you can do this one file at each time, or you can do it by multiple files at the same time; or if you are downloading 1 million websites, you can take advantage of downloading 10 at a time to reduce the total time of downloading. Therefore, learning the basics of parallel computing will help you design code that is more efficient.

Most modern computers use a multicore design, which means that on a single computing component there are multiple independent processing units–the so-called cores–available to do different tasks. For example, on the author's laptop, there is one CPU processor, with six physical cores on it (see Fig. 13.2). Each of them has two logical cores, which will bring the total number of cores to 12 (see the next section when we print out the total number of CPU on your machine).

In Python, there are two basic approaches when you wish to perform parallel computing: you can employ the **multiprocessing** or **threading** library. Let us first take a look of the differences between a process and a thread.

Python Programming and Numerical Methods. https://doi.org/10.1016/B978-0-12-819549-9.00022-1

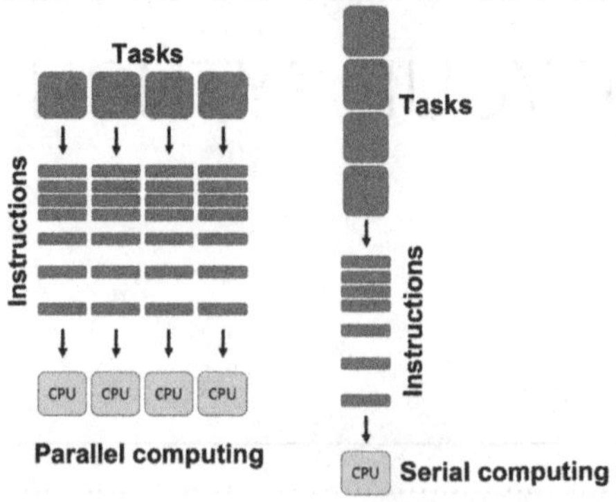

FIGURE 13.1

Parallel computing vs. serial computing.

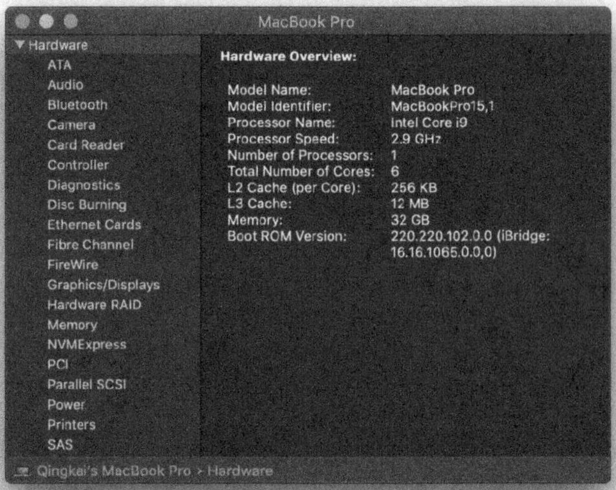

FIGURE 13.2

The hardware on the author's laptop with multicore processor.

13.1.1 PROCESS AND THREAD

A **process** is an instance of a program (such as the Python interpreter, `Jupyter notebook`, etc.). A process is created by the operating system to run a program, and each process has its own memory block. A **thread** is a subprocess that resides within the process. Each process can have multiple threads; these threads will share the same memory block within the process. Therefore, due to the shared memory, the variables or objects for multiple threads in a process are all shared. If you change one variable in one thread, it will change for all the other threads. This is not the case for different processes. If you change one variable in one process, it will not change that variable in other processes. Process and thread both have advantages or disadvantages, and can be used in different tasks to maximize the benefits of each.

13.1.2 PYTHON'S GIL PROBLEM

Python was designed before the multicore processors were available on personal computers (this shows you how old the language is). It contains an inherent limitation called **Global Interpreter Lock** (GIL), whereby only one native thread can run at any time, i.e., it prevents multiple threads from running simultaneously. Even though there are workarounds in Python to do multithreading, only multiprocessing library will be covered below.

13.1.3 DISADVANTAGES OF USING PARALLEL COMPUTING

There are disadvantages when using parallel computing: the code becomes more complicated because of the overheads required to launch and maintain new processes. Thus, if your task is small, using parallel computing will actually take longer since it takes time for the system to initialize new processes and maintain them.

13.2 MULTIPROCESSING

The multiprocessing library, Python's standard library to support its parallel computing using processes, has many different features that are too numerous to discuss herein. We suggest you check out the official documentation.[1] Here, we are introducing only the basics to get you started with parallel computing. Let us start by importing the library and printing out the total number of CPUs on your machine that can be used for parallel computing.

```
In [1]: import multiprocessing as mp

In [2]: print(f"Number of cpu: {mp.cpu_count()}")

Number of cpu: 12
```

The example below shows you how to use multiple cores in one machine to reduce the execution time.

[1] https://docs.python.org/3/library/multiprocessing.html.

EXAMPLE: Generate 10,000,000 random numbers between 0 and 10, and square them. Store the results in a list.

Serial version

```
In [3]: import numpy as np
        import time

        def random_square(seed):
            np.random.seed(seed)
            random_num = np.random.randint(0, 10)
            return random_num**2

In [4]: t0 = time.time()
        results = []
        for i in range(10000000):
            results.append(random_square(i))
        t1 = time.time()
        print(f"Execution time {t1 - t0} s")

Execution time 38.20956087112427 s
```

Parallel version

The simplest way to do parallel computing using the multiprocessing is to use the **pool** class. There are four common methods in this class that are used often: `apply`, `map`, `apply_async`, and `map_async`. Look at the documentation for the differences between them for your own edification. We will only use the `map` function in parallel with the above example. The `map(func, iterable)` function takes in two arguments. Apply the function `func` to each element in the `iterable` and then collect the results.

```
In [5]: t0 = time.time()
        n_cpu = mp.cpu_count()

        pool = mp.Pool(processes=n_cpu)
        results = [pool.map(random_square, range(10000000))]
        t1 = time.time()
        print(f"Execution time {t1 - t0} s")

Execution time 7.130078077316284 s
```

Using the above parallel version of the code reduced the run time from ~38 s to ~7 s. This is a big gain in speed, especially if we were running a code that demanded a lot of computation.

The `pool.apply` function is similar except that it can accept more arguments. The `pool.map` and `pool.apply` will lock the main program until all the processes are finished, which is quite useful if we want to obtain results in a particular order for some applications. In contrast, if we do not need the results in a particular order, we can also use `pool.apply_async` or `pool.map_async`, which will submit all processes at once and retrieve the results as soon as they are finished. Check online to learn more.

13.2.1 VISUALIZE THE EXECUTION TIME

Let us visualize the execution time changes versus the number of data points using both the serial and parallel version. You will see that until a certain point, it is better to use the serial version.

```
In [6]: import matplotlib.pyplot as plt
        plt.style.use("seaborn-poster")
        %matplotlib inline

        def serial(n):
            t0 = time.time()
            results = []
            for i in range(n):
                results.append(random_square(i))
            t1 = time.time()
            exec_time = t1 - t0
            return exec_time

        def parallel(n):
            t0 = time.time()
            n_cpu = mp.cpu_count()

            pool = mp.Pool(processes=n_cpu)
            results = [pool.map(random_square, range(n))]
            t1 = time.time()
            exec_time = t1 - t0
            return exec_time

In [7]: n_run = np.logspace(1, 7, num = 7)

        t_serial = [serial(int(n)) for n in n_run]
        t_parallel = [parallel(int(n)) for n in n_run]

In [8]: plt.figure(figsize = (10, 6))
        plt.plot(n_run, t_serial, "-o", label = "serial")
        plt.plot(n_run, t_parallel, "-o", label = "parallel")
        plt.loglog()
        plt.legend()
        plt.ylabel("Execution time (s)")
```

```
plt.xlabel("Number of random points")
plt.show()
```

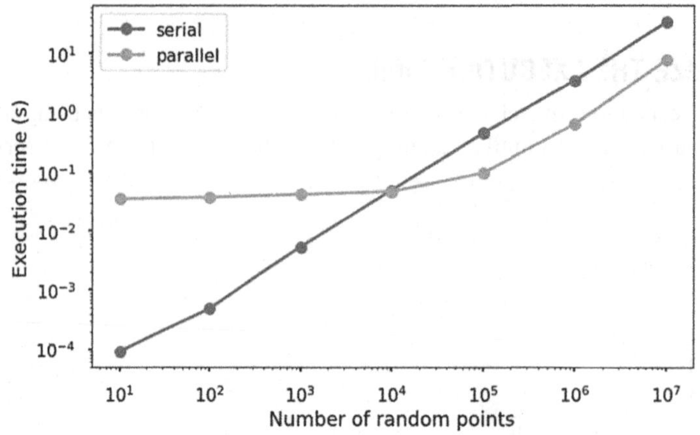

As shown in the figure, when the number of data points is small (below 10000), the execution time for the serial version is faster due to the overhead of the parallel version necessary to launch and maintain the new processes. After that point, the parallel version would be a better choice. For example, when we have 10^7 data points, the parallel version uses less than 10 s to finish the task while the serial version takes roughly 50 s.

13.3 USING JOBLIB

Python also has other third party packages that can make the parallel computing easier, especially for some daily tasks, e.g., package `joblib`[2] provides a simple way to perform parallel computing (and has many other usages as well).

First, install this package by running `pip install joblib`; then run the previous example using this new package.

```
In [1]: from joblib import Parallel, delayed
        import numpy as np

        def random_square(seed):
            np.random.seed(seed)
            random_num = np.random.randint(0, 10)
            return random_num**2
```

2 https://joblib.readthedocs.io/en/latest/index.html.

```
In [2]: results = Parallel(n_jobs=8)\
          (delayed(random_square)(i) for i in range(1000000))
```

Note that the parallel part of the code becomes one line by using the `joblib` package, which is very convenient. `Parallel` is a helper class that essentially provides a convenient interface for the `multiprocessing` module we saw earlier. The `delayed` function is used to capture the arguments of the target function, which in this case is the `random_square`. We ran the above code with eight CPUs. If you want to use all the computational power on your machine, you can use all available CPUs on your machine by setting `n_jobs=-1`. If you set it to -2, all CPUs but one will be used. In addition, turn on the `verbose` argument if you want to output the status messages.

```
In [3]: results = Parallel(n_jobs=-1, verbose=1)\
          (delayed(random_square)(i) for i in range(1000000))

[Parallel(n_jobs=-1)]: Using backend LokyBackend with 12
                        concurrent workers.
[Parallel(n_jobs=-1)]: Done  60 tasks      | elapsed:    0.1s
[Parallel(n_jobs=-1)]: Done 176056 tasks      | elapsed:    3.0s
[Parallel(n_jobs=-1)]: Done 787056 tasks      | elapsed:   12.4s
[Parallel(n_jobs=-1)]: Done 1000000 out of 1000000 |
                        elapsed:   15.5s finished
```

There are multiple backends in `joblib`, which means using different ways to do the parallel computing. If you set the backend as `multiprocessing`, i.e., under the hood, it creates a multiprocessing pool that uses separate Python worker processes to execute tasks concurrently on separate CPUs.

```
In [4]: results = \
    Parallel(n_jobs=-1, backend="multiprocessing", verbose=1)\
    (delayed(random_square)(i) for i in range(1000000))

[Parallel(n_jobs=-1)]: Using backend MultiprocessingBackend
                        with 12 concurrent workers.
[Parallel(n_jobs=-1)]: Done 220 tasks      | elapsed:    0.0s
[Parallel(n_jobs=-1)]: Done 457032 tasks      | elapsed:    1.9s
[Parallel(n_jobs=-1)]: Done 1000000 out of 1000000 |
                        elapsed: 3.8s finished
```

13.4 SUMMARY AND PROBLEMS

13.4.1 SUMMARY

1. Parallel computing can reduce execution time by using multiple cores on your computer.

2. There is a difference between process and thread; it is easier to use a process-based approach in Python to achieve parallelism.
3. Using the multiprocessing package is an easy way to solve your problems when using multiple cores.
4. Using joblib will simplify parallel computing code for many common tasks.

13.4.2 PROBLEMS

1. What is parallel computing?
2. Please specify the difference between process and thread.
3. Find the number of processors on your computer using the `multiprocessing` package.
4. Use the `multiprocessing` package to parallelize the following code and record the running time:

```
def plus_cube(x, y):
    return (x+y)**3

for x, y in zip(range(100), range(100)):
    results.append(plus_cube(x, y))
```

5. Can you provide an example to illustrate the difference of `pool.map` and `pool.map_async`?
6. What is Python's GIL?
7. Use `joblib` to parallelize the above example; use "multiprocessing" as the backend.

PART

2

INTRODUCTION TO NUMERICAL METHODS

CHAPTER

LINEAR ALGEBRA AND SYSTEMS OF LINEAR EQUATIONS

14

CONTENTS

14.1 BASICS OF LINEAR ALGEBRA

First, we introduce some basics of linear algebra, which will be used to describe and solve linear equations. We will just cover the very basics of it in this chapter. If you wish additional guidance, we suggest studying a linear algebra textbook.

14.1.1 SETS

We have discussed the set data structure in Chapter 2 before. Here we take a look at it from a mathematics point of view. In mathematics, a **set** is a collection of objects. As defined earlier, sets are usually denoted by braces $\{\}$. For example, $S = \{\text{orange, apple, banana}\}$ means S is the set containing "orange", "apple", and "banana."

The **empty set** is the set containing no objects and is typically denoted by empty braces such as $\{\}$ or by $\emptyset$. Given two sets, A and B, the **union** of A and B is denoted by $A \cup B$ and is equal to the set containing all the elements of A and B. The **intersection** of A and B is denoted by $A \cap B$ and is equal to the set containing all the elements that belong to both A and B. In set notation, a colon is used

Python Programming and Numerical Methods. https://doi.org/10.1016/B978-0-12-819549-9.00024-5

to mean **"such that."** The usage of these terms will become apparent shortly. The symbol $\in$ is used to denote that an object is contained in a set. For example $a \in A$ means "a is a member of A" or "a is in A." A backslash, \, in set notation means **set minus**. So if $a \in A$ then $A\backslash a$ means "A minus the element, a."

There are several standard sets related to numbers, for example, **natural numbers**, **whole numbers**, **integers**, **rational numbers**, **irrational numbers**, **real numbers**, and **complex numbers**. A description of each set and the symbol used to denote it is shown in the following table.

Set Name	Symbol	Description
Naturals	$\mathbb{N}$	$\mathbb{N} = \{1, 2, 3, 4, \ldots\}$
Wholes	$\mathbb{W}$	$\mathbb{W} = \mathbb{N} \cup \{0\}$
Integers	$\mathbb{Z}$	$\mathbb{Z} = \mathbb{W} \cup \{-1, -2, -3, \ldots\}$
Rationals	$\mathbb{Q}$	$\mathbb{Q} = \{\frac{p}{q} : p \in \mathbb{Z}, q \in \mathbb{Z}\backslash\{0\}\}$
Irrationals	$\mathbb{I}$	$\mathbb{I}$ is the set of real numbers not expressible as a fraction of integers
Reals	$\mathbb{R}$	$\mathbb{R} = \mathbb{Q} \cup \mathbb{I}$
Complex numbers	$\mathbb{C}$	$\mathbb{C} = \{a + bi : a, b \in \mathbb{R}, i = \sqrt{-1}\}$

TRY IT! Let S be the set of all real (x, y) pairs such that $x^2 + y^2 = 1$. Write S using set notation.

$S = \{(x, y) : x, y \in \mathbb{R}, x^2 + y^2 = 1\}$

14.1.2 VECTORS

The set $\mathbb{R}^n$ is the set of all n-tuples of real numbers. In set notation, this is $\mathbb{R}^n = \{(x_1, x_2, x_3, \ldots, x_n) : x_1, x_2, x_3, \ldots, x_n \in \mathbb{R}\}$. For example, the set $\mathbb{R}^3$ represents the set of real triples, (x, y, z) coordinates, in 3D space.

A **vector** in $\mathbb{R}^n$ is an n-tuple, or point, in $\mathbb{R}^n$. Vectors can be written horizontally (i.e., with the elements of the vector next to each other) in a **row vector**, or vertically (i.e., with the elements of the vector on top of each other) in a **column vector**. If the context of a vector is ambiguous, it usually means the vector is a column vector. The ith element of a vector, v, is denoted by v_i. The transpose of a column vector is a row vector of the same length, and the transpose of a row vector is a column vector. In mathematics, the transpose is denoted by a superscript T, or v^T. The **zero vector** is the vector in $\mathbb{R}^n$ containing all zeros.

The **norm** of a vector is a measure of its length. There are many ways of defining the length of a vector depending on the metric used (i.e., the distance formula chosen). The most common is called the L_2 norm, which is computed according to the distance formula. The L_2 **norm** of a vector v is denoted by $\|v\|_2$ and $\|v\|_2 = \sqrt{\sum_i v_i^2}$. This is sometimes also called Euclidean distance and refers to the "physical" length of a vector in 1, 2, or 3D space. The L_1 norm, or "Manhattan distance," is computed as $\|v\|_1 = \sum_i |v_i|$, and is named after the grid-like road structure in New York City. In general, the **p-norm**, L_p, of a vector is $\|v\|_p = \sqrt[p]{(\sum_i v_i^p)}$. The L_∞ **norm** is the p-norm, where $p = \infty$. The L_∞ norm is written as $||v||_\infty$ and is equal to the maximum absolute value in v.

TRY IT! Create a row vector and a column vector, and show their shape.

```
In [1]: import numpy as np
        vector_row = np.array([[1, -5, 3, 2, 4]])
        vector_column = np.array([[1],[2],[3],[4]])
        print(vector_row.shape)
        print(vector_column.shape)

(1, 5)
(4, 1)
```

Note! In Python, the row and column vectors can be tricky. As shown in the example above, in order to get the 1 row and 4 column or 4 row and 1 column vectors, we had to use a list of a list to specify it. You can define `np.array([1,2,3,4])`, but you will soon notice that it does not contain information about row or column.

TRY IT! Transpose the row vector defined above into a column vector and calculate its L_1, L_2, and L_∞ norm. Verify that the L_∞ norm of a vector is equivalent to the maximum value of the elements in the vector.

```
In [2]: from numpy.linalg import norm
        new_vector = vector_row.T
        print(new_vector)
        norm_1 = norm(new_vector, 1)
        norm_2 = norm(new_vector, 2)
        norm_inf = norm(new_vector, np.inf)
        print("L_1 is: %.1f"%norm_1)
        print("L_2 is: %.1f"%norm_2)
        print("L_inf is: %.1f"%norm_inf)

[[ 1]
 [-5]
 [ 3]
 [ 2]
 [ 4]]
L_1 is: 15.0
L_2 is: 7.4
L_inf is: 5.0
```

Vector addition is defined as the pairwise addition of the elements of the added vectors. For example, if v and w are vectors in $\mathbb{R}^n$, then $u = v + w$ is defined as having elements $u_i = v_i + w_i$.

Vector multiplication can be defined in several ways depending on the context. **Scalar multiplication** of a vector is the product of a vector and a **scalar** (i.e., a number in $\mathbb{R}$). Scalar multiplication is defined as the product of each element of the vector by the scalar. More specifically, if α is a scalar and v is a vector, then $u = \alpha v$ is defined as having elements $u_i = \alpha v_i$. Note that this is exactly how Python implements scalar multiplication with a vector.

TRY IT! Show that $a(v + w) = av + aw$ (i.e., scalar multiplication of a vector distributes across vector addition).

By vector addition, $u = v + w$ is the vector with entries $u_i = v_i + w_i$. By definition of scalar multiplication of a vector, $x = \alpha u$ is the vector with elements $x_i = \alpha(v_i + w_i)$. Since α, v_i, and w_i are scalars, multiplication distributes and $x_i = \alpha v_i + \alpha w_i$. Therefore, $a(v + w) = av + aw$.

The **dot-product** of two vectors is the sum of the products of the respective elements and is denoted by $\cdot$, and $v \cdot w$ is read "v dot w." Therefore for $v, w \in \mathbb{R}^n$, $d = v \cdot w$ is defined as $d = \sum_{i=1}^{n} v_i w_i$. The **angle between two vectors**, θ, is defined by the formula:

$$v \cdot w = \|v\|_2 \|w\|_2 \cos\theta.$$

The dot-product is a measure of how similarly directed the two vectors are. For example, the vectors $(1, 1)$ and $(2, 2)$ are parallel. If you compute the angle between them using the dot-product, you will find that $\theta = 0$. If the angle between the vectors is $\theta = \pi/2$, then the vectors are said to be perpendicular or **orthogonal**, and the dot-product is 0.

TRY IT! Compute the angle between the vectors $v = [10, 9, 3]$ and $w = [2, 5, 12]$.

```
In [3]: from numpy import arccos, dot

        v = np.array([[10, 9, 3]])
        w = np.array([[2, 5, 12]])
        theta = arccos(dot(v, w.T)/(norm(v)*norm(w)))
        print(theta)

[[0.97992471]]
```

Finally, the **cross-product** between two vectors, v and w, is written as $v \times w$. It is defined by $v \times w = \|v\|_2 \|w\|_2 \sin(\theta) n$, where θ is the angle between the v and w (which can be computed from the dot-product), and n is a vector perpendicular to both v and w with unit length (i.e., its length is one). The geometric interpretation of the cross-product is a vector perpendicular to both v and w, with the length equal to the area enclosed by the parallelogram created by the two vectors.

TRY IT! Given the vectors $v = [0, 2, 0]$ and $w = [3, 0, 0]$, use the NumPy function `cross` to compute the cross-product of `v` and `w`.

```
In [4]: v = np.array([[0, 2, 0]])
        w = np.array([[3, 0, 0]])
        print(np.cross(v, w))

[[ 0  0 -6]]
```

Assuming that S is a set in which addition and scalar multiplication are defined, a **linear combination** of S is defined as

$$\sum \alpha_i s_i,$$

where α_i is any real number, and s_i is the ith object in S. Sometimes the α_i values are called the **coefficients** of s_i.

Linear combinations can be used to describe numerous things. For example, a grocery bill can be written $\sum c_i n_i$, where c_i is the cost of item i and n_i is the number of items i purchased. Thus, the total cost is a linear combination of the items purchased.

A set is called **linearly independent** if no object in the set can be written as a linear combination of the other objects in the set. For the purposes of this book, we will only consider the linear independence of a set of vectors. A set of vectors that is not linearly independent is **linearly dependent**.

TRY IT! Given the row vectors $v = [0, 3, 2]$, $w = [4, 1, 1]$, and $u = [0, -2, 0]$, write the vector $x = [-8, -1, 4]$ as a linear combination of v, w, and u.

```
In [5]: v = np.array([[0, 3, 2]])
        w = np.array([[4, 1, 1]])
        u = np.array([[0, -2, 0]])
        x = 3*v-2*w+4*u
        print(x)

[[-8 -1  4]]
```

TRY IT! Determine by inspection whether the following set of vectors is linearly independent: $v = [1, 1, 0]$, $w = [1, 0, 0]$, $u = [0, 0, 1]$.

Clearly, u is linearly independent from v and w because only u has a nonzero third element. The vectors v and w are also linearly independent because only v has a nonzero second element. Therefore, v, w, and u are linearly independent.

14.1.3 MATRICES

An $m \times n$ **matrix** is a rectangular table of numbers consisting of m rows and n columns. The norm of a matrix can be considered as a particular kind of vector norm. If we treat the $m \times n$ elements of M as the elements of an mn-dimensional vector, then the p-norm of this vector can be written as

$$\|M\|_p = \sqrt[p]{\sum_i^m \sum_j^n |a_{ij}|^p}.$$

It is possible to calculate the matrix norm using the same `norm` function in `NumPy` as that for a vector.

Matrix addition and scalar multiplication for matrices work the same way as for vectors. However, **matrix multiplication** between two matrices, P and Q, is defined when P is an $m \times p$ matrix and Q is a $p \times n$ matrix. The result of $M = PQ$ is a matrix M that is $m \times n$. The dimension p is called the **inner matrix dimension**, and the inner matrix dimensions must match (i.e., the number of columns in P and the number of rows in Q must be the same) for matrix multiplication to be defined. The dimensions m and n are called the **outer matrix dimensions**. Formally, if P is $m \times p$ and Q is $p \times n$, then $M = PQ$ is defined as

$$M_{ij} = \sum_{k=1}^{p} P_{ik} Q_{kj}.$$

The product of two matrices P and Q in Python is achieved by using the **dot** method in `NumPy`. The **transpose** of a matrix is a reversal of its rows with its columns. The transpose is denoted by a superscript, T, such as M^T is the transpose of matrix M. In Python, the method `T` for a `NumPy` array is used to get the transpose. For example, if `M` is a matrix, then `M.T` is its transpose.

TRY IT!

Let matrices `P` and `Q` be [[1, 7], [2, 3], [5, 0]] and [[2, 6, 3, 1], [1, 2, 3, 4]], respectively. Compute the Python matrix product of `P` and `Q`. Show that the product of `Q` and `P` will produce an error.

```
In [6]: P = np.array([[1, 7], [2, 3], [5, 0]])
        Q = np.array([[2, 6, 3, 1], [1, 2, 3, 4]])
        print(P)
        print(Q)
        print(np.dot(P, Q))
        np.dot(Q, P)

[[1 7]
 [2 3]
 [5 0]]
[[2 6 3 1]
 [1 2 3 4]]
[[ 9 20 24 29]
 [ 7 18 15 14]
 [10 30 15  5]]
```

```
    ---------------------------------------------

    ValueError       Traceback (most recent call last)

    <ipython-input-6-29a4b2da4cb8> in <module>
      4 print(Q)
      5 print(np.dot(P, Q))
----> 6 np.dot(Q, P)

    ValueError: shapes (2,4) and (3,2) not aligned:
                4 (dim 1) != 3 (dim 0)
```

A **square matrix** is an $n \times n$ matrix, that is, it has the same number of rows as columns. The **determinant** is an important property of square matrices. It is a special number that can be calculated directly from a square matrix. The determinant is denoted by `det`, both in mathematics and in `NumPy`'s `linalg` package. Some examples of the use of determinant will be described later.

In the case of a 2×2 matrix, the determinant is

$$|M| = \begin{bmatrix} a & b \\ c & d \end{bmatrix} = ad - bc.$$

Similarly, in the case of a 3×3 matrix, the determinant is:

$$\begin{aligned} |M| = \begin{bmatrix} a & b & c \\ d & e & f \\ g & h & i \end{bmatrix} &= a\begin{bmatrix} \square & \square & \square \\ \square & e & f \\ \square & h & i \end{bmatrix} - b\begin{bmatrix} \square & \square & \square \\ d & \square & f \\ g & \square & i \end{bmatrix} + c\begin{bmatrix} \square & \square & \square \\ d & e & \square \\ g & h & \square \end{bmatrix} \\ &= a\begin{bmatrix} e & f \\ h & i \end{bmatrix} - b\begin{bmatrix} d & f \\ g & i \end{bmatrix} + c\begin{bmatrix} d & e \\ g & h \end{bmatrix} \\ &= aei + bfg + cdh - ceg - bdi - afh. \end{aligned}$$

We can use a similar approach to calculate the determinant for a higher-dimensional matrix, but it is much easier to calculate using Python. See the example below to calculate the determinant in Python.

The **identity matrix** is a square matrix with 1s on the diagonal and 0s elsewhere. The identity matrix is usually denoted by I and is analogous to the real number identity, 1. That is, multiplying any matrix by I (of compatible size) will produce the same matrix.

TRY IT!

Find the determinant of matrix $M = [[0,2,1,3],[3,2,8,1],[1,0,0,3],[0,3,2,1]]$. Use the `np.eye` function to produce a 4×4 identity matrix, I. Multiply M by I to show that the result is M.

```
In [7]: from numpy.linalg import det

        M = np.array([[0,2,1,3],
                      [3,2,8,1],
                      [1,0,0,3],
                      [0,3,2,1]])
        print("M:\n", M)

        print("Determinant: %.1f"%det(M))
        I = np.eye(4)
        print("I:\n", I)
        print("M*I:\n", np.dot(M, I))

M:
 [[0 2 1 3]
 [3 2 8 1]
 [1 0 0 3]
 [0 3 2 1]]
Determinant: -38.0
I:
 [[1. 0. 0. 0.]
 [0. 1. 0. 0.]
 [0. 0. 1. 0.]
 [0. 0. 0. 1.]]
M*I:
 [[0. 2. 1. 3.]
 [3. 2. 8. 1.]
 [1. 0. 0. 3.]
 [0. 3. 2. 1.]]
```

The **inverse** of a square matrix M is a matrix of the same size, N, such that $M \cdot N = I$. The inverse of a matrix is analogous to the inverse of a real number. For example, the inverse of 3 is $\frac{1}{3}$ because $(3)(\frac{1}{3}) = 1$. A matrix is said to be **invertible** if it has an inverse. The inverse of a matrix is unique, that is, for an invertible matrix, there is only one inverse for that matrix. If M is a square matrix, its inverse is denoted by M^{-1} in mathematics, and it can be computed in Python using the function `inv` from `NumPy`'s `linalg` package.

For a 2×2 matrix, the analytical solution of the matrix inverse is

$$M^{-1} = \begin{bmatrix} a & b \\ c & d \end{bmatrix}^{-1} = \frac{1}{|M|}\begin{bmatrix} d & -b \\ -c & a \end{bmatrix}.$$

Calculating the matrix inverse for the analytical solution becomes complicated as the dimension of the matrix increases. There are many other methods which can make things easier, such as Gaussian elimination, Newton's method, eigendecomposition, etc. We will introduce some of these methods after we learn how to solve a system of linear equations (as the process is essentially the same).

Recall that zero has no inverse for multiplication in the setting of real numbers. Similarly, there are matrices that do not have inverses. These matrices are called **singular**. Matrices that do have an inverse are called **nonsingular**.

One way to determine if a matrix is singular is by computing its determinant. If the determinant is 0, then the matrix is singular; if not, the matrix is non-singular.

TRY IT! The matrix M (in the previous example) has a nonzero determinant. Compute the inverse of M. Show that the matrix $P = [[0, 1, 0], [0, 0, 0], [1, 0, 1]]$ has a determinant value of zero, and therefore has no inverse.

```
In [8]: from numpy.linalg import inv

        print("Inv M:\n", inv(M))
        P = np.array([[0,1,0],
                      [0,0,0],
                      [1,0,1]])
        print("det(p):\n", det(P))

Inv M:
 [[-1.57894737 -0.07894737  1.23684211  1.10526316]
 [-0.63157895 -0.13157895  0.39473684  0.84210526]
 [ 0.68421053  0.18421053 -0.55263158 -0.57894737]
 [ 0.52631579  0.02631579 -0.07894737 -0.36842105]]
det(p):
 0.0
```

A matrix that is close to being singular (i.e., the determinant is close to zero) is called **ill-conditioned**. Although ill-conditioned matrices have inverses, they are problematic numerically in the same way that dividing a number by a very, very small number is problematic. That is, it can result in computations that result in overflow, underflow, or numbers small enough to result in significant round-off errors. If you have forgotten any of these concepts, revisit Chapter 9. The **condition number** is a measure of how ill-conditioned a matrix is: it is defined as the norm of the matrix times the norm of the inverse of the matrix, that is, $\|M\|\|M^{-1}\|$. In Python, it can be computed using `NumPy`'s function `cond` from `linalg`. The higher the condition number, the closer the matrix to being singular.

The **rank** of an $m \times n$ matrix A is the number of linearly independent columns or rows of A and is denoted by rank(A). It can be shown that the number of linearly independent rows is always equal to the number of linearly independent columns for any matrix. A matrix has **full rank** if $\text{rank}(A) = \min(m, n)$. The matrix A is also of full rank if all of its columns are linearly independent. An **augmented matrix** is a matrix A concatenated with a vector y and is written $[A, y]$. This is commonly read as "A augmented with y." You can use `np.concatenate` to concatenate. If $\text{rank}([A, y]) = \text{rank}(A) + 1$, then the vector y is "new" information. That is, it cannot be created as a linear combination of the columns in A. Rank is an important characteristic of matrices because of its relationship to solutions of linear equations, which is discussed in the last section of this chapter.

TRY IT! For the matrix $A = [[1, 1, 0], [0, 1, 0], [1, 0, 1]]$, compute the condition number and rank. If $y = [[1], [2], [1]]$, get the augmented matrix $[A, y]$.

```
In [9]: from numpy.linalg import cond, matrix_rank

        A = np.array([[1,1,0],
                      [0,1,0],
                      [1,0,1]])

        print("Condition number:\n", cond(A))
        print("Rank:\n", matrix_rank(A))
        y = np.array([[1], [2], [1]])
        A_y = np.concatenate((A, y), axis = 1)
        print("Augmented matrix:\n", A_y)

Condition number:
 4.048917339522305
Rank:
 3
Augmented matrix:
 [[1 1 0 1]
 [0 1 0 2]
 [1 0 1 1]]
```

14.2 LINEAR TRANSFORMATIONS

For any vectors x and y, and scalars a and b, we say that a function F is a **linear transformation** if

$$F(ax + by) = aF(x) + bF(y).$$

It can be shown that multiplying an $m \times n$ matrix A and an $n \times 1$ vector v of compatible size is a linear transformation of v. Therefore from this point forward, a matrix will be synonymous with a linear transformation function.

TRY IT! Let x be a vector and let $F(x)$ be defined by $F(x) = Ax$, where A is a rectangular matrix of appropriate size. Show that $F(x)$ is a linear transformation.

Proof: Since $F(x) = Ax$, then for vectors v and w, and scalars a and b, $F(av + bw) = A(av + bw)$ (by definition of F)$= aAv + bAw$ (by distributive property of matrix multiplication)$=$ $aF(v) + bF(w)$ (by definition of F).

If A is an $m \times n$ matrix, then there are two important subspaces associated with A: one is $\mathbb{R}^n$, and the other is $\mathbb{R}^m$. The **domain** of A is a subspace of $\mathbb{R}^n$. It is the set of all vectors that can be multiplied by A on the right. The **range** of A is a subspace of $\mathbb{R}^m$. It is the set of all vectors y such that $y = Ax$. It can be denoted as $\mathcal{R}(\mathbf{A})$, where $\mathcal{R}(\mathbf{A}) = \{y \in \mathbb{R}^m : Ax = y\}$. Another way to think about the range of A is as the set of all linear combinations of the columns in A, where x_i is the coefficient of the ith column in A. The **null space** $\mathcal{N}(\mathbf{A}) = \{x \in \mathbb{R}^n : Ax = 0_m\}$ is the subset of vectors x in the domain of A such that $Ax = 0_m$, where 0_m is the **zero vector** (i.e., a vector in $\mathbb{R}^m$ with all zeros).

TRY IT! Let $A = [[1, 0, 0], [0, 1, 0], [0, 0, 0]]$ and let the domain of A be $\mathbb{R}^3$. Characterize the range and nullspace of A.

Let $v = [x, y, z]$ be a vector in $\mathbb{R}^3$; then $u = Av$ is the vector $u = [x, y, 0]$. Since $x, y \in \mathbb{R}$, the range of A is the x–y plane at $z = 0$.

Let $v = [0, 0, z]$ for $z \in \mathbb{R}$. Then $u = Av$ is the vector $u = [0, 0, 0]$. Therefore, the nullspace of A is the z-axis (i.e., the set of vectors $[0, 0, z]$ $z \in \mathbb{R}$).

Therefore, this linear transformation "flattens" any z-component from a vector.

14.3 SYSTEMS OF LINEAR EQUATIONS

A **linear equation** is an equality of the form

$$\sum_{i=1}^{n} a_i x_i = y,$$

where a_i are scalars, x_i are unknown variables in $\mathbb{R}$, and y is a scalar.

TRY IT! Determine which of the following equations is linear and which is not. For the ones that are not linear, can you manipulate them to make them linear?

1. $3x_1 + 4x_2 - 3 = -5x_3$,
2. $\frac{-x_1 + x_2}{x_3} = 2$,
3. $x_1 x_2 + x_3 = 5$.

Equation 1 can be rearranged to be $3x_1 + 4x_2 + 5x_3 = 3$, which clearly has the form of a linear equation. Equation 2 is not linear, but it can be rearranged to be $-x_1 + x_2 - 2x_3 = 0$, which is linear. Equation 3 is not linear.

A **system of linear equations** is a set of linear equations that share the same variables. Consider the following system of linear equations:

$$\begin{array}{rcrcccrcrcl} a_{1,1}x_1 & + & a_{1,2}x_2 & + & \cdots & + & a_{1,n-1}x_{n-1} & + & a_{1,n}x_n & = & y_1, \\ a_{2,1}x_1 & + & a_{2,2}x_2 & + & \cdots & + & a_{2,n-1}x_{n-1} & + & a_{2,n}x_n & = & y_2, \\ & & & & \cdots & & \cdots & & & & \\ a_{m-1,1}x_1 & + & a_{m-1,2}x_2 & + & \cdots & + & a_{m-1,n-1}x_{n-1} & + & a_{m-1,n}x_n & = & y_{m-1}, \\ a_{m,1}x_1 & + & a_{m,2}x_2 & + & \cdots & + & a_{m,n-1}x_{n-1} & + & a_{m,n}x_n & = & y_m, \end{array}$$

where $a_{i,j}$ and y_i are real numbers. The **matrix form** of a system of linear equations is **Ax = y**, where A is an $m \times n$ matrix, $A(i, j) = a_{i,j}$, y is a vector in $\mathbb{R}^m$, and x is an unknown vector in $\mathbb{R}^n$. The matrix form is shown below:

$$\begin{bmatrix} a_{1,1} & a_{1,2} & \cdots & a_{1,n} \\ a_{2,1} & a_{2,2} & \cdots & a_{2,n} \\ \vdots & \vdots & \ddots & \vdots \\ a_{m,1} & a_{m,2} & \cdots & a_{m,n} \end{bmatrix} \begin{bmatrix} x_1 \\ x_2 \\ \vdots \\ x_n \end{bmatrix} = \begin{bmatrix} y_1 \\ y_2 \\ \vdots \\ y_m \end{bmatrix}.$$

If you carry out the matrix multiplication, you will see that you arrive back at the original system of equations.

TRY IT! Put the following system of equations into matrix form:

$$\begin{array}{rcl} 4x + 3y - 5z & = & 2, \\ -2x - 4y + 5z & = & 5, \\ 7x + 8y & = & -3, \\ x + 2z & = & 1, \\ 9 + y - 6z & = & 6, \end{array}$$

$$\begin{bmatrix} 4 & 3 & -5 \\ -2 & -4 & 5 \\ 7 & 8 & 0 \\ 1 & 0 & 2 \\ 9 & 1 & -6 \end{bmatrix} \begin{bmatrix} x \\ y \\ z \end{bmatrix} = \begin{bmatrix} 2 \\ 5 \\ -3 \\ 1 \\ 6 \end{bmatrix}.$$

14.4 SOLUTIONS TO SYSTEMS OF LINEAR EQUATIONS

Consider a system of linear equations in matrix form, $Ax = y$, where A is an $m \times n$ matrix. Recall that this means there are m equations and n unknowns in our system. A **solution** to this system of linear equations is an x in $\mathbb{R}^n$ that satisfies the matrix form equation. Depending on the values that populate A and y, there are three distinct possible solutions for x. Either there is no solution for x, or there is one unique solution for x, or there are infinitely many solutions for x. This fact is not shown in this text.

Case 1: There is no solution for x. If $\text{rank}([A, y]) = \text{rank}(A) + 1$ then y is linearly independent from the columns of A. Therefore, because y is not in the range of A, by definition there cannot be an x that satisfies the equation. Thus, comparing $\text{rank}([A, y])$ and $\text{rank}(A)$ provides an easy way to check if there are no solutions to a system of linear equations.

Case 2: There is a unique solution for x. If $\text{rank}([A, y]) = \text{rank}(A)$, then y can be written as a linear combination of the columns of A, and there is at least one solution for the matrix equation. For there to be only one solution, $\text{rank}(A) = n$ must also be true. In other words, the number of equations must be exactly equal to the number of unknowns. To see why this property results in a unique solution, consider the following three relationships between m and $n : m < n, m = n$, and $m > n$.

- For the case $m < n$, $\text{rank}(A) = n$ cannot possibly be true because this means we have a "fat" matrix with fewer equations than unknowns. Thus, we do not need to consider this subcase.
- When $m = n$ and $\text{rank}(A) = n$, A is square and invertible. Since the inverse of a matrix is unique, the matrix equation $Ax = y$ can be solved by multiplying each side of the equation on the left by A^{-1}. This results in $A^{-1}Ax = A^{-1}y \rightarrow Ix = A^{-1}y \rightarrow x = A^{-1}y$, which gives the unique solution to the equation.
- If $m > n$, then there are more equations than unknowns; however, if $\text{rank}(A) = n$, then it is possible to choose n equations (i.e., rows of A) such that if these equations are satisfied, then the remaining $m - n$ equations will also be satisfied. In other words, they are redundant. If the $m - n$ redundant equations are removed from the system, then the resulting system has an A matrix that is $n \times n$ and invertible. These facts are not proven in this text. The new system then has a unique solution, which is valid for the whole system.

Case 3: There are infinitely many solutions for x. If $\text{rank}([A, y]) = \text{rank}(A)$, then y is in the range of A, and there is at least one solution for the matrix equation; however, if $\text{rank}(A) < n$, then there are infinitely many solutions. Although it is not shown here, if $\text{rank}(A) < n$, then there is at least one nonzero vector, n, that is in the null space of A (Actually, there are infinitely many null space vectors under these conditions.). If n is in the nullspace of A, then $An = 0$ by definition. Now if x^* is a solution to the matrix equation $Ax = y$, then, necessarily, $Ax^* = y$; however, $Ax^* + An = y$ or $A(x^* + n) = y$. Therefore, $x^* + n$ is also a solution for $Ax = y$. In fact, since A is a linear transformation, $x^* + \alpha n$ is a solution for any real number α (you should try to show this on your own). Since there are infinitely many acceptable values for α, there are infinitely many solutions for the matrix equation.

The rest of the chapter will discuss how to solve a system of equations which has a unique solution. First, we will discuss some of the common methods that you will most likely come across in your work and then we will show you how to solve systems in Python.

Let us say we have n equations with n variables, $Ax = y$, as follows:

$$\begin{bmatrix} a_{1,1} & a_{1,2} & \cdots & a_{1,n} \\ a_{2,1} & a_{2,2} & \cdots & a_{2,n} \\ \vdots & \vdots & \ddots & \vdots \\ a_{n,1} & a_{n,2} & \cdots & a_{n,n} \end{bmatrix} \begin{bmatrix} x_1 \\ x_2 \\ \vdots \\ x_n \end{bmatrix} = \begin{bmatrix} y_1 \\ y_2 \\ \vdots \\ y_n \end{bmatrix}.$$

14.4.1 GAUSS ELIMINATION METHOD

The **Gauss elimination** method is a procedure that turns the matrix A into an **upper-triangular** form to solve the system of equations. Let us use a system of four equations and four variables to illustrate the idea. Gauss elimination essentially turns the system of equations into

$$\begin{bmatrix} a_{1,1} & a_{1,2} & a_{1,3} & a_{1,4} \\ 0 & a'_{2,2} & a'_{2,3} & a'_{2,4} \\ 0 & 0 & a'_{3,3} & a'_{3,4} \\ 0 & 0 & 0 & a'_{4,4} \end{bmatrix} \begin{bmatrix} x_1 \\ x_2 \\ x_3 \\ x_4 \end{bmatrix} = \begin{bmatrix} y_1 \\ y'_2 \\ y'_3 \\ y'_4 \end{bmatrix}.$$

By returning to the matrix form using this method, we can see the equations turn into:

$$\begin{aligned} a_{1,1}x_1 + a_{1,2}x_2 + a_{1,3}x_3 + a_{1,4}x_4 &= y_1, \\ a'_{2,2}x_2 + a'_{2,3}x_3 + a'_{2,4}x_4 &= y'_2, \\ a'_{3,3}x_3 + a'_{3,4}x_4 &= y'_3, \\ a'_{4,4}x_4 &= y'_4. \end{aligned}$$

Now, x_4 can be easily solved for by dividing both sides by $a'_{4,4}$, and then by substituting the result into the third equation to solve for x_3. With x_3 and x_4, we can substitute them into the second equation to solve for x_2, and we are now able to solve for all x. We solved the system of equations bottom-up; this is called **backward substitution.** Note that, if A were a lower-triangular matrix, we would solve the system top-down by **forward substitution.**

To illustrate how we solve the equations using Gauss elimination, we use the example below.

TRY IT! Use Gauss elimination to solve the following equations:

$$\begin{aligned} 4x_1 + 3x_2 - 5x_3 &= 2, \\ -2x_1 - 4x_2 + 5x_3 &= 5, \\ 8x_1 + 8x_2 &= -3. \end{aligned}$$

Step 1: Turn these equations to the matrix form $Ax = y$.

$$\begin{bmatrix} 4 & 3 & -5 \\ -2 & -4 & 5 \\ 8 & 8 & 0 \end{bmatrix} \begin{bmatrix} x_1 \\ x_2 \\ x_3 \end{bmatrix} = \begin{bmatrix} 2 \\ 5 \\ -3 \end{bmatrix}.$$

Step 2: Obtain the augmented matrix $[A, y]$ as

$$[A, y] = \begin{bmatrix} 4 & 3 & -5 & 2 \\ -2 & -4 & 5 & 5 \\ 8 & 8 & 0 & -3 \end{bmatrix}.$$

Step 3: Start eliminating the elements in the matrix by choosing a **pivot equation**, which is used to eliminate the elements in other equations. Choose the first equation as the pivot equation and turn the second row's first element to zero by multiplying the first row (pivot equation) by -0.5 and subtracting it from the second row. The multiplier is $m_{2,1} = -0.5$, giving us

$$\begin{bmatrix} 4 & 3 & -5 & 2 \\ 0 & -2.5 & 2.5 & 6 \\ 8 & 8 & 0 & -3 \end{bmatrix}.$$

Step 4: Turn the third row's first element to zero. As was done above, we multiply the first row by 2 and subtract it from the third row. The multiplier is $m_{3,1} = 2$, giving us

$$\begin{bmatrix} 4 & 3 & -5 & 2 \\ 0 & -2.5 & 2.5 & 6 \\ 0 & 2 & 10 & -7 \end{bmatrix}.$$

Step 5: Turn the third row's second element to zero. We multiply the second row by $-4/5$ and subtract it from the third row. The multiplier is $m_{3,2} = -0.8$, giving us

$$\begin{bmatrix} 4 & 3 & -5 & 2 \\ 0 & -2.5 & 2.5 & 6 \\ 0 & 0 & 12 & -2.2 \end{bmatrix}.$$

Step 6: We then obtain $x_3 = -2.2/12 = -0.183$.
Step 7: If we insert x_3 into the second equation, we obtain $x_2 = -2.583$.
Step 8: If we insert x_2 and x_3 into the first equation, we obtain $x_1 = 2.208$.

Note! Sometimes the first element in the first row is zero. When this is the case, switch the first row with a nonzero first element row, then follow the same procedure as outlined above.

We are using the "pivoting" Gauss elimination method here. Note that there is also a "naive" Gauss elimination method which assumes that the pivot values will never be zero.

14.4.2 GAUSS–JORDAN ELIMINATION METHOD

Gauss–Jordan elimination solves systems of equations. It is a procedure to turn A into a diagonal form such that the matrix form of the equations becomes

$$\begin{bmatrix} 1 & 0 & 0 & 0 \\ 0 & 1 & 0 & 0 \\ 0 & 0 & 1 & 0 \\ 0 & 0 & 0 & 1 \end{bmatrix} \begin{bmatrix} x_1 \\ x_2 \\ x_3 \\ x_4 \end{bmatrix} = \begin{bmatrix} y_1' \\ y_2' \\ y_3' \\ y_4' \end{bmatrix}.$$

Essentially, the equations become:

$$\begin{aligned} x_1 + 0 + 0 + 0 &= y_1', \\ 0 + x_2 + 0 + 0 &= y_2', \\ 0 + 0 + x_3 + 0 &= y_3', \\ 0 + 0 + 0 + x_4 &= y_4'. \end{aligned}$$

Let us solve another equation system by using the above example as a blueprint.

TRY IT! Use Gauss–Jordan elimination to solve the following equations:

$$\begin{aligned} 4x_1 + 3x_2 - 5x_3 &= 2, \\ -2x_1 - 4x_2 + 5x_3 &= 5, \\ 8x_1 + 8x_2 &= -3. \end{aligned}$$

Step 1: Construct the augmented matrix $[A, y]$,

$$[A, y] = \begin{bmatrix} 4 & 3 & -5 & 2 \\ -2 & -4 & 5 & 5 \\ 8 & 8 & 0 & -3 \end{bmatrix}.$$

Step 2: The first element in the first row should be 1, so we divide the row by 4:

$$\begin{bmatrix} 1 & 3/4 & -5/4 & 1/2 \\ -2 & -4 & 5 & 5 \\ 8 & 8 & 0 & -3 \end{bmatrix}.$$

Step 3: To eliminate the first element in the second and third rows, we multiply the first row by -2 and 8, respectively, and then subtract it from the second and third rows to get

$$\begin{bmatrix} 1 & 3/4 & -5/4 & 1/2 \\ 0 & -5/2 & 5/2 & 6 \\ 0 & 2 & 10 & -7 \end{bmatrix}.$$

Step 4: To normalize the second element in the second row to 1, we divide both sides of the equation by $-5/2$:

$$\begin{bmatrix} 1 & 3/4 & -5/4 & 1/2 \\ 0 & 1 & -1 & -12/5 \\ 0 & 2 & 10 & -7 \end{bmatrix}.$$

Step 5: To eliminate the second element in the third row, we multiply the second row by 2 and then subtract it from the third row:

$$\begin{bmatrix} 1 & 3/4 & -5/4 & 1/2 \\ 0 & 1 & -1 & -12/5 \\ 0 & 0 & 12 & -11/5 \end{bmatrix}.$$

Step 6: Normalize thê last row by dividing it by 8:

$$\begin{bmatrix} 1 & 3/4 & -5/4 & 1/2 \\ 0 & 1 & -1 & -12/5 \\ 0 & 0 & 1 & -11/60 \end{bmatrix}.$$

Step 7: To eliminate the third element in the second row, multiply the third row by -1 and subtract it from the second row:

$$\begin{bmatrix} 1 & 3/4 & -5/4 & 1/2 \\ 0 & 1 & 0 & -155/60 \\ 0 & 0 & 1 & -11/60 \end{bmatrix}.$$

Step 8: To eliminate the third element in first row, multiply the third row by -5/4 and then subtract it from the first row.

$$\begin{bmatrix} 1 & 3/4 & 0 & 13/48 \\ 0 & 1 & 0 & -2.583 \\ 0 & 0 & 1 & -0.183 \end{bmatrix}.$$

Step 9: To eliminate the second element in first row, multiply the second row by 3/4 and then subtract it from the first row.

$$\begin{bmatrix} 1 & 0 & 0 & 2.208 \\ 0 & 1 & 0 & -2.583 \\ 0 & 0 & 1 & -0.183 \end{bmatrix}.$$

14.4.3 LU DECOMPOSITION METHOD

The two methods shown above involve changing both A and y at the same time while trying to turn A to an upper triangular or diagonal matrix form. Sometimes we may have same set of equations but different sets of y for different experiments. This is actually quite common in the real world, where we have different experiment observations $y_a, y_b, y_c, \ldots$ Therefore, we must solve $Ax = y_a$, $Ax = y_b, \ldots$

many times, since every time the $[A, y]$ will change. Obviously, this is really inefficient. Is there a method by which we only change the left side of A but not the right-hand side y?

The LU decomposition method changes the matrix A only, instead of y. It is ideal for solving the system with the same coefficient matrices A but different constant vectors y. The LU decomposition method aims to turn A into the product of two matrices L and U, where L is a lower triangular matrix while U is an upper triangular matrix. With this decomposition, we convert the system of equations to the following form:

$$LUx = y \rightarrow \begin{bmatrix} l_{1,1} & 0 & 0 & 0 \\ l_{2,1} & l_{2,2} & 0 & 0 \\ l_{3,1} & l_{3,2} & l_{3,3} & 0 \\ l_{4,1} & l_{4,2} & l_{4,3} & l_{4,4} \end{bmatrix} \begin{bmatrix} u_{1,1} & u_{1,2} & u_{1,3} & u_{1,4} \\ 0 & u_{2,2} & u_{2,3} & u_{2,4} \\ 0 & 0 & u_{3,3} & u_{3,4} \\ 0 & 0 & 0 & u_{4,4} \end{bmatrix} \begin{bmatrix} x_1 \\ x_2 \\ x_3 \\ x_4 \end{bmatrix} = \begin{bmatrix} y_1 \\ y_2 \\ y_3 \\ y_4 \end{bmatrix}.$$

If we define $Ux = M$, then the above equations become:

$$\begin{bmatrix} l_{1,1} & 0 & 0 & 0 \\ l_{2,1} & l_{2,2} & 0 & 0 \\ l_{3,1} & l_{3,2} & l_{3,3} & 0 \\ l_{4,1} & l_{4,2} & l_{4,3} & l_{4,4} \end{bmatrix} M = \begin{bmatrix} y_1 \\ y_2 \\ y_3 \\ y_4 \end{bmatrix}.$$

We can easily solve the above problem by forward substitution (the opposite of the backward substitution as we saw in Gauss elimination method). After we solve for M, we can easily solve the rest of the problem using backward substitution:

$$\begin{bmatrix} u_{1,1} & u_{1,2} & u_{1,3} & u_{1,4} \\ 0 & u_{2,2} & u_{2,3} & u_{2,4} \\ 0 & 0 & u_{3,3} & u_{3,4} \\ 0 & 0 & 0 & u_{4,4} \end{bmatrix} \begin{bmatrix} x_1 \\ x_2 \\ x_3 \\ x_4 \end{bmatrix} = \begin{bmatrix} m_1 \\ m_2 \\ m_3 \\ m_4 \end{bmatrix}.$$

But how do we obtain the L and U matrices? There are different ways to obtain the LU decomposition. Below is one example that uses the Gauss elimination method. From the above, we know that we obtain an upper triangular matrix after we conduct the Gauss elimination. At the same time, we also obtain the lower triangular matrix even though it is never explicitly written out. During the Gauss elimination procedure, the matrix A actually turns into the product of two matrices as shown below. The right upper triangular matrix is the one we obtained earlier. The diagonal elements in the left lower triangular matrix are 1, and the elements below the diagonal elements are the multipliers that multiply the pivot equations to eliminate the elements during the calculation:

$$A = \begin{bmatrix} 1 & 0 & 0 & 0 \\ m_{2,1} & 1 & 0 & 0 \\ m_{3,1} & m_{3,2} & 1 & 0 \\ m_{4,1} & m_{4,2} & m_{4,3} & 1 \end{bmatrix} \begin{bmatrix} u_{1,1} & u_{1,2} & u_{1,3} & u_{1,4} \\ 0 & u_{2,2} & u_{2,3} & u_{2,4} \\ 0 & 0 & u_{3,3} & u_{3,4} \\ 0 & 0 & 0 & u_{4,4} \end{bmatrix}.$$

Note that we obtain both L and U at the same time when we perform the Gauss elimination. Using the above example, where U is the one we used before to solve the equations, and L is composed of the multipliers (you can check the examples in the Gauss elimination section), we obtain:

$$L = \begin{bmatrix} 1 & 0 & 0 \\ -0.5 & 1 & 0 \\ 2 & -0.8 & 1 \end{bmatrix},$$

$$U = \begin{bmatrix} 4 & 3 & -5 \\ 0 & -2.5 & 2.5 \\ 0 & 0 & 60 \end{bmatrix}.$$

TRY IT! Verify that the above L and U matrices are the LU decomposition of matrix A. The result should be $A = LU$.

```
In [1]: import numpy as np

        u = np.array([[4, 3, -5],
                      [0, -2.5, 2.5],
                      [0, 0, 12]])
        l = np.array([[1, 0, 0],
                      [-0.5, 1, 0],
                      [2, -0.8, 1]])

        print("LU=", np.dot(l, u))

LU= [[ 4.   3. -5.]
 [-2. -4.   5.]
 [ 8.   8.   0.]]
```

14.4.4 ITERATIVE METHODS – GAUSS–SEIDEL METHOD

The methods introduced above are all direct methods where the solution is computed using a finite number of operations. This section introduces a different class of methods, namely the **iterative methods**, or **indirect methods**. They start with an initial guess of the solution and then repeatedly improve the solution until the change of the solution is below a chosen threshold. In order to use this iterative process, we first need to write the explicit form of a system of equations. If we have a system of linear equations

$$\begin{bmatrix} a_{1,1} & a_{1,2} & \dots & a_{1,n} \\ a_{2,1} & a_{2,2} & \dots & a_{2,n} \\ \vdots & \vdots & \ddots & \vdots \\ a_{m,1} & a_{m,2} & \dots & a_{m,n} \end{bmatrix} \begin{bmatrix} x_1 \\ x_2 \\ \vdots \\ x_n \end{bmatrix} = \begin{bmatrix} y_1 \\ y_2 \\ \vdots \\ y_m \end{bmatrix},$$

we can write its explicit form as

$$x_i = \frac{1}{a_{i,i}}\left[y_i - \sum_{j=1, j\neq i}^{j=n} a_{i,j}x_j\right].$$

This is the basics of the iterative methods; we can assume initial values for all the x, and use it as $x^{(0)}$. In the first iteration, we can substitute $x^{(0)}$ into the right-hand side of the explicit equation above to obtain the first iteration solution $x^{(1)}$. By substituting $x^{(1)}$ into the equation, we obtain $x^{(2)}$, and the iterations continue until the difference between $x^{(k)}$ and $x^{(k-1)}$ is smaller than some predefined value.

Iterative methods require having specific conditions for the solution to converge. A sufficient, but not necessary, condition of the convergence is that the coefficient matrix a is **diagonally dominant**. This means that in each row of the matrix of coefficients a, the absolute value of the diagonal element is greater than the sum of the absolute values of the off-diagonal elements. If the coefficient matrix satisfies this condition, the iterations will converge to the solution. Note that the solution process might still converge even when this condition is not satisfied.

14.4.4.1 Gauss–Seidel Method

The **Gauss–Seidel method** is a specific iterative method that is always using the latest estimated value for each element in x. For example, first assume that the initial values for $x_2, x_3, \ldots, x_n$ (except for x_1) are given and calculate x_1. Using the calculated x_1 and the rest of the x (except for x_2), we can calculate x_2. Continuing in the same manner and calculating all the elements in x will conclude the first iteration. The unique part of the Gauss–Seidel method is the use of the latest value to calculate the next value in x. Such iterations are continued until the value converges. Let us use this method to solve the same problem we just solved above.

EXAMPLE: Solve the following system of linear equations using Gauss–Seidel method using a predefined threshold $\epsilon = 0.01$. Remember to check if the converge condition is satisfied or not.

$$\begin{aligned} 8x_1 + 3x_2 - 3x_3 &= 14, \\ -2x_1 - 8x_2 + 5x_3 &= 5, \\ 3x_1 + 5x_2 + 10x_3 &= -8. \end{aligned}$$

Let us first check if the coefficient matrix is diagonally dominant or not.

```
In [2]: a = [[8, 3, -3], [-2, -8, 5], [3, 5, 10]]

        # Find diagonal coefficients
        diag = np.diag(np.abs(a))

        # Find row sum without diagonal
        off_diag = np.sum(np.abs(a), axis=1) - diag

        if np.all(diag > off_diag):
```

```
        print("matrix is diagonally dominant")
    else:
        print("NOT diagonally dominant")

matrix is diagonally dominant
```

Since it is guaranteed to converge, we can use Gauss–Seidel method to solve the system.

```
In [3]: x1 = 0
        x2 = 0
        x3 = 0
        epsilon = 0.01
        converged = False

        x_old = np.array([x1, x2, x3])

        print("Iteration results")
        print(" k,    x1,    x2,    x3 ")
        for k in range(1, 50):
            x1 = (14-3*x2+3*x3)/8
            x2 = (5+2*x1-5*x3)/(-8)
            x3 = (-8-3*x1-5*x2)/(-5)
            x = np.array([x1, x2, x3])
            # check if it is smaller than threshold
            dx = np.sqrt(np.dot(x-x_old, x-x_old))

            print("%d, %.4f, %.4f, %.4f"%(k, x1, x2, x3))
            if dx < epsilon:
                converged = True
                print("Converged!")
                break

            # assign the latest x value to the old value
            x_old = x

        if not converged:
            print("Not converged, increase the # of iterations")

Iteration results
 k,    x1,    x2,    x3
1, 1.7500, -1.0625, 1.5875
2, 2.7437, -0.3188, 2.9275
3, 2.9673, 0.4629, 3.8433
```

```
4, 3.0177, 1.0226, 4.4332
5, 3.0290, 1.3885, 4.8059
6, 3.0315, 1.6208, 5.0397
7, 3.0321, 1.7668, 5.1861
8, 3.0322, 1.8582, 5.2776
9, 3.0322, 1.9154, 5.3348
10, 3.0323, 1.9512, 5.3705
11, 3.0323, 1.9735, 5.3929
12, 3.0323, 1.9875, 5.4068
13, 3.0323, 1.9962, 5.4156
14, 3.0323, 2.0017, 5.4210
Converged!
```

14.5 SOLVING SYSTEMS OF LINEAR EQUATIONS IN PYTHON

The examples presented above demonstrated the various methods you can use to solve systems of linear equations. This is also very easy to do in Python, as shown below. The easiest way to get a solution is via the `solve` function in `NumPy`.

TRY IT! Use `numpy.linalg.solve` to solve the following equations:

$$\begin{aligned} 4x_1 + 3x_2 - 5x_3 &= 2, \\ -2x_1 - 4x_2 + 5x_3 &= 5, \\ 8x_1 + 8x_2 &= -3. \end{aligned}$$

```
In [1]: import numpy as np

        A = np.array([[4, 3, -5],
                      [-2, -4, 5],
                      [8, 8, 0]])
        y = np.array([2, 5, -3])

        x = np.linalg.solve(A, y)
        print(x)

[ 2.20833333 -2.58333333 -0.18333333]
```

We get the same results as those in the previous section when calculated by hand. Under the "hood," the solver is actually doing an LU decomposition to get the results. If you can check the help of the function, you will see it needs the input matrix to be square and of full rank, i.e., all rows (or, equivalently, columns) must be linearly independent.

TRY IT! Try to solve the above equations using the matrix inversion approach.

```
In [2]: A_inv = np.linalg.inv(A)

        x = np.dot(A_inv, y)
        print(x)

[ 2.20833333 -2.58333333 -0.18333333]
```

We can also obtain the L and U matrices used in the LU decomposition using the `SciPy` package.

TRY IT! Get the L and U for the above matrix A.

```
In [3]: from scipy.linalg import lu

        P, L, U = lu(A)
        print("P:\n", P)
        print("L:\n", L)
        print("U:\n", U)
        print("LU:\n",np.dot(L, U))

P:
 [[0. 0. 1.]
 [0. 1. 0.]
 [1. 0. 0.]]
L:
 [[ 1.    0.    0.  ]
 [-0.25  1.    0.  ]
 [ 0.5   0.5   1.  ]]
U:
 [[ 8.    8.    0. ]
 [ 0.   -2.    5. ]
 [ 0.    0.   -7.5]]
LU:
 [[ 8.  8.  0.]
 [-2. -4.  5.]
 [ 4.  3. -5.]]
```

Why do we obtain different L and U from those calculated by hand in the last section? You will also see that there is a **permutation matrix** P that is returned by the `lu` function. This permutation matrix records how it changes the order of the equations for easier calculation purposes. For example, if the first element in the first row is zero, it cannot be the pivot equation since you cannot turn the first elements in other rows to zero; therefore, we need to switch the order of the equations to get a new pivot equation. If you multiply P and A, you will see that this permutation matrix reverses the order of the equations for this case.

TRY IT! Multiply P and A and see what is the effect of the permutation matrix on A.

```
In [4]: print(np.dot(P, A))

[[ 8.  8.  0.]
 [-2. -4.  5.]
 [ 4.  3. -5.]]
```

14.6 MATRIX INVERSION

We defined the inverse of a square matrix M as a matrix of the same size, M^{-1}, such that $M \cdot M^{-1} = M^{-1} \cdot M = I$. If the dimension of the matrix is high, the analytical solution for the matrix inversion will be complicated. Therefore, we need some other efficient ways to obtain the inverse of the matrix.

Let us use a 4×4 matrix for illustration. Suppose we have

$$M = \begin{bmatrix} m_{1,1} & m_{1,2} & m_{1,3} & m_{1,4} \\ m_{2,1} & m_{2,2} & m_{2,3} & m_{2,4} \\ m_{3,1} & m_{3,2} & m_{3,3} & m_{3,4} \\ m_{4,1} & m_{4,2} & m_{4,3} & m_{4,4} \end{bmatrix},$$

and the inverse of M is

$$X = \begin{bmatrix} x_{1,1} & x_{1,2} & x_{1,3} & x_{1,4} \\ x_{2,1} & x_{2,2} & x_{2,3} & x_{2,4} \\ x_{3,1} & x_{3,2} & x_{3,3} & x_{3,4} \\ x_{4,1} & x_{4,2} & x_{4,3} & x_{4,4} \end{bmatrix}.$$

Therefore, we will have:

$$M \cdot X = \begin{bmatrix} m_{1,1} & m_{1,2} & m_{1,3} & m_{1,4} \\ m_{2,1} & m_{2,2} & m_{2,3} & m_{2,4} \\ m_{3,1} & m_{3,2} & m_{3,3} & m_{3,4} \\ m_{4,1} & m_{4,2} & m_{4,3} & m_{4,4} \end{bmatrix} \begin{bmatrix} x_{1,1} & x_{1,2} & x_{1,3} & x_{1,4} \\ x_{2,1} & x_{2,2} & x_{2,3} & x_{2,4} \\ x_{3,1} & x_{3,2} & x_{3,3} & x_{3,4} \\ x_{4,1} & x_{4,2} & x_{4,3} & x_{4,4} \end{bmatrix} = \begin{bmatrix} 1 & 0 & 0 & 0 \\ 0 & 1 & 0 & 0 \\ 0 & 0 & 1 & 0 \\ 0 & 0 & 0 & 1 \end{bmatrix}.$$

We can rewrite the above equation as four separate equations, i.e.,

$$\begin{bmatrix} m_{1,1} & m_{1,2} & m_{1,3} & m_{1,4} \\ m_{2,1} & m_{2,2} & m_{2,3} & m_{2,4} \\ m_{3,1} & m_{3,2} & m_{3,3} & m_{3,4} \\ m_{4,1} & m_{4,2} & m_{4,3} & m_{4,4} \end{bmatrix} \begin{bmatrix} x_{1,1} \\ x_{2,1} \\ x_{3,1} \\ x_{4,1} \end{bmatrix} = \begin{bmatrix} 1 \\ 0 \\ 0 \\ 0 \end{bmatrix},$$

$$\begin{bmatrix} m_{1,1} & m_{1,2} & m_{1,3} & m_{1,4} \\ m_{2,1} & m_{2,2} & m_{2,3} & m_{2,4} \\ m_{3,1} & m_{3,2} & m_{3,3} & m_{3,4} \\ m_{4,1} & m_{4,2} & m_{4,3} & m_{4,4} \end{bmatrix} \begin{bmatrix} x_{1,2} \\ x_{2,2} \\ x_{3,2} \\ x_{4,2} \end{bmatrix} = \begin{bmatrix} 0 \\ 1 \\ 0 \\ 0 \end{bmatrix},$$

$$\begin{bmatrix} m_{1,1} & m_{1,2} & m_{1,3} & m_{1,4} \\ m_{2,1} & m_{2,2} & m_{2,3} & m_{2,4} \\ m_{3,1} & m_{3,2} & m_{3,3} & m_{3,4} \\ m_{4,1} & m_{4,2} & m_{4,3} & m_{4,4} \end{bmatrix} \begin{bmatrix} x_{1,3} \\ x_{2,3} \\ x_{3,3} \\ x_{4,3} \end{bmatrix} = \begin{bmatrix} 0 \\ 0 \\ 1 \\ 0 \end{bmatrix},$$

$$\begin{bmatrix} m_{1,1} & m_{1,2} & m_{1,3} & m_{1,4} \\ m_{2,1} & m_{2,2} & m_{2,3} & m_{2,4} \\ m_{3,1} & m_{3,2} & m_{3,3} & m_{3,4} \\ m_{4,1} & m_{4,2} & m_{4,3} & m_{4,4} \end{bmatrix} \begin{bmatrix} x_{1,4} \\ x_{2,4} \\ x_{3,4} \\ x_{4,4} \end{bmatrix} = \begin{bmatrix} 0 \\ 0 \\ 0 \\ 1 \end{bmatrix}.$$

Solving the above four systems of equations will provide the inverse of the matrix. We can use any method introduced previously to solve these equations (such as Gauss elimination, Gauss–Jordan, and LU decomposition). Below is an example of matrix inversion using the Gauss–Jordan method.

Recall that in the Gauss–Jordan method, we convert our problem from

$$\begin{bmatrix} m_{1,1} & m_{1,2} & m_{1,3} & m_{1,4} \\ m_{2,1} & m_{2,2} & m_{2,3} & m_{2,4} \\ m_{3,1} & m_{3,2} & m_{3,3} & m_{3,4} \\ m_{4,1} & m_{4,2} & m_{4,3} & m_{4,4} \end{bmatrix} \begin{bmatrix} x_1 \\ x_2 \\ x_3 \\ x_4 \end{bmatrix} = \begin{bmatrix} y_1 \\ y_2 \\ y_3 \\ y_4 \end{bmatrix}$$

to

$$\begin{bmatrix} 1 & 0 & 0 & 0 \\ 0 & 1 & 0 & 0 \\ 0 & 0 & 1 & 0 \\ 0 & 0 & 0 & 1 \end{bmatrix} \begin{bmatrix} x_1 \\ x_2 \\ x_3 \\ x_4 \end{bmatrix} = \begin{bmatrix} y'_1 \\ y'_2 \\ y'_3 \\ y'_4 \end{bmatrix}$$

to obtain the solution. Essentially, we are converting

$$\begin{bmatrix} m_{1,1} & m_{1,2} & m_{1,3} & m_{1,4} & y_1 \\ m_{2,1} & m_{2,2} & m_{2,3} & m_{2,4} & y_2 \\ m_{3,1} & m_{3,2} & m_{3,3} & m_{3,4} & y_3 \\ m_{4,1} & m_{4,2} & m_{4,3} & m_{4,4} & y_4 \end{bmatrix}$$

to

$$\begin{bmatrix} 1 & 0 & 0 & 0 & y'_1 \\ 0 & 1 & 0 & 0 & y'_2 \\ 0 & 0 & 1 & 0 & y'_3 \\ 0 & 0 & 0 & 1 & y'_4 \end{bmatrix}.$$

In summary, all we need to do is to convert

$$\begin{bmatrix} m_{1,1} & m_{1,2} & m_{1,3} & m_{1,4} & 1 & 0 & 0 & 0 \\ m_{2,1} & m_{2,2} & m_{2,3} & m_{2,4} & 0 & 1 & 0 & 0 \\ m_{3,1} & m_{3,2} & m_{3,3} & m_{3,4} & 0 & 0 & 1 & 0 \\ m_{4,1} & m_{4,2} & m_{4,3} & m_{4,4} & 0 & 0 & 0 & 1 \end{bmatrix}$$

to

$$\begin{bmatrix} 1 & 0 & 0 & 0 & m'_{1,1} & m'_{1,2} & m'_{1,3} & m'_{1,4} \\ 0 & 1 & 0 & 0 & m'_{2,1} & m'_{2,2} & m'_{2,3} & m'_{2,4} \\ 0 & 0 & 1 & 0 & m'_{3,1} & m'_{3,2} & m'_{3,3} & m'_{1,4} \\ 0 & 0 & 0 & 1 & m'_{4,1} & m'_{4,2} & m'_{4,3} & m'_{1,4} \end{bmatrix}.$$

Then the matrix

$$\begin{bmatrix} m'_{1,1} & m'_{1,2} & m'_{1,3} & m'_{1,4} \\ m'_{2,1} & m'_{2,2} & m'_{2,3} & m'_{2,4} \\ m'_{3,1} & m'_{3,2} & m'_{3,3} & m'_{1,4} \\ m'_{4,1} & m'_{4,2} & m'_{4,3} & m'_{1,4} \end{bmatrix}$$

is the inverse of M we are looking for.

Can you explain how to use LU decomposition to get the inverse of a matrix?

14.7 SUMMARY AND PROBLEMS

14.7.1 SUMMARY

1. Linear algebra is the foundation of many engineering fields.
2. Vectors can be considered as points in $\mathbb{R}^n$; addition and multiplication are defined, although this is not necessarily the case for scalars.
3. A set of vectors is linearly independent if none of the vectors can be written as a linear combination of the others.
4. Matrices are tables of numbers. They have several important characteristics including the determinant, rank, and inverse.
5. A system of linear equations can be represented by the matrix equation $Ax = y$.
6. The number of solutions to a system of linear equations is related to rank(A) and rank($[A, y]$). It can be zero, one, or infinity.
7. We can solve the equations using Gauss elimination, Gauss–Jordan elimination, LU decomposition, and the Gauss–Seidel method.
8. We introduced methods to find the matrix inverse.

14.7.2 PROBLEMS

1. It is strongly recommended that you read a book on linear algebra, which will give you greater mastery of the contents of this chapter. We strongly recommend reading the first part of book **Optimization Models** by Giuseppe Calafiore and Laurent El Ghaoui to get you started.

2. Show that matrix multiplication distributes over matrix addition: show $A(B+C) = AB + AC$ assuming that A, B, and C are matrices of compatible size.

3. Write a function `my_is_orthogonal(v1,v2,tol)` where `v1` and `v2` are column vectors of the same size, and `tol` is a scalar value strictly larger than zero. The output should be 1 if the angle between `v1` and `v2` is within tol of $\pi/2$, that is, $|\pi/2 - \theta| <$ `tol`, and zero otherwise. You may assume that `v1` and `v2` are column vectors of the same size, and that `tol` is a positive scalar.

```
In [ ]: # Test cases for Problem 3
        a = np.array([[1], [0.001]])
        b = np.array([[0.001], [1]])
        # output: 1
        my_is_orthogonal(a,b, 0.01)

        # output: 0
        my_is_orthogonal(a,b, 0.001)

        # output: 0
        a = np.array([[1], [0.001]])
        b = np.array([[1], [1]])
        my_is_orthogonal(a,b, 0.01)

        # output: 1
        a = np.array([[1], [1]])
        b = np.array([[-1], [1]])
        my_is_orthogonal(a,b, 1e-10)
```

3. Write a function `my_is_similar(s1,s2,tol)` where `s1` and `s2` are strings, not necessarily of the same size, and `tol` is a scalar value strictly larger than zero. From `s1` and `s2`, the function should construct two vectors, `v1` and `v2`, where `v1[0]` is the number of a's in `s1`, `v1[1]` is the number b's in `s1`, and so on until `v1[25]`, which is the number of z's in v1. The vector `v2` should be similarly constructed from `s2`. The output should be 1 if the absolute value of the angle between `v1` and `v2` is less than `tol`, that is, $|\theta| <$ `tol`.

4. Write a function `my_make_lin_ind(A)` where A and B are matrices. Let rank$(A) = n$. Then B should be a matrix containing the first n columns of A that are all linearly independent. Note that this implies that B has full rank.

```
In [ ]: ## Test cases for Problem 4


        A = np.array([[12,24,0,11,-24,18,15],
                      [19,38,0,10,-31,25,9],
```

```
        [1,2,0,21,-5,3,20],
        [6,12,0,13,-10,8,5],
        [22,44,0,2,-12,17,23]])

B = my_make_lin_ind(A)

# B = [[12,11,-24,15],
#      [19,10,-31,9],
#      [1,21,-5,20],
#      [6,13,-10,5],
#      [22,2,-12,23]]
```

5. Cramer's rule is a method of computing the determinant of a matrix. Consider an $n \times n$ square matrix M. Let $M(i, j)$ be the element of M in the ith row and jth column of M, and let $m_{i,j}$ be the minor of M created by removing the ith row and jth column from M. Cramer's rule says that

$$\det(M) = \sum_{i=1}^{n} (-1)^{i-1} M(1,i) \det(m_{i,j}).$$

Write a function `my_rec_det(M)` where the output is $\det(M)$. Use Cramer's rule to compute the determinant, not `NumPy`'s function.

6. What is the complexity of `my_rec_det` in the previous problem? Do you think this is an effective way of determining if a matrix is singular or not?

7. Let p be a vector with length L containing the coefficients of a polynomial of order $L-1$. For example, the vector $p = [1, 0, 2]$ is a representation of the polynomial $f(x) = 1x^2 + 0x + 2$. Write a function `my_poly_der_mat(p)` where p is the aforementioned vector, and the output D is the matrix that will return the coefficients of the derivative of p when p is left multiplied by D. For example, the derivative of $f(x)$ is $f'(x) = 2x$; therefore, $d = Dp$ should yield $d = [2, 0]$. Note this implies that the dimension of D is $L - 1 \times L$. The point of this problem is to show that differentiating polynomials is actually a linear transformation.

8. Use the Gauss elimination method to solve the following equations:

$$\begin{aligned} 3x_1 - x_2 + 4x_3 &= 2, \\ 17x_1 + 2x_2 + x_3 &= 14, \\ x_1 + 12x_2 - 7z &= 54. \end{aligned}$$

9. Use the Gauss–Jordan elimination method to solve the equations in Problem 8.

10. Obtain the lower triangular matrix L and upper triangular matrix U from the equations in Problem 8.

11. Show that the dot-product distributes across vector addition, that is, show that $u \cdot (v + w) = u \cdot v + u \cdot w$.

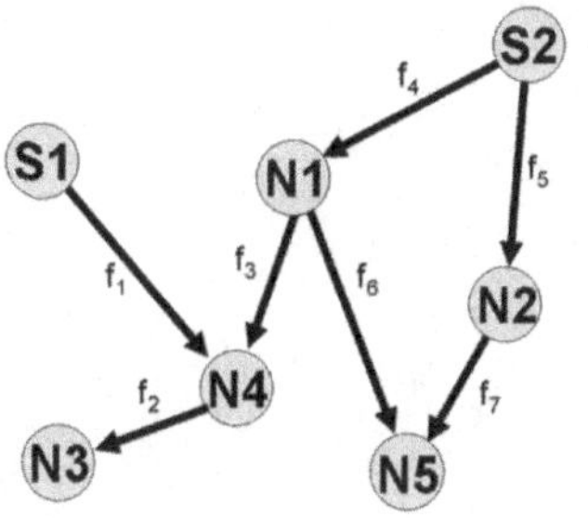

FIGURE 14.1

Graph for Problem 12.

12. Consider the following network shown in Fig. 14.1 consisting of two power supply stations denoted by `S1` and `S2` and five power recipient nodes denoted by `N1` to `N5`. The nodes are connected by power lines, which are denoted by arrows, and power can flow between nodes along these lines in both directions.
Let d_i be a positive scalar denoting the power demand for node i; assume that this demand must be met exactly. The capacity of the power supply stations is denoted by S. Power supply stations must run at capacity. For each arrow, let f_j be the power flow along that arrow. Negative flow implies that power is running in the opposite direction of the arrow.
Write a function `my_flow_calculator(S,d)` where `S` is a 1×2 vector representing the capacity of each power supply station, and `d` is a 1×5 row vector representing the demands at each node (i.e., $d[0]$ is the demand at node 1). The output argument, `f`, should be a 1×7 row vector denoting the flows in the network (i.e., `f[0]` = f_1 in the diagram). The flows contained in `f` should satisfy all the constraints of the system, like power generation and demands. Note that there may be more than one solution to the system of equations.
The total flow into a node must equal the total flow out of the node plus the demand; that is, for each node i, $f_{\text{inflow}} = f_{\text{outflow}} + d_i$. You may assume that $\sum S_j = \sum d_i$.

```
In [ ]: ## Test cases for Problem 12

        s = np.array([[10, 10]])
        d = np.array([[4, 4, 4, 4, 4]])

        # f = [[10.0, 4.0, -2.0, 4.5, 5.5, 2.5, 1.5]]
        f = my_flow_calculator(s, d)

        s = np.array([[10, 10]])
        d = np.array([[3, 4, 5, 4, 4]])
        # f = [[10.0, 5.0, -1.0, 4.5, 5.5, 2.5, 1.5]]
        f = my_flow_calculator(s, d)
```

CHAPTER

EIGENVALUES AND EIGENVECTORS

15

CONTENTS

15.1 EIGENVALUES AND EIGENVECTORS PROBLEM STATEMENT

15.1.1 EIGENVALUES AND EIGENVECTORS

We learned from the previous chapter that matrix A applied to a column vector x, that is, Ax, is a linear transformation of x. There is a special transform in the following form:

$$Ax = \lambda x,$$

where A is an $n \times n$ matrix, x is an $n \times 1$ column vector ($x \neq 0$), and λ is a scalar. Any λ that satisfies the above equation is known as an **eigenvalue** of the matrix A, while the associated vector x is called an **eigenvector** corresponding to λ.

15.1.2 THE MOTIVATION BEHIND EIGENVALUES AND EIGENVECTORS

The motivation behind eigenvalues and eigenvectors is that if we understand the characteristics of the linear transformation, it will help simplify the solutions to our problem. Say, we can multiply a vector A to another vector x, i.e., Ax. It essentially transforms the vector x into another vector, whereby the transformation represents a scale of the length of the vector and/or the rotation of the vector. The above equation points out that for some vectors, the effect of transformation Ax is only scaling (stretching, compressing, and flipping). The eigenvectors are the vectors having this property and the eigenvalues λ are the scale factors. Let us look at the following example.

Python Programming and Numerical Methods. https://doi.org/10.1016/B978-0-12-819549-9.00025-7

TRY IT! Plot the vector $x = [[1], [1]]$ and the vector $b = Ax$, where $A = [[2, 0], [0, 1]]$

```
In [1]: import numpy as np
        import matplotlib.pyplot as plt

        plt.style.use("seaborn-poster")

        %matplotlib inline

        def plot_vect(x, b, xlim, ylim):
            """
            function to plot two vectors,
            x - the original vector
            b - the transformed vector
            xlim - the limit for x
            ylim - the limit for y
            """
            plt.figure(figsize = (10, 6))
            plt.quiver(0,0,x[0],x[1],\
                color="k",angles="xy",\
                scale_units="xy",scale=1,\
                label="Original vector")
            plt.quiver(0,0,b[0],b[1],\
                color="g",angles="xy",\
                scale_units="xy",scale=1,\
                label ="Transformed vector")
            plt.xlim(xlim)
            plt.ylim(ylim)
            plt.xlabel("X")
            plt.ylabel("Y")
            plt.legend()
            plt.show()

In [2]: A = np.array([[2, 0],[0, 1]])

        x = np.array([[1],[1]])
        b = np.dot(A, x)
        plot_vect(x,b,(0,3),(0,2))
```

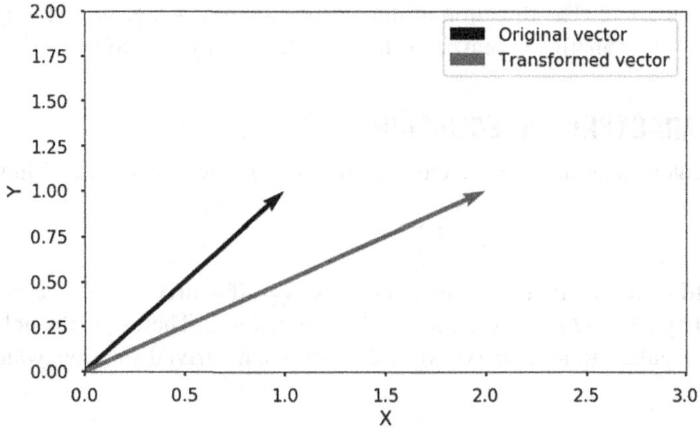

We can see from the generated figure that the original vector x is rotated and stretched longer after being transformed by A. The vector [[1], [1]] is transformed to [[2], [1]]. Let us try to do the same exercise with a different vector [[1], [0]].

TRY IT! Plot the vector x = [[1], [0]] and the vector $b = Ax$ where A = [[2, 0], [0, 1]]

```
In [3]: x = np.array([[1], [0]])
        b = np.dot(A, x)

        plot_vect(x,b,(0,3),(-0.5,0.5))
```

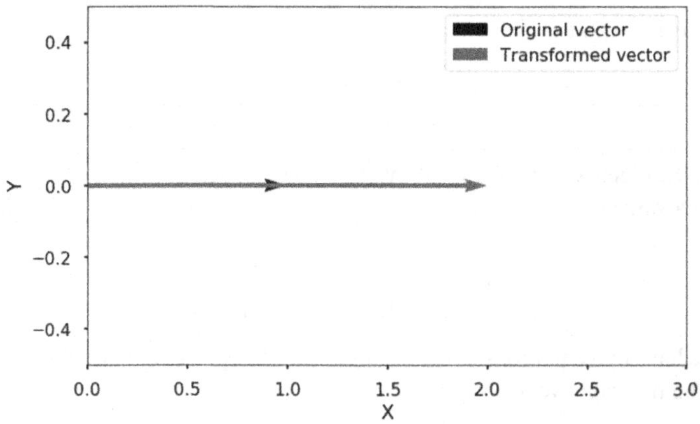

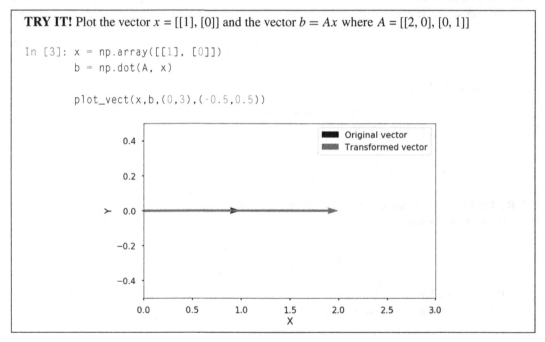

With this new vector, the only change after the transformation is the length of the vector; it is stretched. The new vector is [[2], [0]], therefore, the transform is

$$Ax = 2x$$

with x = [[1], [0]] and $\lambda = 2$. The direction of the vector does not change at all (no rotation). You can also check that [[0], [1]] is another eigenvector; try to verify this by yourself.

15.1.3 THE CHARACTERISTIC EQUATION

In order to get the eigenvalues and eigenvectors, from $Ax = \lambda x$, we can use the following form:

$$(A - \lambda I)x = 0,$$

where I is the identify matrix with the same dimensions as A. If matrix $A - \lambda I$ has an inverse and both sides are multiplied by $(A - \lambda I)^{-1}$, we get a trivial solution $x = 0$. Therefore, the only interesting case is when $A - \lambda I$ is singular (no inverse exists), and we have a nontrivial solution, which means that the determinant is zero:

$$\det(A - \lambda I) = 0.$$

This equation is called the **characteristic equation** that will lead to a polynomial equation for λ, which we can solve for the eigenvalues; see the example below.

TRY IT! Obtain the eigenvalues for matrix [[0, 2], [2, 3]]

The characteristic equation gives us

$$\begin{vmatrix} 0-\lambda & 2 \\ 2 & 3-\lambda \end{vmatrix} = 0.$$

Therefore, we have

$$-\lambda(3-\lambda) - 4 = 0 \Rightarrow \lambda^2 - 3\lambda - 4 = 0.$$

We obtain two eigenvalues:

$$\lambda_1 = 4, \quad \lambda_2 = -1.$$

TRY IT! Obtain the eigenvectors for the above two eigenvalues.

If the first eigenvalue is $\lambda_1 = 4$, we can simply insert it back to $A - \lambda I = 0$, where we have:

$$\begin{bmatrix} -4 & 2 \\ 2 & -1 \end{bmatrix} \begin{bmatrix} x_1 \\ x_2 \end{bmatrix} = \begin{bmatrix} 0 \\ 0 \end{bmatrix}.$$

Therefore, we have two equations $-4x_1 + 2x_2 = 0$ and $2x_1 - x_2 = 0$, both indicating that $x_2 = 2x_1$. Therefore, the first eigenvector is

$$x_1 = k_1 \begin{bmatrix} 1 \\ 2 \end{bmatrix}.$$

The symbol k_1 is a scalar ($k_1 \neq 0$); as long as we have the ratio between x_2 and x_1 as 2, it will be an eigenvector. We can verify that the vector [[1], [2]] is an eigenvector by inserting it back:

$$\begin{bmatrix} 0 & 2 \\ 2 & 3 \end{bmatrix} \begin{bmatrix} 1 \\ 2 \end{bmatrix} = \begin{bmatrix} 4 \\ 8 \end{bmatrix} = 4 \begin{bmatrix} 1 \\ 2 \end{bmatrix}.$$

By inserting $\lambda_2 = -1$ as above, we obtain the other eigenvector as follows where $k_2 \neq 0$:

$$x_2 = k_2 \begin{bmatrix} -2 \\ 1 \end{bmatrix}.$$

The above example demonstrates how we can obtain the eigenvalues and eigenvectors from a matrix A; the choice of the eigenvectors for a system is not unique. When you have a larger matrix A and try to solve the nth order polynomial characteristic equation, the solution becomes more complicated. Luckily, many different numerical methods have been developed to solve the eigenvalue problems for larger matrices (with a few hundred to thousands of dimensions). We will introduce the power method and the QR method in the next two sections.

15.2 THE POWER METHOD

15.2.1 FINDING THE LARGEST EIGENVALUE

Some problems only require finding the largest dominant eigenvalue and its corresponding eigenvector. In this case, we can use the **power method**, which is an iterative method that will converge to the largest eigenvalue; see the example below.

Consider an $n \times n$ matrix A that has n real eigenvalues $\lambda_1, \lambda_2, \ldots, \lambda_n$ and the corresponding linearly independent eigenvectors $v_1, v_2, \ldots, v_n$. Since the eigenvalues are scalars, we can rank them so that

$$|\lambda_1| > |\lambda_2| \geq \cdots \geq |\lambda_n|.$$

Note that we only require $|\lambda_1| > |\lambda_2|$; the other eigenvalues may be equal to each other.

Because the eigenvectors are assumed linearly independent, they are a set of basis vectors, which means that any vector that is in the same space can be written as a linear combination of the basis vectors. That is, for any vector x_0 and can be written as:

$$x_0 = c_1 v_1 + c_2 v_2 + \cdots + c_n v_n$$

where $c_1 \neq 0$ is the constraint. If it is zero, then we need to choose another initial vector so that $c_1 \neq 0$.

Now let us multiply both sides by A:

$$Ax_0 = c_1 A v_1 + c_2 A v_2 + \cdots + c_n A v_n.$$

Since $Av_i = \lambda_i v_i$,

$$Ax_0 = c_1 \lambda_1 v_1 + c_2 \lambda_2 v_2 + \cdots + c_n \lambda_n v_n.$$

We can change the above equation to:

$$Ax_0 = c_1\lambda_1\left[v_1 + \frac{c_2}{c_1}\frac{\lambda_2}{\lambda_1}v_2 + \cdots + \frac{c_n}{c_1}\frac{\lambda_n}{\lambda_1}v_n\right] = c_1\lambda_1 x_1,$$

where x_1 is a new vector, $x_1 = v_1 + \frac{c_2}{c_1}\frac{\lambda_2}{\lambda_1}v_2 + \cdots + \frac{c_n}{c_1}\frac{\lambda_n}{\lambda_1}v_n$.

This finishes the first iteration. For the second iteration, we apply A to x_1:

$$Ax_1 = \lambda_1 v_1 + \frac{c_2}{c_1}\frac{\lambda_2^2}{\lambda_1}v_2 + \cdots + \frac{c_n}{c_1}\frac{\lambda_n^2}{\lambda_1}v_n.$$

Similarly, we can rearrange the above equation to get

$$Ax_1 = \lambda_1\left[v_1 + \frac{c_2}{c_1}\frac{\lambda_2^2}{\lambda_1^2}v_2 + \cdots + \frac{c_n}{c_1}\frac{\lambda_n^2}{\lambda_1^2}v_n\right] = \lambda_1 x_2,$$

where x_2 is a new vector, $x_2 = v_1 + \frac{c_2}{c_1}\frac{\lambda_2^2}{\lambda_1^2}v_2 + \cdots + \frac{c_n}{c_1}\frac{\lambda_n^2}{\lambda_1^2}v_n$.

If we continue applying A to the new vector, we obtain from each iteration k:

$$Ax_{k-1} = \lambda_1\left[v_1 + \frac{c_2}{c_1}\frac{\lambda_2^k}{\lambda_1^k}v_2 + \cdots + \frac{c_n}{c_1}\frac{\lambda_n^k}{\lambda_1^k}v_n\right] = \lambda_1 x_k.$$

Because λ_1 is the largest eigenvalue, the ratio $\frac{\lambda_i}{\lambda_1} < 1$ for all $i > 1$. When k is sufficiently large, the factor $(\frac{\lambda_n}{\lambda_1})^k$ will be close to zero, so that all terms that contain this factor can be neglected as k increases:

$$Ax_{k-1} \sim \lambda_1 v_1.$$

Essentially, if k is large enough, we will obtain the largest eigenvalue and its corresponding eigenvector. When implementing this power method, the resulting vector in each iteration is usually normalized. This can be done by factoring out the largest element in the vector, which will make the largest element in the vector equal to 1. This normalization will provide the largest eigenvalue and its corresponding eigenvector at the same time; see the example below.

When should we stop the iteration? The basic stopping criterion should be one of these the following: (1) the difference between eigenvalues is less than some specified tolerance; (2) the angle between eigenvectors is smaller than a threshold; or (3) the norm of the residual vector is small enough.

TRY IT! We know from the last section that the largest eigenvalue is 4 for the matrix $A = \begin{bmatrix} 0 & 2 \\ 2 & 3 \end{bmatrix}$. Use the power method to find the largest eigenvalue and the associated eigenvector. You can use the initial vector [1, 1] to start the iteration.

First iteration:

$$\begin{bmatrix} 0 & 2 \\ 2 & 3 \end{bmatrix} \begin{bmatrix} 1 \\ 1 \end{bmatrix} = \begin{bmatrix} 2 \\ 5 \end{bmatrix} = 5 \begin{bmatrix} 0.4 \\ 1 \end{bmatrix}.$$

Second iteration:

$$\begin{bmatrix} 0 & 2 \\ 2 & 3 \end{bmatrix} \begin{bmatrix} 0.4 \\ 1 \end{bmatrix} = \begin{bmatrix} 2 \\ 3.8 \end{bmatrix} = 3.8 \begin{bmatrix} 0.5263 \\ 1 \end{bmatrix}.$$

Third iteration:

$$\begin{bmatrix} 0 & 2 \\ 2 & 3 \end{bmatrix} \begin{bmatrix} 0.5263 \\ 1 \end{bmatrix} = \begin{bmatrix} 2 \\ 4.0526 \end{bmatrix} = 4.0526 \begin{bmatrix} 0.4935 \\ 1 \end{bmatrix}.$$

Fourth iteration:

$$\begin{bmatrix} 0 & 2 \\ 2 & 3 \end{bmatrix} \begin{bmatrix} 0.4935 \\ 1 \end{bmatrix} = \begin{bmatrix} 2 \\ 3.987 \end{bmatrix} = 3.987 \begin{bmatrix} 0.5016 \\ 1 \end{bmatrix}.$$

Fifth iteration:

$$\begin{bmatrix} 0 & 2 \\ 2 & 3 \end{bmatrix} \begin{bmatrix} 0.5016 \\ 1 \end{bmatrix} = \begin{bmatrix} 2 \\ 4.0032 \end{bmatrix} = 4.0032 \begin{bmatrix} 0.4996 \\ 1 \end{bmatrix}.$$

Sixth iteration:

$$\begin{bmatrix} 0 & 2 \\ 2 & 3 \end{bmatrix} \begin{bmatrix} 0.4996 \\ 1 \end{bmatrix} = \begin{bmatrix} 2 \\ 3.9992 \end{bmatrix} = 3.9992 \begin{bmatrix} 0.5001 \\ 1 \end{bmatrix}.$$

Seventh iteration:

$$\begin{bmatrix} 0 & 2 \\ 2 & 3 \end{bmatrix} \begin{bmatrix} 0.5001 \\ 1 \end{bmatrix} = \begin{bmatrix} 2 \\ 4.0002 \end{bmatrix} = 4.0002 \begin{bmatrix} 0.5000 \\ 1 \end{bmatrix}.$$

After seven iterations, the eigenvalue has converged to four, with [0.5, 1] as the corresponding eigenvector.

TRY IT! Implement the power method in Python.

```
In [1]: import numpy as np

In [2]: def normalize(x):
            fac = abs(x).max()
            x_n = x / x.max()
            return fac, x_n
```

```
In [3]: x = np.array([1, 1])
        a = np.array([[0, 2],
                      [2, 3]])

        for i in range(8):
            x = np.dot(a, x)
            lambda_1, x = normalize(x)

        print("Eigenvalue:", lambda_1)
        print("Eigenvector:", x)

Eigenvalue: 3.999949137887188
Eigenvector: [0.50000636 1.]
```

15.2.2 THE INVERSE POWER METHOD

The eigenvalues of the inverse matrix A^{-1} are the reciprocals of the eigenvalues of A. By taking advantage of this feature, as well as the power method, we are able to obtain the smallest eigenvalue of A; this will be basis of the **inverse power method**. The steps are very simple: instead of applying A as described above, we just apply A^{-1} for our iteration to find the largest value of $\frac{1}{\lambda_1}$, which will be the smallest value of the eigenvalues for A. In practice, we can use the methods we covered in the previous chapter to calculate the inverse of the matrix. We will not go to greater detail here, but we present an example below.

TRY IT! Find the smallest eigenvalue and eigenvector for $A = \begin{bmatrix} 0 & 2 \\ 2 & 3 \end{bmatrix}$.

```
In [4]: from numpy.linalg import inv

In [5]: a_inv = inv(a)

        for i in range(8):
            x = np.dot(a_inv, x)
            lambda_1, x = normalize(x)

        print("Eigenvalue:", lambda_1)
        print("Eigenvector:", x)

Eigenvalue: 0.20000000000003912
Eigenvector: [1. 1.]
```

15.2.3 THE SHIFTED POWER METHOD

In some cases, it is necessary to find all the eigenvalues and eigenvectors instead of just the largest and smallest. One simple, but inefficient way is to use the **shifted power method**; we will introduce you a more efficient method in the next section.

Given $Ax = \lambda_{1x}$, and λ_1 being the largest eigenvalue obtained by the power method, we have

$$[A - \lambda_1 I]x = \alpha x,$$

where α's are the eigenvalues of the shifted matrix $A - \lambda_1 I$, which will be $0, \lambda_2 - \lambda_1, \lambda_3 - \lambda_1, \ldots, \lambda_n - \lambda_1$.

Now if we apply the power method to the shifted matrix, we can determine the largest eigenvalue of the shifted matrix, i.e., α_k. Since $\alpha_k = \lambda_k - \lambda_1$, we can obtain the eigenvalue λ_k easily. Repeating this process many times will find the all the other eigenvalues, but you can see it is very labor intensive. A better method for finding all the eigenvalues is to use the QR method, which we will introduced next.

15.3 THE QR METHOD

The **QR method** is the preferred iterative method to find all the eigenvalues of a matrix (but not the eigenvectors at the same time). The idea is based on the following two concepts:

1. Similar matrices will have the same eigenvalues and associated eigenvectors. Two square matrices A and B are similar if

   $$A = C^{-1}BC$$

   where C is an invertible matrix.

2. The QR method is a way to decompose a matrix into two matrices Q and R, where Q is an orthogonal matrix, and R is an upper triangular matrix. An orthogonal matrix satisfies $Q^{-1} = Q^T$, which means $Q^{-1}Q = Q^TQ = I$.

How do we link these two concepts to find the eigenvalues? Say, we have a matrix A_0 whose eigenvalues must be determined. At the kth step (starting with $k = 0$), we can perform the QR decomposition and obtain

$$A_k = Q_k R_k$$

where Q_k is an orthogonal matrix, and R_k is an upper triangular matrix. We then form

$$A_{k+1} = R_k Q_k$$

to obtain

$$A_{k+1} = R_k Q_k = Q_k^{-1} Q_k R_k Q_k = Q_k^{-1} A_k Q_k.$$

Because all the A_k are similar, as we discussed above, they all have the same eigenvalues.

As the iteration continues, we will eventually converge to an upper triangular matrix with the form:

$$A_k = R_k Q_k = \begin{bmatrix} \lambda_1 & X & \dots & X \\ 0 & \lambda_2 & \dots & X \\ \vdots & \vdots & \ddots & \vdots \\ 0 & 0 & \dots & \lambda_n \end{bmatrix},$$

where the diagonal values are the eigenvalues of the matrix. In each iteration of the QR method, factoring a matrix into an orthogonal and an upper triangular matrix can be done by using a special matrix called **Householder matrix**. We will not go into the mathematical details how you get the Q and R from the matrix. Instead, we will use the Python function to obtain the two matrices directly.

TRY IT! Use the `qr` function in `numpy.linalg` to decompose matrix $A = \begin{bmatrix} 0 & 2 \\ 2 & 3 \end{bmatrix}$. Verify the results.

```
In [1]: import numpy as np
        from numpy.linalg import qr

In [2]: a = np.array([[0, 2],
                      [2, 3]])

        q, r = qr(a)

        print("Q:", q)
        print("R:", r)

        b = np.dot(q, r)
        print("QR:", b)

Q: [[ 0. -1.]
 [-1.  0.]]
R: [[-2. -3.]
 [ 0. -2.]]
QR: [[0. 2.]
 [2. 3.]]
```

TRY IT! Use the QR method to get the eigenvalues of matrix $A = \begin{bmatrix} 0 & 2 \\ 2 & 3 \end{bmatrix}$. Do 20 iterations, and print out the first, fifth, 10th, and 20th iteration.

```
In [3]: a = np.array([[0, 2],
                      [2, 3]])
        p = [1, 5, 10, 20]
        for i in range(20):
            q, r = qr(a)
            a = np.dot(r, q)
            if i+1 in p:
                print(f"Iteration {i+1}:")
                print(a)

Iteration 1:
[[3. 2.]
 [2. 0.]]
Iteration 5:
[[ 3.99998093  0.00976559]
 [ 0.00976559 -0.99998093]]
Iteration 10:
[[ 4.00000000e+00  9.53674316e-06]
 [ 9.53674316e-06 -1.00000000e+00]]
Iteration 20:
[[ 4.00000000e+00  9.09484250e-12]
 [ 9.09494702e-12 -1.00000000e+00]]
```

Note that after the fifth iteration, the eigenvalues are converged to the correct ones. The next section will demonstrate how to obtain the eigenvalues and eigenvectors in Python using the built-in function.

15.4 EIGENVALUES AND EIGENVECTORS IN PYTHON

The methods introduced above are fairly complicated to execute. The calculation of eigenvalues and eigenvectors in Python is fairly easy. The main built-in function in Python to solve the eigenvalue/eigenvector problem for a square array is the `eig` function in `numpy.linalg`; see below for an example in how to execute it.

TRY IT! Calculate the eigenvalues and eigenvectors for matrix $A = \begin{bmatrix} 0 & 2 \\ 2 & 3 \end{bmatrix}$.

```
In [1]: import numpy as np
        from numpy.linalg import eig

In [2]: a = np.array([[0, 2],
                      [2, 3]])
```

```
        w,v=eig(a)
        print("E-value:", w)
        print("E-vector", v)

E-value: [-1.  4.]
E-vector [[-0.89442719 -0.4472136 ]
 [ 0.4472136  -0.89442719]]
```

TRY IT! Compute the eigenvalues and eigenvectors for the matrix $A = \begin{bmatrix} 2 & 2 & 4 \\ 1 & 3 & 5 \\ 2 & 3 & 4 \end{bmatrix}$.

```
In [3]: a = np.array([[2, 2, 4],
                      [1, 3, 5],
                      [2, 3, 4]])
        w,v=eig(a)
        print("E-value:", w)
        print("E-vector", v)

E-value: [ 8.80916362  0.92620912 -0.73537273]
E-vector [[-0.52799324 -0.77557092 -0.36272811]
 [-0.604391    0.62277013 -0.7103262 ]
 [-0.59660259 -0.10318482  0.60321224]]
```

15.5 SUMMARY AND PROBLEMS

15.5.1 SUMMARY

1. Eigenvalues and eigenvectors help us understand the characteristics of a linear transformation.
2. Eigenvectors of a matrix are the vectors that can only be scaled lengthwise without rotation after applying the matrix transformation; the eigenvalues are the factors of the scaling.
3. We can use power method to get the largest eigenvalue and corresponding eigenvector of a matrix.
4. The inverse power method can help us get the smallest eigenvalue and corresponding eigenvector of a matrix.
5. The shifted power method can get all the other eigenvectors/eigenvectors of a matrix.
6. The preferred method to get all the eigenvalues is the QR method.

15.5.2 PROBLEMS

1. Write down the characteristic equation for matrix $A = \begin{bmatrix} 3 & 2 \\ 5 & 3 \end{bmatrix}$.
2. Use the above characteristic equation to solve for eigenvalues and eigenvectors of matrix A.

3. Use the first eigenvector derived from Problem 2 to verify that $Ax = \lambda x$.
4. Use the power method to obtain the largest eigenvalue and eigenvector for the matrix $A = \begin{bmatrix} 2 & 1 & 2 \\ 1 & 3 & 2 \\ 2 & 4 & 1 \end{bmatrix}$. Start with initial vector [1, 1, 1] and see the results after eight iterations.
5. Using the inverse power method to get the smallest eigenvalue and eigenvector for the matrix in Problem 4, see how many iterations are needed for it to converge to the smallest eigenvalue.
6. Perform a QR decomposition for matrix A in Problem 4. Verify that $A = QR$ and Q is an orthogonal matrix.
7. Use the QR method to get all the eigenvalues for matrix A in Problem 4.
8. Obtain the eigenvalues and eigenvectors for matrix A in Problem 4 using the Python built-in function.

CHAPTER

LEAST SQUARES REGRESSION 16

CONTENTS

16.1 LEAST SQUARES REGRESSION PROBLEM STATEMENT

Given a set of independent data points x_i and dependent data points $y_i, i = 1, \ldots, m$, we would like to find an **estimation function**, $\hat{y}(x)$, that describes the data as accurately as possible. Note that $\hat{y}$ can be a function of several variables, but for the sake of this discussion, we restrict the domain of $\hat{y}$ to a single variable. In least squares regression, the estimation function must be a linear combination of **basis functions**, $f_i(x)$. That is, the estimation function must be of the form

$$\hat{y}(x) = \sum_{i=1}^{n} \alpha_i f_i(x).$$

The scalars α_i are referred to as the **parameters** of the estimation function, and each basis function must be linearly independent from the others. In other words, in the proper "functional space," no basis function should be expressible as a linear combination of the other functions. Note that, in general, there are significantly more data points, m, than basis functions, n (i.e., $m \gg n$).

Python Programming and Numerical Methods. https://doi.org/10.1016/B978-0-12-819549-9.00026-9

TRY IT! Create an estimation function for the force–displacement relationship of a linear spring. Identify the basis function(s) and model parameters.

The relationship between the force, F, and the displacement, x, can be described by the function $F(x) = kx$, where k is the spring stiffness. The only basis function is the function $f_1(x) = x$ and the model parameter to determine is $\alpha_1 = k$.

The goal of **least squares regression** is to find the parameters of the estimation function that minimize the **total squared error**, E, defined by $E = \sum_{i=1}^{m}(\hat{y} - y_i)^2$. The **individual errors** or **residuals** are defined as $e_i = (\hat{y} - y_i)$. If e is the vector containing all the individual errors, then we are also trying to minimize $E = \|e\|_2^2$, which is the L_2 norm defined in the previous chapter.

In the next two sections, we will derive the least squares method of finding the desired parameters. The first derivation comes from linear algebra, and the second comes from multivariate calculus. Although they are different derivations, they lead to the same least squares formula. You are free to focus on the section with which you are most comfortable.

16.2 LEAST SQUARES REGRESSION DERIVATION (LINEAR ALGEBRA)

First, we enumerate the estimation of the data at each data point x_i as

$$\hat{y}(x_1) = \alpha_1 f_1(x_1) + \alpha_2 f_2(x_1) + \cdots + \alpha_n f_n(x_1),$$
$$\hat{y}(x_2) = \alpha_1 f_1(x_2) + \alpha_2 f_2(x_2) + \cdots + \alpha_n f_n(x_2),$$
$$\cdots$$
$$\hat{y}(x_m) = \alpha_1 f_1(x_m) + \alpha_2 f_2(x_m) + \cdots + \alpha_n f_n(x_m).$$

Let $X \in \mathbb{R}^n$ be a column vector such that the ith element of X contains the value of the ith x-data point, x_i, $\hat{Y}$ be a column vector with elements, $\hat{Y}_i = \hat{y}(x_i)$, β be a column vector such that $\beta_i = \alpha_i$, $F_i(x)$ be a function that returns a column vector of $f_i(x)$ computed on every element of x, and A be an $m \times n$ matrix such that the ith column of A is $F_i(x)$. Given this notation, the previous system of equations becomes $\hat{Y} = A\beta$.

Now if Y is a column vector such that $Y_i = y_i$, the total squared error is given by $E = \|\hat{Y} - Y\|_2^2$. Verify this by substituting the definition of the L_2 norm. Since we want to make E as small as possible and norms are a measure of distance, this previous expression is equivalent to saying that we want $\hat{Y}$ and Y to be as "close as possible." Note that, in general, Y will not be in the range of A and therefore $E > 0$.

Consider the following simplified depiction of the range of A; see Fig. 16.1. Note this is *not* a plot of the data points (x_i, y_i).

From observation, the vector in the range of A, namely $\hat{Y}$ which is closest to Y, is that which can point perpendicularly to Y; therefore, we want a vector $Y - \hat{Y}$ that is perpendicular to the vector $\hat{Y}$.

Recall from linear algebra that two vectors are perpendicular, or orthogonal, if their dot-product is 0. Noting that the dot-product between two vectors, v and w, can be written as $\text{dot}(v, w) = v^T w$, we can state that $\hat{Y}$ and $Y - \hat{Y}$ are perpendicular if $\text{dot}(\hat{Y}, Y - \hat{Y}) = 0$; therefore, $\hat{Y}^T(Y - \hat{Y}) = 0$, which is equivalent to $(A\beta)^T(Y - A\beta) = 0$.

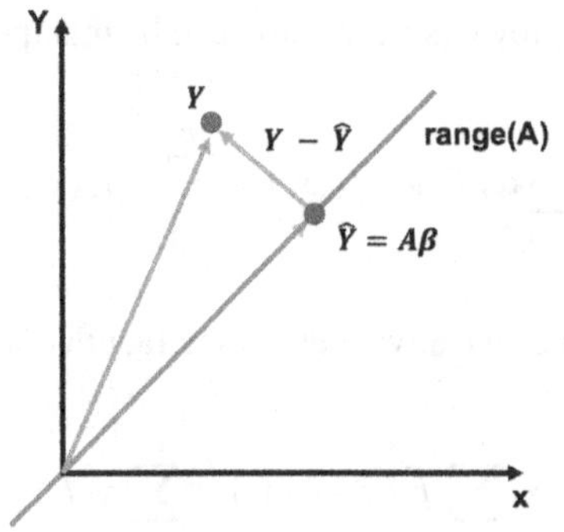

FIGURE 16.1

Illustration of the L_2 projection of Y on the range of A.

Noting that for the two matrices A and B we have $(AB)^T = B^T A^T$ and using distributive properties of vector multiplication, this is equivalent to $\beta^T A^T Y - \beta^T A^T A\beta = \beta^T(A^T Y - A^T A\beta) = 0$. The solution, $\beta = \mathbf{0}$, is a trivial solution, so we use $A^T Y - A^T A\beta = 0$ to find a more interesting solution. Solving this equation for β gives the **least squares regression formula**:

$$\beta = (A^T A)^{-1} A^T Y.$$

Note that $(A^T A)^{-1} A^T$ is called the **pseudo-inverse** of A and it exists when $m > n$ and A has linearly independent columns. Proving the invertibility of $(A^T A)$ is outside the scope of this book, but it is always invertible except for some specific cases.

16.3 LEAST SQUARES REGRESSION DERIVATION (MULTIVARIATE CALCULUS)

Recall that the total error for m data points and n basis functions is:

$$E = \sum_{i=1}^{m} e_i^2 = \sum_{i=1}^{m} (\hat{y}(x_i) - y_i)^2 = \sum_{i=1}^{m} \left(\sum_{j=1}^{n} \alpha_j f_j(x_i) - y_i \right)^2,$$

which is an n-dimensional paraboloid in α_k. From calculus, we know that the minimum of a paraboloid is where all the partial derivatives equal zero. Taking the partial derivative of E with respect to the variable α_k (remember that in this case the parameters are our variables) and setting these derivatives equal to 0, and then solving the system of equations for the α_k's should give the correct results.

Computing the partial derivative with respect to α_k and setting it equal to zero yields

$$\frac{\partial E}{\partial \alpha_k} = \sum_{i=1}^{m} 2 \left(\sum_{j=1}^{n} \alpha_j f_j(x_i) - y_i \right) f_k(x_i) = 0.$$

With some rearrangement, the previous expression can be manipulated as follows:

$$\sum_{i=1}^{m}\sum_{j=1}^{n}\alpha_j f_j(x_i) f_k(x_i) - \sum_{i=1}^{m} y_i f_k(x_i) = 0,$$

and upon further rearrangement (we take advantage of the fact that addition commutes) the result is

$$\sum_{j=1}^{n}\alpha_j \sum_{i=1}^{m} f_j(x_i) f_k(x_i) = \sum_{i=1}^{m} y_i f_k(x_i).$$

Now let X be a column vector such that the ith element of X is x_i and Y similarly constructed, and let $F_j(X)$ be a column vector such that the ith element of $F_j(X)$ is $f_j(x_i)$. Using this notation, the previous expression can be rewritten in vector notation as

$$\left[F_k^T(X)F_1(X), F_k^T(X)F_2(X), \ldots, F_k^T(X)F_j(X), \ldots, F_k^T(X)F_n(X)\right] \begin{bmatrix} \alpha_1 \\ \alpha_2 \\ \vdots \\ \alpha_j \\ \vdots \\ \alpha_n \end{bmatrix} = F_k^T(X)Y.$$

If we repeat this equation for every k, we get the following system of linear equations in matrix form:

$$\begin{bmatrix} F_1^T(X)F_1(X), F_1^T(X)F_2(X), \ldots, F_1^T(X)F_j(X), \ldots, F_1^T(X)F_n(X) \\ F_2^T(X)F_1(X), F_2^T(X)F_2(X), \ldots, F_2^T(X)F_j(X), \ldots, F_2^T(X)F_n(X) \\ \vdots \\ F_n^T(X)F_1(X), F_n^T(X)F_2(X), \ldots, F_n^T(X)F_j(X), \ldots, F_n^T(X)F_n(X) \end{bmatrix} \begin{bmatrix} \alpha_1 \\ \alpha_2 \\ \vdots \\ \alpha_j \\ \vdots \\ \alpha_n \end{bmatrix} = \begin{bmatrix} F_1^T(X)Y \\ F_2^T(X)Y \\ \vdots \\ F_n^T(X)Y \end{bmatrix}.$$

If we let $A = [F_1(X), F_2(X), \ldots, F_j(X), \ldots, F_n(X)]$ and β be a column vector such that the jth element of β is α_j, then the previous system of equations becomes

$$A^T A\beta = A^T Y,$$

and then solving this matrix equation for β gives $\beta = (A^T A)^{-1} A^T Y$, which is exactly the same formula as the previous derivation.

16.4 LEAST SQUARES REGRESSION IN PYTHON

Recall that enumerating the estimation of the data at each data point, x_i, will give us the following system of equations:

$$\hat{y}(x_1) = \alpha_1 f_1(x_1) + \alpha_2 f_2(x_1) + \cdots + \alpha_n f_n(x_1),$$
$$\hat{y}(x_2) = \alpha_1 f_1(x_2) + \alpha_2 f_2(x_2) + \cdots + \alpha_n f_n(x_2),$$
$$\cdots$$
$$\hat{y}(x_m) = \alpha_1 f_1(x_m) + \alpha_2 f_2(x_m) + \cdots + \alpha_n f_n(x_m).$$

If the data were absolutely perfect (i.e., no noise), then the estimation function would go through all the data points, resulting in the following system of equations:

$$y_1 = \alpha_1 f_1(x_1) + \alpha_2 f_2(x_1) + \cdots + \alpha_n f_n(x_1),$$
$$y_2 = \alpha_1 f_1(x_2) + \alpha_2 f_2(x_2) + \cdots + \alpha_n f_n(x_2),$$
$$\cdots$$
$$y_m = \alpha_1 f_1(x_m) + \alpha_2 f_2(x_m) + \cdots + \alpha_n f_n(x_m).$$

If we were to take A as defined previously, this would result in the matrix equation

$$Y = A\beta.$$

Because the data is not perfect, there will not be an estimation function that can go through all the data points, and this system will have *no solution*. Therefore, we need to use the least square regression that we derived in the previous two sections to obtain a solution:

$$\beta = (A^T A)^{-1} A^T Y.$$

TRY IT! Consider the artificial data created by `x = np.linspace(0, 1, 101)` and `y = 1 + x + x * np.random.random(len(x))`. Do a least squares regression with an estimation function defined by $\hat{y} = \alpha_1 x + \alpha_2$. Plot the data points along with the least squares regression. Note that we expect $\alpha_1 = 1.5$ and $\alpha_2 = 1.0$ based on this data. Due to the random noise we added into the data, your results maybe slightly different. In the next few subsections, we will see how we solve this problem using different approaches.

16.4.1 USING THE DIRECT INVERSE METHOD

```
In [1]: import numpy as np
        from scipy import optimize
        import matplotlib.pyplot as plt

        plt.style.use("seaborn-poster")
```

```
In [2]: # generate x and y
        x = np.linspace(0, 1, 101)
        y = 1 + x + x * np.random.random(len(x))

In [3]: # assemble matrix A
        A = np.vstack([x, np.ones(len(x))]).T

        # turn y into a column vector
        y = y[:, np.newaxis]

In [4]: # Direct least squares regression
        alpha = np.dot((np.dot(np.linalg.inv(np.dot(A.T,A)),A.T)),y)
        print(alpha)

[[1.459573  ]
 [1.02952189]]


In [5]: # plot the results
        plt.figure(figsize = (10,8))
        plt.plot(x, y, "b.")
        plt.plot(x, alpha[0]*x + alpha[1], "r")
        plt.xlabel("x")
        plt.ylabel("y")
        plt.show()
```

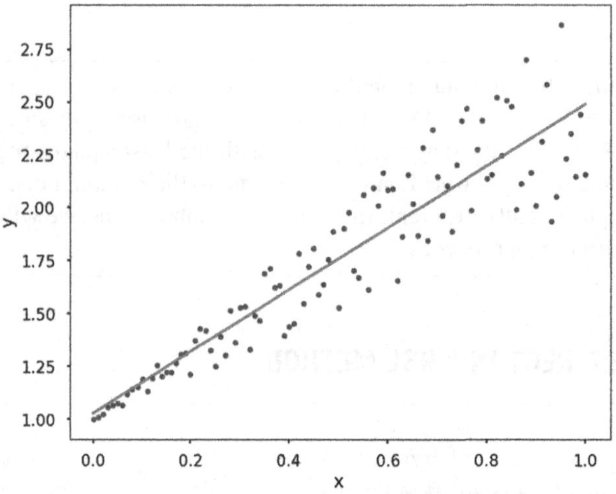

Python has several packages and functions that can perform a least squares regression. These include `NumPy`, `SciPy`, `statsmodels`, and `sklearn`. Below are several examples of such applications. Feel free to choose the one you like.

16.4.2 USING THE PSEUDO-INVERSE

We mentioned earlier that the matrix $(A^T A)^{-1} A^T$ is called the pseudo-inverse, therefore, we can use the `pinv` function in `NumPy` to calculate it directly.

```
In [6]: pinv = np.linalg.pinv(A)
        alpha = pinv.dot(y)
        print(alpha)

[[1.459573  ]
 [1.02952189]]
```

16.4.3 USING `NUMPY.LINALG.LSTSQ`

`NumPy` has already implemented the least squares methods, so we can just call the function to get a solution. The function will return more data than the solution itself; please check the documentation for details.

```
In [7]: alpha = np.linalg.lstsq(A, y, rcond=None)[0]
        print(alpha)

[[1.459573  ]
 [1.02952189]]
```

16.4.4 USING `OPTIMIZE.CURVE_FIT` FROM `SCIPY`

This `SciPy` function is very powerful. It is not only suitable for linear functions, but many different function forms as well, such as nonlinear functions. Here we will only show the linear example from above. Note that, when using this function, we do not need to turn y into a column vector.

```
In [8]: # generate x and y
        x = np.linspace(0, 1, 101)
        y = 1 + x + x * np.random.random(len(x))

In [9]: def func(x, a, b):
            y = a*x + b
            return y

        alpha=optimize.curve_fit(func, xdata=x, ydata=y)[0]
        print(alpha)

[1.44331612 1.0396133 ]
```

16.5 LEAST SQUARES REGRESSION FOR NONLINEAR FUNCTIONS

A least squares regression requires that the estimation function be a linear combination of basis functions. There are some functions that cannot be put in this form, but where a least squares regression is still appropriate.

Introduced below are several ways to deal with nonlinear functions.

- We can accomplish this by taking advantage of the properties of logarithms and transform the nonlinear function into a linear function.
- We can use the `curve_fit` function from `SciPy` to estimate directly the parameters for the nonlinear function using least square.

16.5.1 LOG TRICKS FOR EXPONENTIAL FUNCTIONS

Assume you have a function in the form $\hat{y}(x) = \alpha e^{\beta x}$ and data for x and y, and that you want to perform least squares regression to find α and β. Clearly, the previous set of basis functions (linear) would be inappropriate to describe $\hat{y}(x)$; however, if we take the log of both sides, we get $\log(\hat{y}(x)) = \log(\alpha) + \beta x$. Now, we see that if $\tilde{y}(x) = \log(\hat{y}(x))$ and $\tilde{\alpha} = \log(\alpha)$, then $\tilde{y}(x) = \tilde{\alpha} + \beta x$. Thus, we can perform a least squares regression on the linearized expression to find $\tilde{y}(x)$, $\tilde{\alpha}$, and β, and then recover α by using the expression $\alpha = e^{\tilde{\alpha}}$.

For the example below, we will generate data using $\alpha = 0.1$ and $\beta = 0.3$.

```
In [1]: import numpy as np
        from scipy import optimize
        import matplotlib.pyplot as plt
        plt.style.use("seaborn-poster")

In [2]: # let's generate x and y, and add some noise into y
        x = np.linspace(0, 10, 101)
        y = 0.1*np.exp(0.3*x) + 0.1*np.random.random(len(x))

In [3]: # Let's have a look of the data
        plt.figure(figsize = (10,8))
        plt.plot(x, y, "b.")
        plt.xlabel("x")
        plt.ylabel("y")
        plt.show()
```

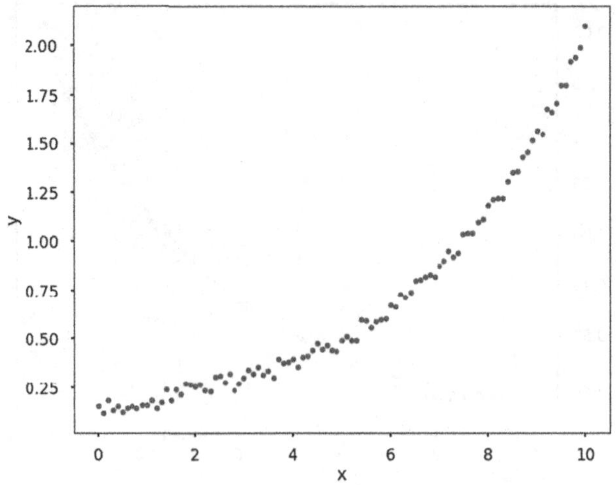

Once the log trick has been applied, we can fit the data.

```
In [4]: A = np.vstack([x, np.ones(len(x))]).T
        beta, log_alpha = np.linalg.lstsq(A, np.log(y), rcond = None)[0]
        alpha = np.exp(log_alpha)
        print(f"alpha={alpha}, beta={beta}")

alpha=0.13973103064296616, beta=0.26307478591152406

In [5]: # Let's have a look of the data
        plt.figure(figsize = (10,8))
        plt.plot(x, y, "b.")
        plt.plot(x, alpha*np.exp(beta*x), "r")
        plt.xlabel("x")
        plt.ylabel("y")
        plt.show()
```

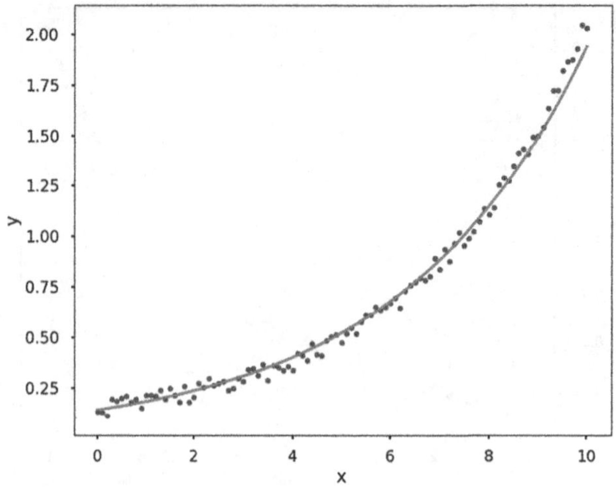

16.5.2 LOG TRICKS FOR POWER FUNCTIONS

The power function case is very similar. Assume we have a function in the form $\hat{y}(x) = bx^m$ and data for x and y. Then we can turn this function into a linear form by taking log to both sides: $\log(\hat{y}(x)) = m\log(x) + \log b$, solving this function as a linear regression. Since it is very similar to the above example, we will not devote any more time on this.

16.5.3 POLYNOMIAL REGRESSION

We can also use polynomial and least squares to fit a nonlinear function. Previously, we had our functions all in linear form, that is, $y = ax + b$. But "polynomials" are functions with the following form:

$$f(x) = a_n x^n + a_{n-1}x^{n-1} + \cdots + a_2 x^2 + a_1 x^1 + a_0$$

where $a_n, a_{n-1}, \ldots, a_2, a_1, a_0$ are the real number coefficients, and n, a non-negative integer, is the **order** or **degree** of the polynomial. If we have a set of data points, we can use different orders of polynomials to fit it. The coefficients of the polynomials can be estimated using the least squares method as before, i.e., minimizing the error between the real data and the polynomial fitting results.

In Python, we can use `numpy.polyfit` to obtain the coefficients of different order polynomials with the least squares. With the coefficients, we can get the specific values using `numpy.polyval`. Below is an example of how to perform this in Python.

```
In [6]:  x_d = np.array([0, 1, 2, 3, 4, 5, 6, 7, 8])
         y_d=np.array([0,0.8,0.9,0.1,-0.6,-0.8,-1,-0.9,-0.4])

         plt.figure(figsize = (12, 8))
         for i in range(1, 7):
```

```
        # get the polynomial coefficients
        y_est = np.polyfit(x_d, y_d, i)
        plt.subplot(2,3,i)
        plt.plot(x_d, y_d, "o")
        # evaluate the values for a polynomial
        plt.plot(x_d, np.polyval(y_est, x_d))
        plt.title(f"Polynomial order {i}")
    plt.tight_layout()
    plt.show()
```

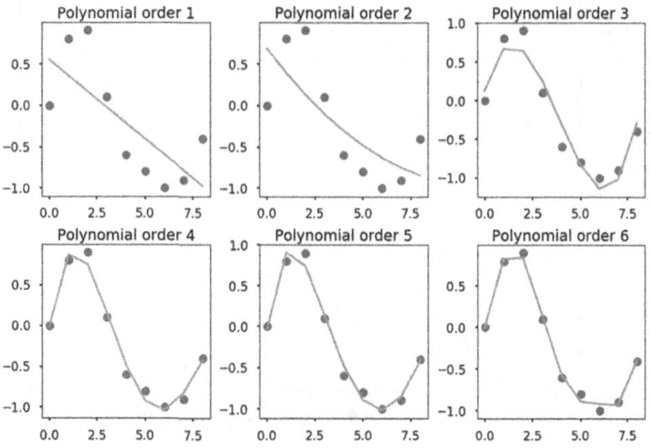

The figure above shows that we can use different orders of polynomials to fit the same data. The higher the order, the more flexible the data curve required to fit the data. But what order to use is not a simple question, it depends on the specific problems in science and engineering.

16.5.4 USING `OPTIMIZE.CURVE_FIT` FROM `SCIPY`

We can use the `curve_fit` function to fit any form function and estimate its parameters. We can solve the above problem using the `curve_fit` function as follows:

```
In [7]: # let's define the function form
        def func(x, a, b):
            y = a*np.exp(b*x)
            return y
        alpha, beta = optimize.curve_fit(func, xdata = x, ydata = y)[0]
        print(f"alpha={alpha}, beta={beta}")

alpha=0.12663549356730994, beta=0.27760076897453045
```

```
In [8]: # Let's have a look of the data
        plt.figure(figsize = (10,8))
        plt.plot(x, y, "b.")
        plt.plot(x, alpha*np.exp(beta*x), "r")
        plt.xlabel("x")
        plt.ylabel("y")
        plt.show()
```

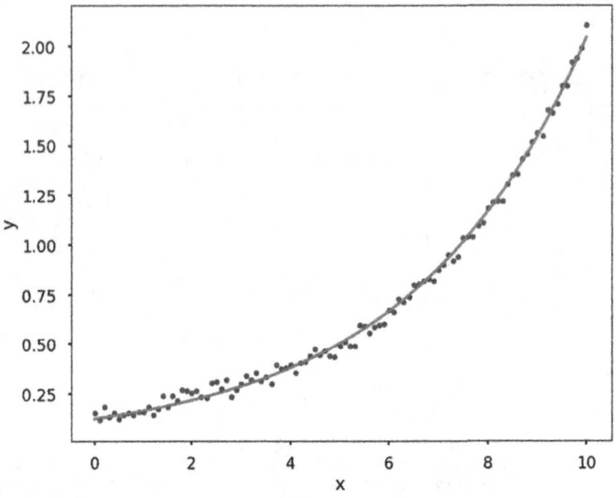

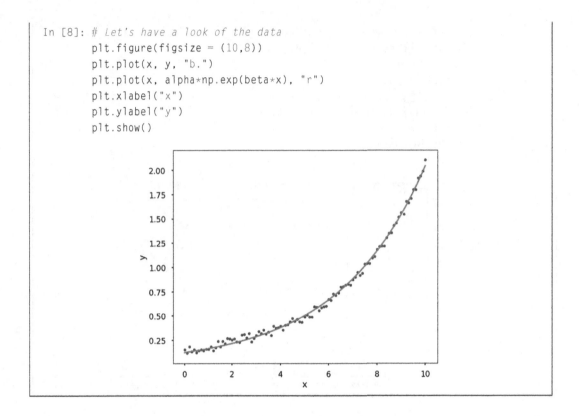

16.6 SUMMARY AND PROBLEMS

16.6.1 SUMMARY

1. Mathematical models are used to understand, predict, and control engineering systems. These models consist of parameters that govern the way the model behaves.
2. Given a set of experimental data, a least squares regression is a method of finding a set of model parameters that fits the data well. That is, it minimizes the squared error between the model, or estimation function, and the data points.
3. In a linear least squares regression, the estimation function must be a linear combination of linearly independent basis functions.
4. The set of parameters β can be determined by the least squares equation $\beta = (A^T A)^{-1} A^T Y$, where the jth column of A is the jth basis function evaluated at each X data point.
5. To estimate a nonlinear function, we transform it into a linear estimation function or use directly a least squares regression to solve the nonlinear function using `curve_fit` from `SciPy`.

16.6.2 PROBLEMS

1. Repeat the multivariate calculus derivation of the least squares regression formula for an estimation function $\hat{y}(x) = ax^2 + bx + c$, where a, b, and c are the parameters.

2. Write a function `my_ls_params(f, x, y)` where `x` and `y` are arrays of the same size containing experimental data, and `f` is a list with each element a function object to a basis vector of the estimation function. The output argument, `beta`, should be an array of the parameters of the least squares regression for `x`, `y`, and `f`.

3. Write a function `my_func_fit (x,y)` where `x` and `y` are column vectors of the same size containing experimental data, and the function returns `alpha` and `beta` are the parameters of the estimation function $\hat{y}(x) = \alpha x^{\beta}$.

4. Given four data points (x_i, y_i) and the parameters for a cubic polynomial $\hat{y}(x) = ax^3 + bx^2 + cx + d$, what will be the total error associated with the estimation function $\hat{y}(x)$? Can we place another data point `(x,y)` such that no additional error is incurred for the estimation function?

5. Write a function `my_lin_regression(f, x, y)` where `f` is a list containing function objects to basis functions, and `x` and `y` are arrays containing noisy data. Assume that `x` and `y` are the same size. Let an estimation function for the data contained in `x` and `y` be defined as $\hat{y}(x) = \beta(1) \cdot f_1(x) + \beta(2) \cdot f_2(x) + \cdots + \beta(n) \cdot f_n(x)$, where `n` is the length of `f`. Your function should compute *beta* according to the least squares regression formula.
Test Case: Note that your solution may vary by a little bit, depending on the random numbers generated.

```
x = np.linspace(0, 2*np.pi, 1000)
y = 3*np.sin(x) - 2*np.cos(x) + np.random.random(len(x))
f = [np.sin, np.cos]
beta = my_lin_regression(f, x, y)

plt.figure(figsize = (10,8))
plt.plot(x,y,"b.", label = "data")
plt.plot(x, beta[0]*f[0](x)+beta[1]*f[1](x)+beta[2], "r", label="regression")
plt.xlabel("x")
plt.ylabel("y")
plt.title("Least Square Regression Example")
plt.legend()
plt.show()
```

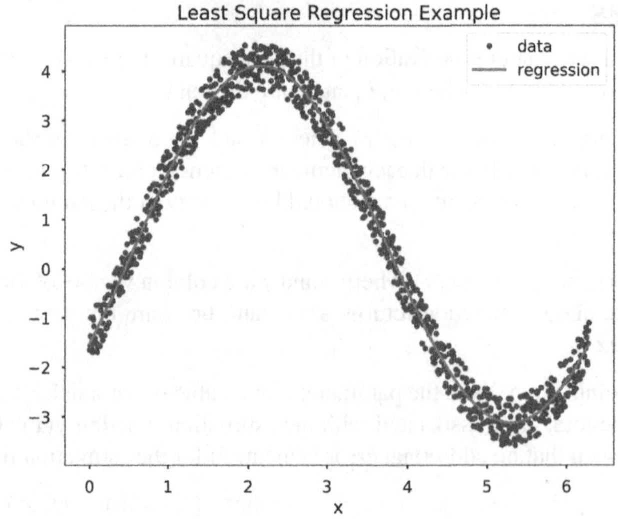

6. Write a function `my_exp_regression (x,y)` where `x` and `y` are arrays of the same size. Let an estimation function for the data contained in `x` and `y` be defined as $\hat{y}(x) = \alpha e^{\beta x}$. Your function should compute α and β to solve the least squares regression formula.
Test Cases: Note that your solution may vary slightly from the test case, depending on the random numbers generated.

```
x = np.linspace(0, 1, 1000)
y = 2*np.exp(-0.5*x) + 0.25*np.random.random(len(x))

alpha, beta = my_exp_regression(x, y)

plt.figure(figsize = (10,8))
plt.plot(x,y,"b.", label = "data")
plt.plot(x, alpha*np.exp(beta*x), "r", label="regression")
plt.xlabel("x")
plt.ylabel("y")
plt.title("Least Square Regression on Exponential Model")
plt.legend()
plt.show()
```

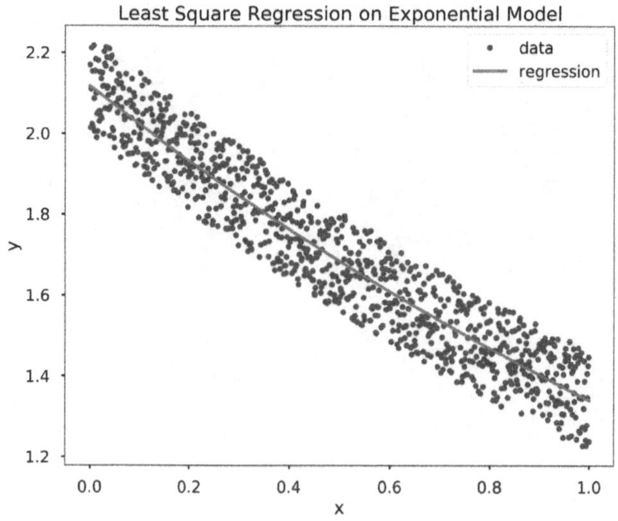
Least Square Regression on Exponential Model
data
regression
2.2
2.0
1.8
1.6
1.4
1.2
y
0.0
0.2
0.4
0.6
0.8
1.0
x

CHAPTER

INTERPOLATION

17

CONTENTS

17.1 INTERPOLATION PROBLEM STATEMENT

Assume we have a dataset consisting of independent data values, x_i, and dependent data values, y_i, where $i = 1, \ldots, n$. We would like to find an estimation function $\hat{y}(x)$ such that $\hat{y}(x_i) = y_i$ for every point in our dataset. This means the estimation function goes through our data points. Given a new x^*, we can **interpolate** its function value using $\hat{y}(x^*)$. In this context, $\hat{y}(x)$ is called an **interpolation function**. Fig. 17.1 shows the interpolation problem statement.

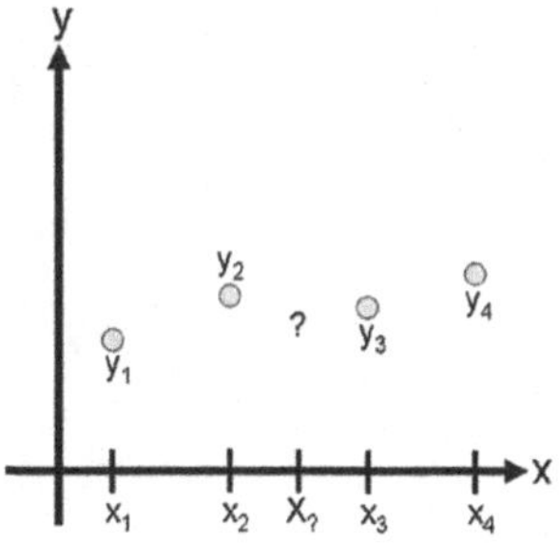

FIGURE 17.1

Illustration of the interpolation problem: estimate the value of a function in-between data points.

Unlike regression, interpolation does not require the user to have an underlying model for the data, especially when there are many reliable data points. However, the processes that underly the data must still inform the user about the quality of the interpolation. For example, our data may consist of (x, y)

Python Programming and Numerical Methods. https://doi.org/10.1016/B978-0-12-819549-9.00027-0

coordinates of a car over time. Since motion is restricted to the maneuvering physics of the car, we can expect that the points between the (x, y) coordinates in our set will be "smooth" rather than jagged.

The following sections will present several common interpolation methods.

17.2 LINEAR INTERPOLATION

In **linear interpolation**, the estimated point is assumed to lie on the line joining the nearest points to the left and right. Assume, without loss of generality, that the x-data points are in ascending order, that is, $x_i < x_{i+1}$, and let x be a point such that $x_i < x < x_{i+1}$. Then the linear interpolation at x is

$$\hat{y}(x) = y_i + \frac{(y_{i+1} - y_i)(x - x_i)}{(x_{i+1} - x_i)}.$$

TRY IT! Find the linear interpolation at $x = 1.5$ based on the data `x = [0, 1, 2]`, `y = [1, 3, 2]`. Verify the result using `SciPy`'s function `interp1d`.

Since $1 < x < 2$, we use the second and third data points to compute the linear interpolation. Plugging in the corresponding values gives

$$\hat{y}(x) = y_i + \frac{(y_{i+1} - y_i)(x - x_i)}{(x_{i+1} - x_i)} = 3 + \frac{(2-3)(1.5-1)}{(2-1)} = 2.5.$$

```
In [1]: from scipy.interpolate import interp1d
        import matplotlib.pyplot as plt

        plt.style.use("seaborn-poster")

In [2]: x = [0, 1, 2]
        y = [1, 3, 2]

        f = interp1d(x, y)
        y_hat = f(1.5)
        print(y_hat)

2.5

In [3]: plt.figure(figsize = (10,8))
        plt.plot(x, y, "-ob")
        plt.plot(1.5, y_hat, "ro")
        plt.title("Linear Interpolation at x = 1.5")
        plt.xlabel("x")
```

```
plt.ylabel("y")
plt.show()
```

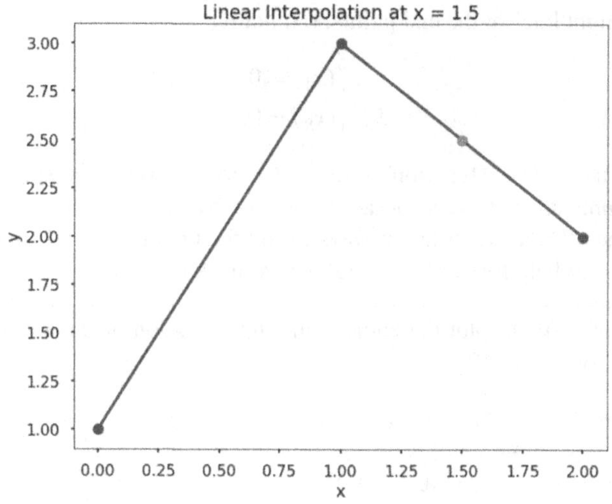

17.3 CUBIC SPLINE INTERPOLATION

In **cubic spline interpolation** (as shown in the following figure), the interpolation function is a set of piecewise cubic functions. Specifically, we assume that the points (x_i, y_i) and (x_{i+1}, y_{i+1}) are joined by a cubic polynomial $S_i(x) = a_i x^3 + b_i x^2 + c_i x + d_i$ that is valid for $x_i \leq x \leq x_{i+1}$ for $i = 1, \ldots, n-1$. To find the interpolation function, we must first determine the coefficients a_i, b_i, c_i, d_i for each of the cubic functions. For n points, there are $n-1$ cubic functions to find, and each cubic function requires four coefficients. Therefore we have a total of $4(n-1)$ unknowns, and so we need $4(n-1)$ independent equations to find all the coefficients.

First, the cubic functions must intersect the data the points on the left and the right:

$$S_i(x_i) = y_i, \qquad i = 1, \ldots, n-1,$$
$$S_i(x_{i+1}) = y_{i+1}, \qquad i = 1, \ldots, n-1,$$

which gives us $2(n-1)$ equations. Next, we want each cubic function to join as smoothly with its neighbors as possible, so we constrain the splines to have continuous first and second derivatives at the data points $i = 2, \ldots, n-1$:

$$S_i'(x_{i+1}) = S_{i+1}'(x_{i+1}), \quad i = 1, \ldots, n-2,$$
$$S_i''(x_{i+1}) = S_{i+1}''(x_{i+1}), \quad i = 1, \ldots, n-2,$$

which gives us $2(n-2)$ equations.

Two more equations are required to compute the coefficients of $S_i(x)$. These last two constraints are arbitrary; they can be chosen to fit the circumstances of the interpolation being performed. A common set of final constraints is to assume that the second derivatives are zero at the endpoints. This means that the curve is a "straight line" at the end points. Explicitly,

$$S_1''(x_1) = 0,$$
$$S_{n-1}''(x_n) = 0.$$

In Python, we can use SciPy's function CubicSpline to perform cubic spline interpolation. Note that the above constraints are not the same as those used by SciPy's CubicSpline as a default for performing cubic splines. There are different ways to add the final two constraints in SciPy by setting the bc_type argument (see help for CubicSpline to learn more).

TRY IT! Use CubicSpline to plot the cubic spline interpolation of the dataset x = [0, 1, 2] and y = [1, 3, 2] for $0 \le x \le 2$.

```
In [1]: from scipy.interpolate import CubicSpline
        import numpy as np
        import matplotlib.pyplot as plt

        plt.style.use("seaborn-poster")

In [2]: x = [0, 1, 2]
        y = [1, 3, 2]

        # use bc_type = "natural" adds the constraints
        f = CubicSpline(x, y, bc_type="natural")
        x_new = np.linspace(0, 2, 100)
        y_new = f(x_new)

In [3]: plt.figure(figsize = (10,8))
        plt.plot(x_new, y_new, "b")
        plt.plot(x, y, "ro")
        plt.title("Cubic Spline Interpolation")
        plt.xlabel("x")
        plt.ylabel("y")
        plt.show()
```

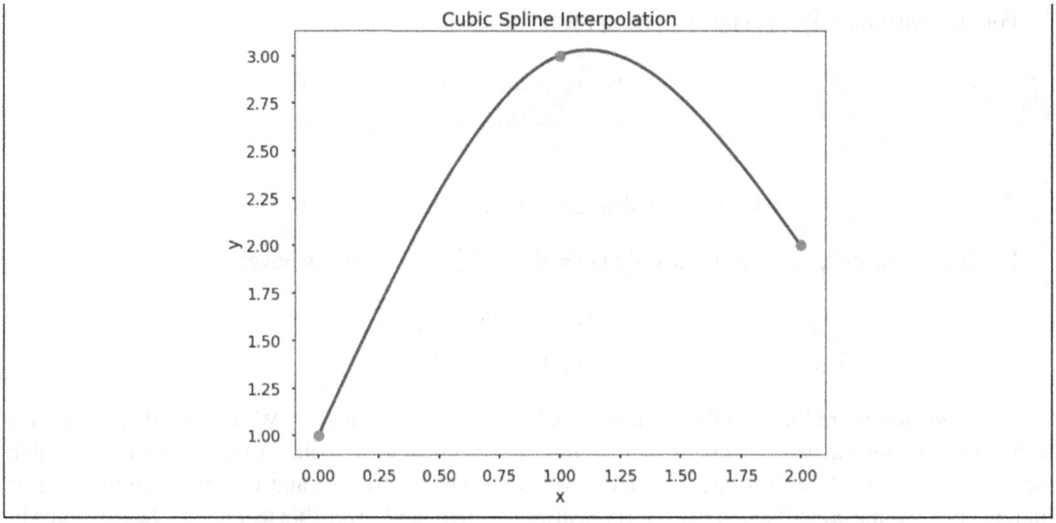

To determine the coefficients of each cubic function, we write down the constraints explicitly as a system of linear equations with $4(n-1)$ unknowns. For n data points, the unknowns are the coefficients a_i, b_i, c_i, d_i of the cubic spline, S_i, joining the points x_i and x_{i+1}.

For the constraints $S_i(x_i) = y_i$ we have:

$$a_1x_1^3 + b_1x_1^2 + c_1x_1 + d_1 = y_1,$$
$$a_2x_2^3 + b_2x_2^2 + c_2x_2 + d_2 = y_2,$$
$$\ldots$$
$$a_{n-1}x_{n-1}^3 + b_{n-1}x_{n-1}^2 + c_{n-1}x_{n-1} + d_{n-1} = y_{n-1}.$$

For the constraints $S_i(x_{i+1}) = y_{i+1}$ we have:

$$a_1x_2^3 + b_1x_2^2 + c_1x_2 + d_1 = y_2,$$
$$a_2x_3^3 + b_2x_3^2 + c_2x_3 + d_2 = y_3,$$
$$\ldots$$
$$a_{n-1}x_n^3 + b_{n-1}x_n^2 + c_{n-1}x_n + d_{n-1} = y_n.$$

For the constraints $S_i'(x_{i+1}) = S_{i+1}'(x_{i+1})$ we have:

$$3a_1x_2^2 + 2b_1x_2 + c_1 - 3a_2x_2^2 - 2b_2x_2 - c_2 = 0,$$
$$3a_2x_3^2 + 2b_2x_3 + c_2 - 3a_3x_3^2 - 2b_3x_3 - c_3 = 0,$$
$$\ldots$$
$$3a_{n-2}x_{n-1}^2 + 2b_{n-2}x_{n-1} + c_{n-2} - 3a_{n-1}x_{n-1}^2 - 2b_{n-1}x_{n-1} - c_{n-1} = 0.$$

For the constraints $S''_i(x_{i+1}) = S''_{i+1}(x_{i+1})$ we have:

$$6a_1x_2 + 2b_1 - 6a_2x_2 - 2b_2 = 0,$$
$$6a_2x_3 + 2b_2 - 6a_3x_3 - 2b_3 = 0,$$
$$\dots$$
$$6a_{n-2}x_{n-1} + 2b_{n-2} - 6a_{n-1}x_{n-1} - 2b_{n-1} = 0.$$

Finally, for the endpoint constraints $S''_1(x_1) = 0$ and $S''_{n-1}(x_n) = 0$, we have:

$$6a_1x_1 + 2b_1 = 0,$$
$$6a_{n-1}x_n + 2b_{n-1} = 0.$$

These equations are linear in the unknown coefficients a_i, b_i, c_i, and d_i. We can put them in matrix form and solve for the coefficients of each spline by left division. Remember that whenever we solve the matrix equation $Ax = b$ for x, we must make sure that A is square and invertible. In the case of finding cubic spline equations, the A matrix is always square and invertible as long as the x_i values in the dataset are unique.

TRY IT! Find the cubic spline interpolation at `x = 1.5` based on the data `x = [0, 1, 2]`, `y = [1, 3, 2]`.

First, we create the appropriate system of equations and find the coefficients of the cubic splines by solving the system in matrix form.

The matrix form of the system of equations is

$$\begin{bmatrix} 0 & 0 & 0 & 1 & 0 & 0 & 0 & 0 \\ 0 & 0 & 0 & 0 & 1 & 1 & 1 & 1 \\ 1 & 1 & 1 & 1 & 0 & 0 & 0 & 0 \\ 0 & 0 & 0 & 0 & 8 & 4 & 2 & 1 \\ 3 & 2 & 1 & 0 & -3 & -2 & -1 & 0 \\ 6 & 2 & 0 & 0 & -6 & -2 & 0 & 0 \\ 0 & 2 & 0 & 0 & 0 & 0 & 0 & 0 \\ 0 & 0 & 0 & 0 & 12 & 2 & 0 & 0 \end{bmatrix} \begin{bmatrix} a_1 \\ b_1 \\ c_1 \\ d_1 \\ a_2 \\ b_2 \\ c_2 \\ d_2 \end{bmatrix} = \begin{bmatrix} 1 \\ 3 \\ 3 \\ 2 \\ 0 \\ 0 \\ 0 \\ 0 \end{bmatrix}.$$

```
In [4]: b = np.array([1, 3, 3, 2, 0, 0, 0, 0])
        b = b[:, np.newaxis]
        A = np.array([[0, 0, 0, 1, 0, 0, 0, 0],
                      [0, 0, 0, 0, 1, 1, 1, 1],
                      [1, 1, 1, 1, 0, 0, 0, 0],
                      [0, 0, 0, 0, 8, 4, 2, 1],
                      [3, 2, 1, 0, -3, -2, -1, 0],
                      [6, 2, 0, 0, -6, -2, 0, 0],
                      [0, 2, 0, 0, 0, 0, 0, 0],
                      [0, 0, 0, 0, 12, 2, 0, 0]])
```

```
In [5]: np.dot(np.linalg.inv(A), b)

Out[5]: array([[-0.75],
               [ 0.  ],
               [ 2.75],
               [ 1.  ],
               [ 0.75],
               [-4.5 ],
               [ 7.25],
               [-0.5 ]])
```

The two cubic polynomials are

$$S_1(x) = -0.75x^3 + 2.75x + 1 \quad \text{for} \quad 0 \leq x \leq 1, \text{ and} \tag{17.1}$$

$$S_2(x) = 0.75x^3 - 4.5x^2 + 7.25x - 0.5 \quad \text{for} \quad 1 \leq x \leq 2. \tag{17.2}$$

So for $x = 1.5$, we evaluate $S_2(1.5)$ and obtain an estimated value of 2.7813.

17.4 LAGRANGE POLYNOMIAL INTERPOLATION

Rather than finding cubic polynomials between subsequent pairs of data points, **Lagrange polynomial interpolation** finds a single polynomial that goes through all the data points. This polynomial is referred to as a **Lagrange polynomial**, $L(x)$. As an interpolation function, it should have the property $L(x_i) = y_i$ for every point in the dataset. When computing Lagrange polynomials, it is useful to write them as a linear combination of **Lagrange basis polynomials**, $P_i(x)$, where

$$P_i(x) = \prod_{j=1, j\neq i}^{n} \frac{x - x_j}{x_i - x_j},$$

and

$$L(x) = \sum_{i=1}^{n} y_i P_i(x).$$

Here, $\prod$ means "the product of" or "multiply out." Note that, by construction, $P_i(x)$ has the property that $P_i(x_j) = 1$ when $i = j$ and $P_i(x_j) = 0$ when $i \neq j$. Since $L(x)$ is a sum of these polynomials, observe that $L(x_i) = y_i$ for every point, exactly as desired.

TRY IT! Find the Lagrange basis polynomials for the dataset $x = [0, 1, 2]$ and $y = [1, 3, 2]$. Plot each polynomial and verify the property that $P_i(x_j) = 1$ when $i = j$ and $P_i(x_j) = 0$ when $i \neq j$.

$$P_1(x) = \frac{(x - x_2)(x - x_3)}{(x_1 - x_2)(x_1 - x_3)} = \frac{(x - 1)(x - 2)}{(0 - 1)(0 - 2)} = \frac{1}{2}(x^2 - 3x + 2),$$
$$P_2(x) = \frac{(x - x_1)(x - x_3)}{(x_2 - x_1)(x_2 - x_3)} = \frac{(x - 0)(x - 2)}{(1 - 0)(1 - 2)} = -x^2 + 2x,$$
$$P_3(x) = \frac{(x - x_1)(x - x_2)}{(x_3 - x_1)(x_3 - x_2)} = \frac{(x - 0)(x - 1)}{(2 - 0)(2 - 1)} = \frac{1}{2}(x^2 - x).$$

```
In [1]: import numpy as np
        import numpy.polynomial.polynomial as poly
        import matplotlib.pyplot as plt

        plt.style.use("seaborn-poster")

In [2]: x = [0, 1, 2]
        y = [1, 3, 2]
        P1_coeff = [1,-1.5,.5]
        P2_coeff = [0, 2,-1]
        P3_coeff = [0,-.5,.5]

        # get the polynomial function
        P1 = poly.Polynomial(P1_coeff)
        P2 = poly.Polynomial(P2_coeff)
        P3 = poly.Polynomial(P3_coeff)

        x_new = np.arange(-1.0, 3.1, 0.1)

        fig = plt.figure(figsize = (10,8))
        plt.plot(x_new, P1(x_new), "b", label = "P1")
        plt.plot(x_new, P2(x_new), "r", label = "P2")
        plt.plot(x_new, P3(x_new), "g", label = "P3")

        plt.plot(x, np.ones(len(x)), "ko", x,np.zeros(len(x)), "ko")
        plt.title("Lagrange Basis Polynomials")
        plt.xlabel("x")
        plt.ylabel("y")
        plt.grid()
        plt.legend()
        plt.show()
```

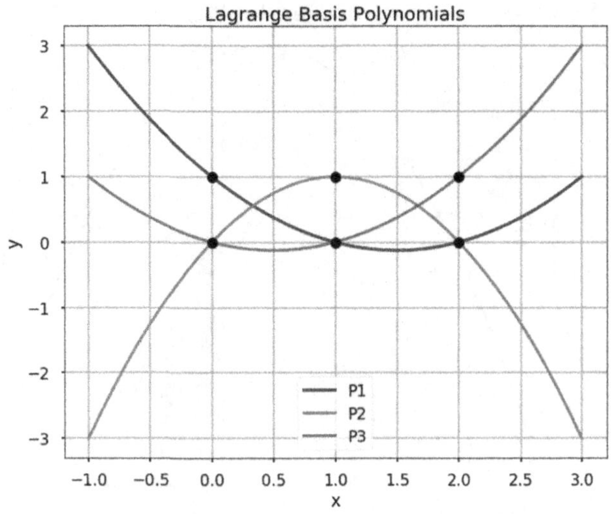

TRY IT! Using the previous example, compute and plot the Lagrange polynomial. Verify that it goes through each of the data points.

```
In [3]: L = P1 + 3*P2 + 2*P3

        fig = plt.figure(figsize = (10,8))
        plt.plot(x_new, L(x_new), "b", x, y, "ro")
        plt.title("Lagrange Polynomial")
        plt.grid()
        plt.xlabel("x")
        plt.ylabel("y")
        plt.show()
```

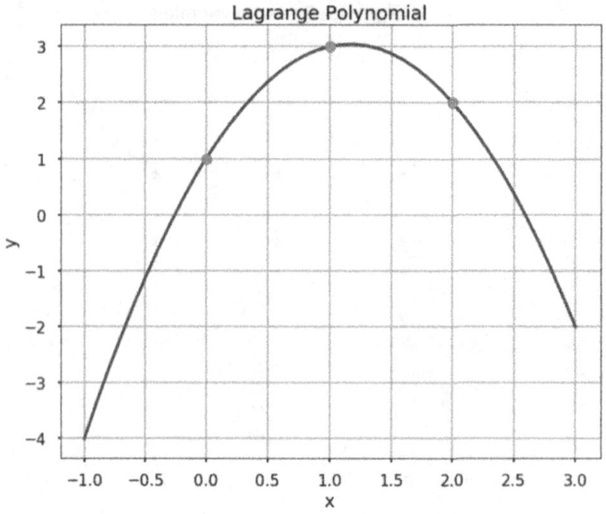

WARNING! Lagrange interpolation polynomials are defined outside the area of interpolation, that is, outside of the interval $[x_1, x_n]$, and will grow very fast and unbounded outside this region. This is not a desirable feature because, in general, this does not agree with the behavior of the underlying data. Thus, Lagrange interpolation should be used with caution outside the region of interest.

17.4.1 USING THE LAGRANGE FUNCTION FROM `SCIPY`

Instead of calculating everything from scratch, in `SciPy` we can use the `lagrange` function directly to interpolate the data. Let us use the above example to illustrate our point.

```
In [4]: from scipy.interpolate import lagrange

In [5]: f = lagrange(x, y)

In [6]: fig = plt.figure(figsize = (10,8))
        plt.plot(x_new, f(x_new), "b", x, y, "ro")
        plt.title("Lagrange Polynomial")
        plt.grid()
        plt.xlabel("x")
        plt.ylabel("y")
        plt.show()
```

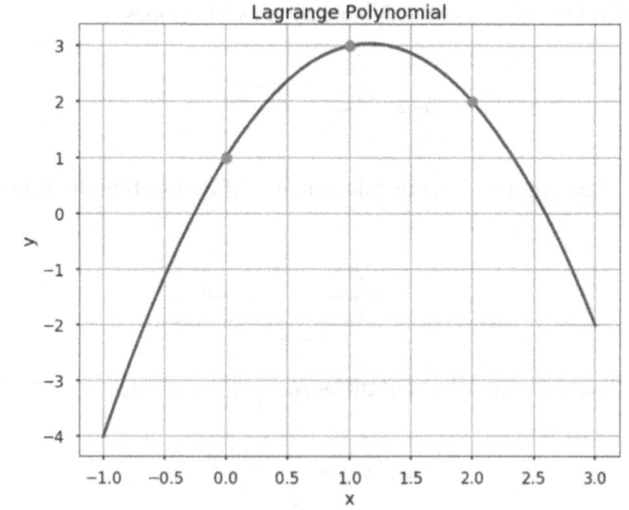

17.5 NEWTON'S POLYNOMIAL INTERPOLATION

Newton's polynomial interpolation is another popular way to exactly fit a set of data points. The general form of the $(n-1)$th order Newton's polynomial that goes through n points is

$$f(x) = a_0 + a_1(x - x_0) + a_2(x - x_0)(x - x_1) + \cdots + a_n(x - x_0)(x - x_1)\cdots(x - x_n),$$

which can be rewritten as

$$f(x) = \sum_{i=0}^{n} a_i n_i(x)$$

where

$$n_i(x) = \prod_{j=0}^{i-1}(x - x_j).$$

The special feature of the Newton's polynomial is that the coefficients a_i can be determined using a very simple mathematical procedure. For example, since the polynomial goes through each data point, for a data point (x_i, y_i), we will have $f(x_i) = y_i$, thus we have

$$f(x_0) = a_0 = y_0$$

and $f(x_1) = a_0 + a_1(x_1 - x_0) = y_1$. If we rearrange it to obtain a_1, we have

$$a_1 = \frac{y_1 - y_0}{x_1 - x_0}.$$

If we insert data point (x_2, y_2), we can calculate a_2, and it becomes

$$a_2 = \frac{\frac{y_2 - y_1}{x_2 - x_1} - \frac{y_1 - y_0}{x_1 - x_0}}{x_2 - x_0}.$$

Let us do one more data point (x_3, y_3) to calculate a_3. After inserting the data point into the equation, we obtain

$$a_3 = \frac{\frac{\frac{y_3 - y_2}{x_3 - x_2} - \frac{y_2 - y_1}{x_2 - x_1}}{x_3 - x_1} - \frac{\frac{y_2 - y_1}{x_2 - x_1} - \frac{y_1 - y_0}{x_1 - x_0}}{x_2 - x_0}}{x_3 - x_0}.$$

See the patterns? These are called **divided differences**. If we define

$$f[x_1, x_0] = \frac{y_1 - y_0}{x_1 - x_0},$$

then

$$f[x_2, x_1, x_0] = \frac{\frac{y_2 - y_1}{x_2 - x_1} - \frac{y_1 - y_0}{x_1 - x_0}}{x_2 - x_0} = \frac{f[x_2, x_1] - f[x_1, x_0]}{x_2 - x_1}.$$

If we continue write this out, we will obtain the following iteration equation:

$$f[x_k, x_{k-1}, \ldots, x_1, x_0] = \frac{f[x_k, x_{k-1}, \ldots, x_2, x_2] - f[x_{k-1}, x_{k-2}, \ldots, x_1, x_0]}{x_k - x_0}.$$

The advantage of using this method is that once the coefficients are determined, adding new data points will not change the previously calculated coefficient; we only need to calculate the higher differences in the same manner. The whole procedure for finding these coefficients can be summarized into a divided difference table. An example using five data points is shown below:

x_0	y_0				
		$f[x_1, x_0]$			
x_1	y_1		$f[x_2, x_1, x_0]$		
		$f[x_2, x_1]$		$f[x_3, x_2, x_1, x_0]$	
x_2	y_2		$f[x_3, x_2, x_1]$		$f[x_4, x_3, x_2, x_1, x_0]$
		$f[x_3, x_2]$		$f[x_4, x_3, x_2, x_1]$	
x_3	y_3		$f[x_4, x_3, x_2]$		
		$f[x_4, x_3]$			
x_4	y_4				

Each element in the table can be calculated using the two previous elements (to the left). In reality, we can calculate all elements and store them in a diagonal matrix–that is, as the coefficient matrix–

which can be written as

$$\begin{matrix} y_0 & f[x_1,x_0] & f[x_2,x_1,x_0] & f[x_3,x_2,x_1,x_0] & f[x_4,x_3,x_2,x_1,x_0] \\ y_1 & f[x_2,x_1] & f[x_3,x_2,x_1] & f[x_4,x_3,x_2,x_1] & 0 \\ y_2 & f[x_3,x_2] & f[x_4,x_3,x_2] & 0 & 0 \\ y_3 & f[x_4,x_3] & 0 & 0 & 0 \\ y_4 & 0 & 0 & 0 & 0 \end{matrix}$$

Note that the first row in the matrix is actually all the coefficients that we need, i.e., a_0, a_1, a_2, a_3, and a_4. Shown below is an example of how to do this.

TRY IT! Calculate the divided difference table for `x = [-5, -1, 0, 2]`, `y = [-2, 6, 1, 3]`.

```
In [1]: import numpy as np
        import matplotlib.pyplot as plt

        plt.style.use("seaborn-poster")

        %matplotlib inline

In [2]: def divided_diff(x, y):
            """
            function to calculate the divided
            difference table
            """
            n = len(y)
            coef = np.zeros([n, n])
            # the first column is y
            coef[:,0] = y

            for j in range(1,n):
                for i in range(n-j):
                    coef[i][j] = (coef[i+1][j-1]-coef[i][j-1])/(x[i+j]-x[i])

            return coef

        def newton_poly(coef, x_data, x):
            """
            evaluate the Newton polynomial
            at x
            """
            n = len(x_data) - 1
            p = coef[n]
```

```
        for k in range(1,n+1):
            p = coef[n-k] + (x -x_data[n-k])*p
        return p

In [3]: x = np.array([-5, -1, 0, 2])
        y = np.array([-2, 6, 1, 3])
        # get the divided difference coef
        a_s = divided_diff(x, y)[0, :]

        # evaluate on new data points
        x_new = np.arange(-5, 2.1, .1)
        y_new = newton_poly(a_s, x, x_new)

        plt.figure(figsize = (12, 8))
        plt.plot(x, y, "bo")
        plt.plot(x_new, y_new)

Out[3]: [<matplotlib.lines.Line2D at 0x11bd4e630>]
```

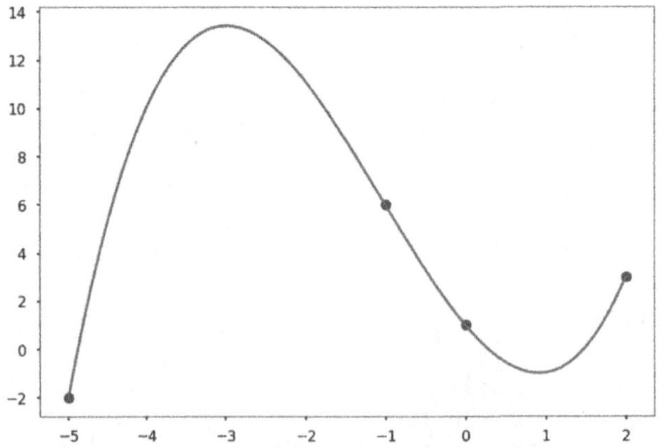

We can see that the Newton's polynomial goes through all the data points and fits the data.

17.6 SUMMARY AND PROBLEMS

17.6.1 SUMMARY

1. Given a set of reliable data points, interpolation is a method of estimating dependent variable values for independent variable values not in the dataset.
2. Linear, Cubic Spline, Lagrange, and Newton's polynomial interpolation are common interpolation methods.

17.6.2 PROBLEMS

1. Write a function `my_lin_interp(x, y, X)` where `x` and `y` are arrays that contain experimental data points, and `X` is an array. Assume that `x` and `X` are in ascending order and have unique elements. The output argument, `Y`, should be an array the same size as `X`, where `Y[i]` is the linear interpolation of `X[i]`. Do not use `interp` from `NumPy` or `interp1d` from `SciPy`.

2. Write a function `my_cubic_spline(x, y, X)` where `x` and `y` are arrays that contain experimental data points, and `X` is an array. Assume that `x` and `X` are in ascending order and have unique elements. The output argument, `Y`, should be an array the same size as `X`, where `Y[i]` is cubic spline interpolation of `X[i]`. Do not use `interp1d` or `CubicSpline`.

3. Write a function `my_nearest_neighbor(x, y, X)` where `x` and `y` are arrays that contain experimental data points, and `X` is an array. Assume that `x` and `X` are in ascending order and have unique elements. The output argument, `Y`, should be an array the same size as `X`, where `Y[i]` is the nearest neighbor interpolation of `X[i]`. That is, `Y[i]` should be the `y[j]` where `x[j]` is the closest independent data point of `X[i]`. Do not use `interp1d` from `SciPy`.

4. Think of a circumstance where using the nearest neighbor interpolation would be superior to cubic spline interpolation.

5. Write a function `my_cubic_spline_flat(x, y, X)` where `x` and `y` are arrays that contain experimental data points, and `X` is an array. Assume that `x` and `X` are in ascending order and have unique elements. The output argument, `Y`, should be an array the same size as `X`, where `Y[i]` is the cubic spline interpolation of `X[i]`. Instead of the constraints introduced previously, use $S_1'(x_1) = 0$ and $S_{n-1}'(x_n) = 0$.

6. Write a function `my_quintic_spline(x, y, X)` where `x` and `y` are arrays that contain experimental data points, and `X` is an array. Assume that `x` and `X` are in ascending order and have unique elements. The output argument, `Y`, should be an array the same size as `X`, where `Y[i]` is the quintic spline interpolation of `X[i]`. You will need to use additional endpoint constraints to come up with enough constraints. You may use endpoint constraints at your discretion.

7. Write a function `my_interp_plotter(x, y, X, option)` where `x` and `y` are arrays containing experimental data points, and `X` is an array that contains the coordinates for which an interpolation is desired. The input argument option should be a string, either "linear," "spline," or "nearest." Your function should produce a plot of the data points (x, y) marked as red circles. The points (X, Y), where `X` is the input and `Y` is the interpolation at the points contained in `X` defined by the input argument specified by option. The points (X, Y) should be connected by a blue line. Be sure to include the title, axis labels, and a legend. Hint: You should use `interp1d` from `SciPy`, and checkout the `kind` option.
   Test cases:

```
x = np.array([0, .1, .15, .35, .6, .7, .95, 1])
y=np.array([1,0.8187,0.7408,0.4966,0.3012,0.2466,0.1496,0.1353])

my_interp_plotter(x, y, np.linspace(0, 1, 101), "nearest")
```

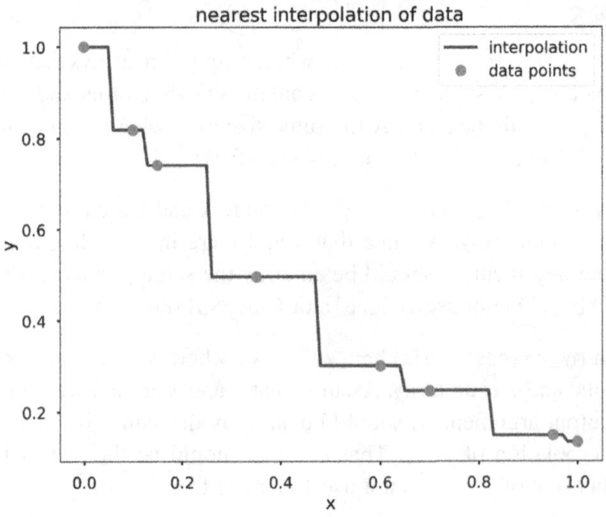


```
my_interp_plotter(x, y, np.linspace(0, 1, 101), "linear")
```

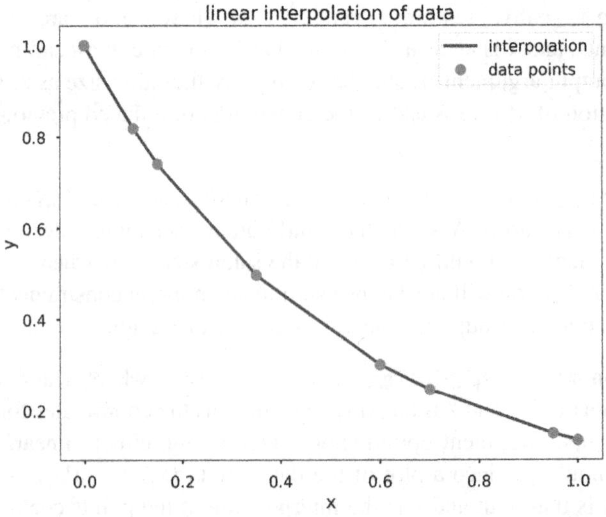


```
my_interp_plotter(x, y, np.linspace(0, 1, 101), "cubic")
```

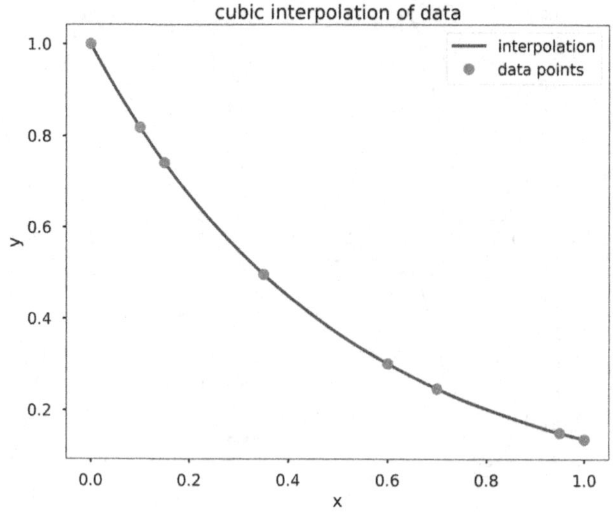

8. Write a function `my_D_cubic_spline(x, y, X, D)`, where the output `Y` is the cubic spline interpolation at `X` taken from the data points contained in `x` and `y`. Instead of the standard pinned endpoint conditions (i.e., $S_1''(x_1) = 0$ and $S_{n-1}''(x_n) = 0$), use the endpoint conditions $S_1'(x_1) = D$ and $S_{n-1}'(x_n) = D$ (i.e., the slopes of the interpolating polynomials at the endpoints are D).
Test cases:

```
x = [0, 1, 2, 3, 4]
y = [0, 0, 1, 0, 0]
X = np.linspace(0, 4, 101)

# Solution: Y = 0.54017857
Y = my_D_cubic_spline(x, y, 1.5, 1)

plt.figure(figsize = (10, 8))
plt.subplot(221)
plt.plot(x, y, "ro", X, my_D_cubic_spline(x, y, X, 0), "b")
plt.subplot(222)
plt.plot(x, y, "ro", X, my_D_cubic_spline(x, y, X, 1), "b")
plt.subplot(223)
plt.plot(x, y, "ro", X, my_D_cubic_spline(x, y, X, -1), "b")
plt.subplot(224)
plt.plot(x, y, "ro", X, my_D_cubic_spline(x, y, X, 4), "b")
plt.tight_layout()
plt.show()
```

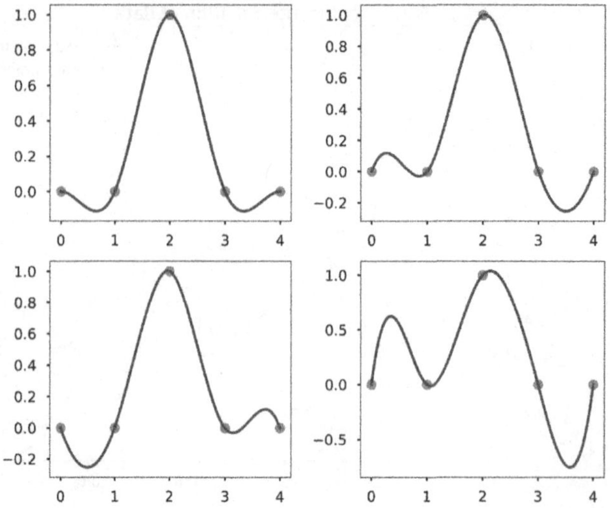

9. Write a function `my_lagrange(x, y, X)` where the output `Y` is the Lagrange interpolation of the data points contained in `x` and `y` computed at `X`. Hint: Use a nested for-loop, where the inner for-loop computes the product for the Lagrange basis polynomial and the outer loop computes the sum for the Lagrange polynomial. Don't use the existing `lagrange` function from `SciPy`.
Test cases:

```
x = [0, 1, 2, 3, 4]
y = [2, 1, 3, 5, 1]

X = np.linspace(0, 4, 101)

plt.figure(figsize = (10,8 ))
plt.plot(X, my_lagrange(x, y, X), "b", label = "interpolation")
plt.plot(x, y, "ro", label = "data points")

plt.xlabel("x")
plt.ylabel("y")

plt.title(f"Lagrange Interpolation of Data Points")
plt.legend()
plt.show()
```

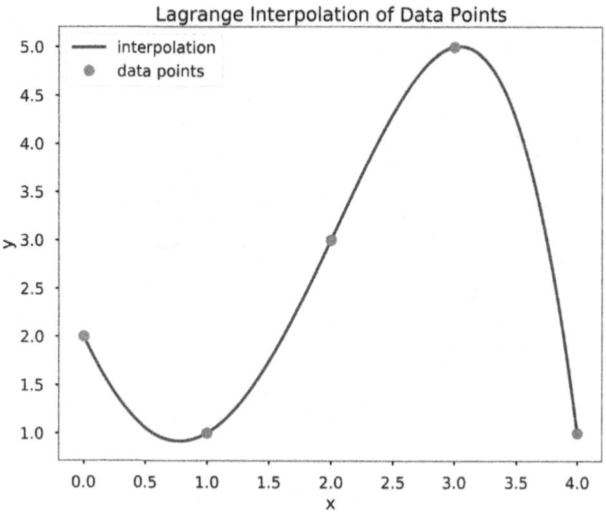

10. Fit the data x = [0, 1, 2, 3, 4], y = [2, 1, 3, 5, 1] using Newton's polynomial interpolation.

CHAPTER

TAYLOR SERIES

18

CONTENTS

18.1 EXPRESSING FUNCTIONS USING A TAYLOR SERIES

A **sequence** is an ordered set of numbers denoted by the list of numbers inside parentheses. For example, $s = (s_1, s_2, s_3, \ldots)$ means s is the sequence $s_1, s_2, s_3, \ldots$, and so on. In this context, "ordered" means that s_1 comes *before* s_2, not that $s_1 < s_2$. Many sequences have a more complicated structure. For example, $s = (n^2, n \in N)$ is the sequence 0, 1, 4, 9, ... A **series** is the sum of a sequence up to a certain element. An **infinite sequence** is a sequence with an infinite number of terms, and an **infinite series** is the sum of an infinite sequence.

A **Taylor series expansion** is a representation of a function by an infinite series of polynomials around a point. Mathematically, the Taylor series of a function, $f(x)$, is defined as

$$f(x) = \sum_{n=0}^{\infty} \frac{f^{(n)}(a)(x-a)^n}{n!},$$

where $f^{(n)}$ is the nth derivative of f and $f^{(0)}$ is the function f.

> **TRY IT!** Compute a Taylor series expansion for $f(x) = 5x^2 + 3x + 5$ around $a = 0$, and $a = 1$. Verify that f and its Taylor series expansions are identical.

Python Programming and Numerical Methods. https://doi.org/10.1016/B978-0-12-819549-9.00028-2

First compute derivatives analytically:

$$\begin{aligned} f(x) &= 5x^2 + 3x + 5, \\ f'(x) &= 10x + 3, \\ f''(x) &= 10. \end{aligned}$$

Around $a = 0$:

$$f(x) = \frac{5x^0}{0!} + \frac{3x^1}{1!} + \frac{10x^2}{2!} + 0 + 0 + \cdots = 5x^2 + 3x + 5.$$

Around $a = 1$:

$$\begin{aligned} f(x) &= \frac{13(x-1)^0}{0!} + \frac{13(x-1)^1}{1!} + \frac{10(x-1)^2}{2!} + 0 + \cdots \\ &= 13 + 13x - 13 + 5x^2 - 10x + 5 = 5x^2 + 3x + 5. \end{aligned}$$

Note that a Taylor series expansion of any polynomial has finitely many terms because the nth derivative of any polynomial is zero when n is large enough.

TRY IT! Write Taylor series for $\sin(x)$ around the point $a = 0$.

Let $f(x) = \sin(x)$. According to Taylor series expansion,

$$f(x) = \frac{\sin(0)}{0!}x^0 + \frac{\cos(0)}{1!}x^1 + \frac{-\sin(0)}{2!}x^2 + \frac{-\cos(0)}{3!}x^3 + \frac{\sin(0)}{4!}x^4 + \frac{\cos(0)}{5!}x^5 + \cdots.$$

The expansion can be written compactly by the formula

$$f(x) = \sum_{n=0}^{\infty} \frac{(-1)^n x^{2n+1}}{(2n+1)!},$$

which ignores the terms that contain $\sin(0)$ (i.e., the even terms). Because these terms are ignored, the terms in this series and the proper Taylor series expansion agree after renumbering. For example, the $n = 0$ term in the formula is the $n = 1$ term in the Taylor series, and the $n = 1$ term in the formula is the $n = 3$ term in the Taylor series.

18.2 APPROXIMATIONS USING TAYLOR SERIES

Clearly, it is not useful to express functions as infinite sums because we cannot compute them. That said, it is often useful to approximate functions by using an N**th order Taylor series approximation** of a function, which is a truncation of its Taylor expansion at some $n = N$. This technique is especially powerful especially when there is a point around which we have knowledge about a function and all its derivatives. For example, if we take the Taylor expansion of e^x around $a = 0$, then $f^{(n)}(a) = 1$ for all n, and we do not have to compute the derivatives in the Taylor expansion to approximate e^x!

TRY IT! Use Python to plot the sin function along with the first, third, fifth, and seventh order Taylor series approximations. Note that this involves the *zeroth* to third terms in the formula given earlier.

```
In [1]: import numpy as np
        import matplotlib.pyplot as plt
        plt.style.use("seaborn-poster")

In [2]: x = np.linspace(-np.pi, np.pi, 200)
        y = np.zeros(len(x))

        labels = ["First Order", "Third Order",
                  "Fifth Order", "Seventh Order"]

        plt.figure(figsize = (10,8))
        for n, label in zip(range(4), labels):
            y=y+((-1)**n*(x)**(2*n+1))/np.math.factorial(2*n+1)
            plt.plot(x,y, label = label)

        plt.plot(x, np.sin(x), "k", label = "Analytic")
        plt.grid()
        plt.title("Taylor Series Approximations of Various Orders")
        plt.xlabel("x")
        plt.ylabel("y")
        plt.legend()
        plt.show()
```

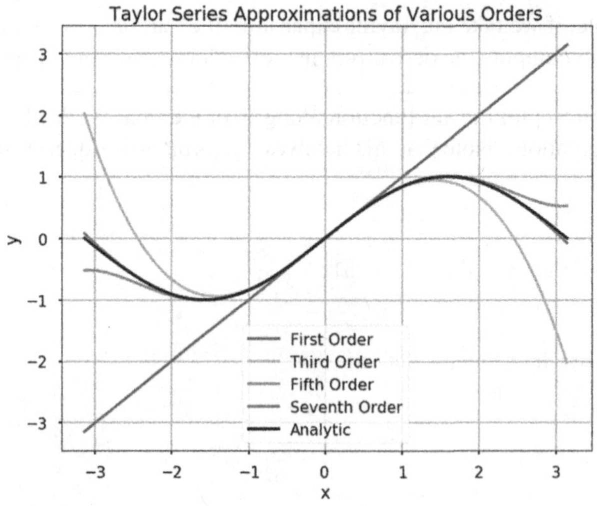

Obviously, the approximation approaches the analytic function quickly, even for x not near $a = 0$. Note that in the above code, we also used a new function `zip`, which allows us to loop through two parameters `range(4)` and `labels` to use in our plot.

TRY IT! Compute the seventh order Taylor series approximation for $\sin(x)$ around $a = 0$ at $x = \pi/2$. Compare the value to the correct value, 1.

```
In [3]: x = np.pi/2
        y = 0

        for n in range(4):
            y=y+((-1)**n *(x)**(2*n+1))/np.math.factorial(2*n+1)

        print(y)

0.9998431013994987
```

The seventh order Taylor series approximation is very close to the theoretical value of the function even if it is computed far from the point around which the Taylor series was computed (i.e., $x = \pi/2$ and $a = 0$).

The most common Taylor series approximation is the first order approximation, or **linear approximation**. Intuitively, for "smooth" functions the linear approximation of the function around a point, a, is legitimate provided you stay sufficiently close to a. In other words, "smooth" functions look more and more like a line the more you zoom into any point. In the figure below this has been plotted in successive levels of zoom using a smooth function to illustrate the linear nature of functions locally. Linear approximations are useful tools when analyzing "complicated" functions locally.

```
In [4]: x = np.linspace(0, 3, 30)
        y = np.exp(x)

        plt.figure(figsize = (14, 4.5))
        plt.subplot(1, 3, 1)
        plt.plot(x, y)
        plt.grid()
        plt.subplot(1, 3, 2)
        plt.plot(x, y)
        plt.grid()
        plt.xlim(1.7, 2.3)
        plt.ylim(5, 10)
        plt.subplot(1, 3, 3)
        plt.plot(x, y)
        plt.grid()
        plt.xlim(1.92, 2.08)
        plt.ylim(6.6, 8.2)
        plt.tight_layout()
        plt.show()
```

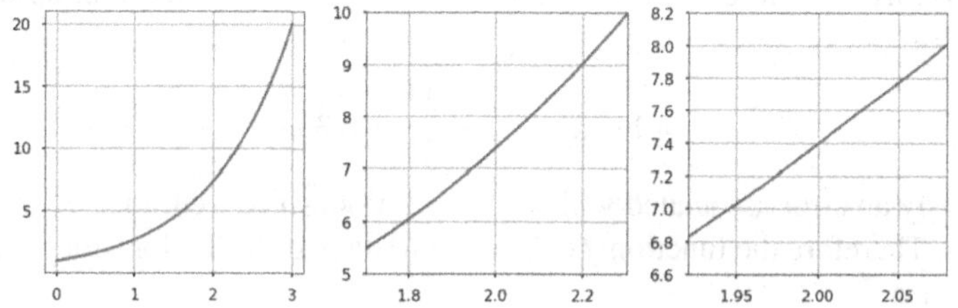

TRY IT! Take the linear approximation for e^x around point $a = 0$. Use the linear approximation for e^x to approximate the value of e^1 and $e^{0.01}$. Use the `NumPy`'s function `exp` to compute `exp(1)` and `exp(0.01)` and compare the results.

The linear approximation of e^x around $a = 0$ is $1 + x$.

The `NumPy`'s `exp` function gives the following:

```
In [5]: np.exp(1)

Out[5]: 2.718281828459045

In [6]: np.exp(0.01)

Out[6]: 1.010050167084168
```

The linear approximation of e^1 is 2, which is inaccurate, and the linear approximation of $e^{0.01}$ is 1.01, which is very good. This example illustrates how the linear approximation becomes close to the point from which the approximation is taken.

18.3 DISCUSSION ABOUT ERRORS

18.3.1 TRUNCATION ERRORS FOR TAYLOR SERIES

In numerical analysis, there are usually two sources of error, **round-off** and **truncation**. The round-off errors are due to the inexactness in the representation of real numbers on a computer and the arithmetic operations performed with them. Refer to Chapter 9 for additional discussion on this type of error. The truncation errors are due to the approximate nature of the method used and usually arise when using an approximation in place of an exact mathematical procedure, e.g., when we used Taylor series to approximate a function. For example, if we use Taylor series to approximate the function e^x, we obtain

$$e^x = 1 + x + \frac{x^2}{2!} + \frac{x^3}{3!} + \frac{x^4}{4!} + \cdots.$$

Since it takes an infinite sequence to approximate the function, using only a few terms will result in an approximation (or truncation) error. For example, if we only use the first four terms to approximate e^2, we obtain

$$e^2 \approx 1 + 2 + \frac{2^2}{2!} + \frac{2^3}{3!} = 6.3333.$$

There is obviously an error associated with such a solution, since we truncated the rest of the terms in the Taylor series. Therefore the function $f(x)$ can be written as the Taylor series approximation plus a truncation error term

$$f(x) = f_n(x) + E_n(x).$$

The more terms we use, the closer the approximation will be to the exact value. Let us use Python to calculate the above example.

TRY IT! Approximate e^2 using different order Taylor series and print out the results.

```
In [1]: import numpy as np

In [2]: exp = 0
        x = 2
        for i in range(10):
            exp = exp + (x**i)/np.math.factorial(i)
            print(f"Using {i}-term, {exp}")
```

```
        print(f"The true e^2 is: \n{np.exp(2)}")

Using 0-term, 1.0
Using 1-term, 3.0
Using 2-term, 5.0
Using 3-term, 6.333333333333333
Using 4-term, 7.0
Using 5-term, 7.266666666666667
Using 6-term, 7.355555555555555
Using 7-term, 7.3809523809523805
Using 8-term, 7.387301587301587
Using 9-term, 7.3887125220458545
The true e^2 is:
7.38905609893065
```

18.3.2 ESTIMATING TRUNCATION ERRORS

We can see that the higher the order used to approximate the function at a point, the closer we are to the true value. For each order we choose, there is an error associated with it, and the approximation is only useful if we have an idea of how accurate the approximation is. This is the motivation that we need to understand more about the errors.

Using the Taylor series, if we use only the first n terms, we can see that

$$f(x) = f_n(x) + E_n(x) = \sum_{k=0}^{n} \frac{f^{(k)}(a)(x-a)^k}{k!} + E_n(x).$$

The $E_n(x)$ is the remainder of the Taylor series, or the truncation error that measures how far off the approximation $f_n(x)$ is from $f(x)$. We can estimate the error using the **Taylor Remainder Estimation Theorem**, which states:

If the function $f(x)$ has $n+1$ derivatives for all x in an interval I containing a, then for each x in I there exists a z between x and a such that

$$E_n(x) = \frac{f^{(n+1)}(z)(x-a)^{n+1}}{(n+1)!}.$$

If we know that M is the maximum value of $|f^{(n+1)}|$ in the interval, then we obtain

$$|E_n(x)| \leq \frac{M|x-a|^{n+1}}{(n+1)!}.$$

This provides us with a bound for the truncation error using this theorem. See the example below.

TRY IT! Estimate the remainder bound for the approximation using Taylor series for e^2 using $n = 9$.

To understand the basis of this error, when we use $n = 9$, we know that $(e^x)' = e^x$, and $a = 0$; therefore, the error related to $x = 2$ is

$$E_n(x) = \frac{f^{(9+1)}(z)(x)^{(9+1)}}{(9+1)!} = \frac{e^z 2^{10}}{10!}.$$

Recall that $0 \le z \le 2$, and $e < 3$; therefore,

$$|E_n(x)| \le \frac{3^2 2^{10}}{10!} = 0.00254.$$

If we use Taylor series with $n = 9$ to approximate e^2, our absolute error should be less than 0.00254. We verify this below.

```
In [3]: abs(7.3887125220458545-np.exp(2))

Out[3]: 0.0003435768847959153
```

18.3.3 ROUND-OFF ERRORS FOR TAYLOR SERIES

Numerically, when adding many terms in a sum, we should be mindful of numerical accumulation of errors that is due to floating point round-off errors; see the following example.

EXAMPLE: Approximate e^{-30} using different order Taylor series, and print out the results.

```
In [4]: exp = 0
x = -30
for i in range(200):
    exp = exp + (x**i)/np.math.factorial(i)

print(f"Using {i}-term, our result is {exp}")
print(f"The true e^2 is: {np.exp(x)}")

Using 199-term, our result is -8.553016433669241e-05
The true e^2 is: 9.357622968840175e-14
```

From the above example, it is clear that our estimation using Taylor series is not close to the true value anymore, no matter how many terms we include into the calculation, which is due to the round-off errors we discussed earlier. To obtain a small result, when using negative large arguments, the Taylor series requires alternating large numbers to cancel out. We need many digits for precision in the series to capture both the large and small numbers with enough remaining digits to get the result in the desired output precision. This is why the program threw an error message in the example above.

18.4 SUMMARY AND PROBLEMS

18.4.1 SUMMARY

1. Some functions can be perfectly represented by a Taylor series, which is an infinite sum of polynomials.
2. Functions that have a Taylor series expansion can be approximated by truncating its Taylor series.
3. The linear approximation is a common local approximation for functions.
4. The truncation error can be estimated using the Taylor Remainder Estimation Theorem.
5. Be mindful of the round-off error in the Taylor series.

18.4.2 PROBLEMS

1. Use Taylor series expansions to show that $e^{ix} = \cos(x) + i\sin(x)$, where $i = \sqrt{-1}$.
2. Use the linear approximation of $\sin(x)$ around $a = 0$ to show that $\frac{\sin(x)}{x} \approx 1$ for small x.
3. Write the Taylor series expansion for e^{x^2} around $a = 0$. Write a function `my_double_exp(x, n)`, which computes an approximation of e^{x^2} using the first n terms of the Taylor series expansion. Be sure that `my_double_exp` can take array inputs.
4. Write a function that gives the Taylor series approximation to the `np.exp` function around 0 for an order 1 through 7. Calculate the truncation error bound for order 7.
5. Compute the fourth order Taylor expansion for $\sin(x)$ and $\cos(x)$, and $\sin(x)\cos(x)$ around 0, which produces a smaller error for $x = \pi/2$. Which is correct: computing separately Taylor expansion for sin and cos and then multiplying the result together, or computing the Taylor expansion for the product first and then plugging in x?
6. Use the fourth order Taylor series to approximate $\cos(0.2)$ and determine the truncation error bound.
7. Write a function `my_cosh_approximator(x, n)` where output is the nth order Taylor series approximation for $\cosh(x)$, the hyperbolic cosine of x taken around $a = 0$. You may assume that x is an array, and n is a positive integer (including zero). Recall that

   $$\cosh(x) = (e^x + e^{-x})/2.$$

   Warning: The approximations for $n = 0$ and $n = 1$ will be equivalent, and the approximations for $n = 2$ and $n = 3$ will be equivalent, etc.

CHAPTER

ROOT FINDING

19

CONTENTS

19.1 ROOT FINDING PROBLEM STATEMENT

The **root** or **zero** of a function, $f(x)$, is an x_r such that $f(x_r) = 0$. For functions such as $f(x) = x^2 - 9$, the roots are clearly 3 and -3. However, for other functions, such as $f(x) = \cos(x) - x$, determining an **analytic** or exact solution for the roots of functions can be difficult. For these cases, it is useful to generate numerical approximations of the roots of f and understand their limitations.

TRY IT! Use the `fsolve` function from `SciPy` to compute the root of $f(x) = \cos(x) - x$ near -2. Verify that the solution is a root (or close enough).

```
In [1]: import numpy as np
        from scipy import optimize

        f = lambda x: np.cos(x) - x
        r = optimize.fsolve(f, -2)
        print("r =", r)

        # Verify the solution is a root
        result = f(r)
        print("result=", result)

r = [0.73908513]
result= [0.]
```

Python Programming and Numerical Methods. https://doi.org/10.1016/B978-0-12-819549-9.00029-4

TRY IT! The function $f(x) = \frac{1}{x}$ has no root. Use the `fsolve` function to try to compute the root of $f(x) = \frac{1}{x}$. Turn on the `full_output` to see what is going on. Check the documentation for details.

```
In [2]: f = lambda x: 1/x

        r, infodict, ier, mesg =
          optimize.fsolve(f, -2, full_output=True)
        print("r =", r)

        result = f(r)
        print("result=", result)

        print(mesg)

r = [-3.52047359e+83]
result= [-2.84052692e-84]
The number of calls to function has reached maxfev = 400.
```

The value r that was returned is not a root, even though the value of $f(r)$ is a very small number. Since we turned on the `full_output`, we can see more information. A message would have been returned if no solution was found; we can see `mesg` details for the cause of failure: "The number of calls to function has reached `maxfev = 400`."

19.2 TOLERANCE

In engineering and science, **error** is a deviation from an expected or computed value. **Tolerance** is the level of error that is acceptable for an engineering application. We say that a computer program has **converged** to a solution when it has found a solution with an error smaller than the tolerance. When computing roots numerically, or conducting any other kind of numerical analysis, it is important to establish both a metric for error and a tolerance that is suitable for a given engineering/science application.

For computing roots, we want an x_r such that $f(x_r)$ is very close to zero. Therefore $|f(x)|$ is a possible choice for the measure of error since the smaller it is, the likelier we are to a root. Also, if we assume that x_i is the ith guess of an algorithm for finding a root, then $|x_{i+1} - x_i|$ is another possible choice for measuring the error since we expect improvement between subsequent guesses to diminish as it approaches a solution. As will be demonstrated in the following examples, these different choices have their advantages and disadvantages.

TRY IT! Let the error be measured by $e = |f(x)|$ and tol be the acceptable level of error. The function $f(x) = x^2 + \text{tol}/2$ has no real roots. Because $|f(0)| = \text{tol}/2$, it is acceptable as a solution for a root finding program.

TRY IT! Let the error be measured by $e = |x_{i+1} - x_i|$ and tol be the acceptable level of error. The function $f(x) = 1/x$ has no real roots, but the guesses $x_i = -\text{tol}/4$ and $x_{i+1} = \text{tol}/4$ have an error of $e = \text{tol}/2$ and are an acceptable solution for a computer program.

Based on these observations, the use of tolerance and convergence criteria must be done very carefully and in the context of the program that uses them.

19.3 BISECTION METHOD

The **Intermediate Value Theorem** says that if $f(x)$ is a continuous function between a and b, and $\text{sign}(f(a)) \neq \text{sign}(f(b))$, then there must be a c, such that $a < c < b$ and $f(c) = 0$. This is illustrated in Fig. 19.1.

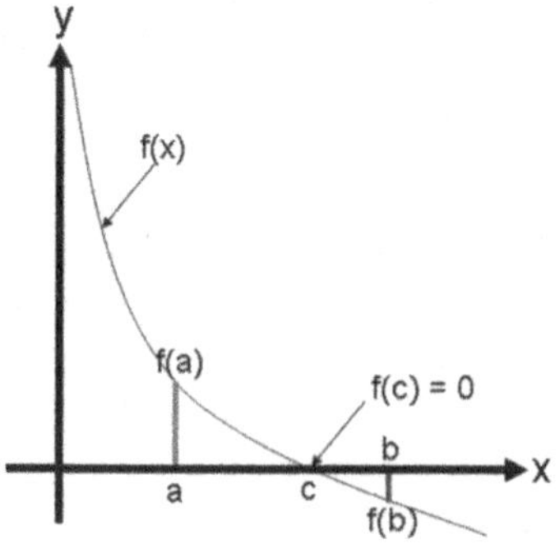

FIGURE 19.1

Illustration of the intermediate value theorem. If $\text{sign}(f(a))$ and $\text{sign}(f(b))$ are not equal, then there exists a c in (a, b) such that $f(c) = 0$.

The **bisection method** uses the intermediate value theorem iteratively to find roots. Let $f(x)$ be a continuous function, and a and b be real scalar values such that $a < b$. Assume, without loss of generality, that $f(a) > 0$ and $f(b) < 0$. Then, by the intermediate value theorem, there must be a root in the open interval (a, b). Now let $m = \frac{b+a}{2}$ be the midpoint between and a and b. If $f(m) = 0$ or is close enough, then m is a root. If $f(m) > 0$, then m is an improvement on the left bound, a, and it is guaranteed that there is a root in the open interval (m, b). If $f(m) < 0$, then m is an improvement on the right bound, b, it is guaranteed that there is a root in the open interval (a, m). This scenario is depicted in Fig. 19.2.

The process of updating a and b can be repeated until the error is acceptably low.

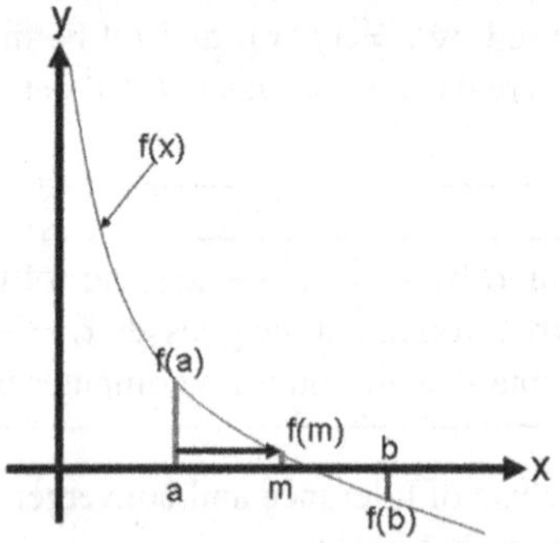

FIGURE 19.2

Illustration of the bisection method. The sign of $f(m)$ is checked to determine if the root is contained in the interval (a, m) or (m, b). This new interval is used in the next iteration of the bisection method; in the case depicted in the figure, the root is in the interval (m, b).

TRY IT! Program a function `my_bisection(f,a,b,tol)` that approximates a root `r` of `f`, bounded by `a` and `b` to within $|f(\frac{a+b}{2})| <$ `tol`.

```
In [1]: import numpy as np

        def my_bisection(f, a, b, tol):
            # approximates a root, R, of f bounded
            # by a and b to within tolerance
            # | f(m) | < tol with m being the midpoint
            # between a and b. Recursive implementation

            # check if a and b bound a root
            if np.sign(f(a)) == np.sign(f(b)):
                raise Exception(
                 "The scalars a and b do not bound a root")

            # get midpoint
            m = (a + b)/2

            if np.abs(f(m)) < tol:
                # stopping condition, report m as root
                return m
            elif np.sign(f(a)) == np.sign(f(m)):
                # case where m is an improvement on a.
                # Make recursive call with a = m
                return my_bisection(f, m, b, tol)
```

```
        elif np.sign(f(b)) == np.sign(f(m)):
            # case where m is an improvement on b.
            # Make recursive call with b = m
            return my_bisection(f, a, m, tol)
```

TRY IT! The $\sqrt{2}$ can be computed as the root of the function $f(x) = x^2 - 2$. Starting at $a = 0$ and $b = 2$, use `my_bisection` to approximate the $\sqrt{2}$ to a tolerance of $|f(x)| < 0.1$ and $|f(x)| < 0.01$. Verify that the results are close to a root by plugging the root back into the function.

```
In [2]: f = lambda x: x**2 - 2

        r1 = my_bisection(f, 0, 2, 0.1)
        print("r1 =", r1)
        r01 = my_bisection(f, 0, 2, 0.01)
        print("r01 =", r01)

        print("f(r1) =", f(r1))
        print("f(r01) =", f(r01))

r1 = 1.4375
r01 = 1.4140625
f(r1) = 0.06640625
f(r01) = -0.00042724609375
```

TRY IT! See what happens if you use $a = 2$ and $b = 4$ for the above function.

```
In [3]: my_bisection(f, 2, 4, 0.01)


        ---------------------------------------------------------


        Exception                 Traceback (most recent call last)

        <ipython-input-3-4158b7a9ae67> in <module>
    ----> 1 my_bisection(f, 2, 4, 0.01)


        <ipython-input-1-36f06123e87c> in my_bisection(f,a,b,tol)
         10     if np.sign(f(a)) == np.sign(f(b)):
         11         raise Exception(
    ---> 12          "The scalars a and b do not bound a root")
         13
```

```
# get midpoint

Exception: The scalars a and b do not bound a root
```

19.4 NEWTON–RAPHSON METHOD

Let $f(x)$ be a smooth function, and x_r be an unknown root of $f(x)$. Assume that x_0 is a guess for x_r. Unless x_0 is a very lucky guess, $f(x_0)$ will not be a root. Given this scenario, we want to find an x_1 that is an improvement on x_0 (i.e., closer to x_r than x_0). If we assume that x_0 is "close enough" to x_r, then we can improve upon it by taking the linear approximation of $f(x)$ around x_0, which is a line, and finding the intersection of this line with the x-axis. Written out, the linear approximation of $f(x)$ around x_0 is $f(x) \approx f(x_0) + f'(x_0)(x - x_0)$. Using this approximation, we find x_1 such that $f(x_1) = 0$. Plugging these values into the linear approximation results in the following equation:

$$0 = f(x_0) + f'(x_0)(x_1 - x_0),$$

which when solved for x_1 yields

$$x_1 = x_0 - \frac{f(x_0)}{f'(x_0)}.$$

An illustration of how this linear approximation improves an initial guess is shown in Fig. 19.3.

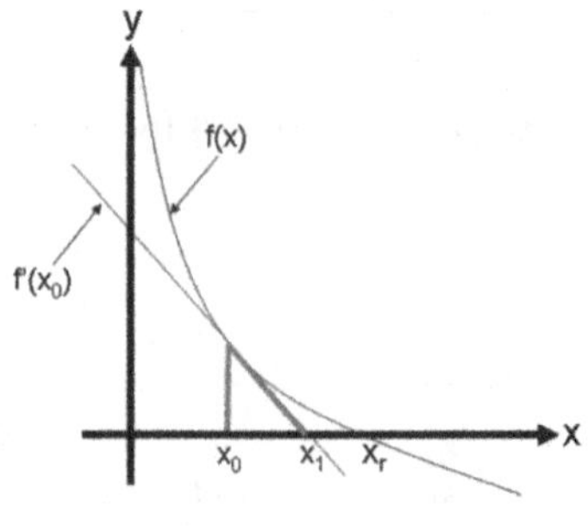

FIGURE 19.3

Illustration of Newton step for a smooth function, $g(x)$.

Written generally, a **Newton step** computes an improved guess, x_i, using a previous guess, x_{i-1}, and is given by the equation

$$x_i = x_{i-1} - \frac{g(x_{i-1})}{g'(x_{i-1})}.$$

The **Newton–Raphson Method** of finding roots iterates Newton steps from x_0 until the error is less than the tolerance.

TRY IT! Again, the $\sqrt{2}$ is the root of the function $f(x) = x^2 - 2$. Using $x_0 = 1.4$ as a starting point, use the previous equation to estimate $\sqrt{2}$. Compare this approximation with the value computed by Python's `sqrt` function.

$$x = 1.4 - \frac{1.4^2 - 2}{2(1.4)} = 1.4142857142857144.$$

```
In [1]: import numpy as np

        f = lambda x: x**2 - 2
        f_prime = lambda x: 2*x
        newton_raphson = 1.4 - (f(1.4))/(f_prime(1.4))

        print("newton_raphson =", newton_raphson)
        print("sqrt(2) =", np.sqrt(2))

newton_raphson = 1.4142857142857144
sqrt(2) = 1.4142135623730951
```

TRY IT! Write a function `my_newton(f,df,x0,tol)` where the output is an estimate of the root of `f`, `f` is a function object $f(x)$, `df` is a function object $f'(x)$, `x0` is an initial guess, and `tol` is the error tolerance. The error measurement should be $|f(x)|$.

```
In [2]: def my_newton(f, df, x0, tol):
            # output is an estimation of the root of f
            # using the Newton-Raphson method
            # recursive implementation
            if abs(f(x0)) < tol:
                return x0
            else:
                return my_newton(f, df, x0 - f(x0)/df(x0), tol)
```

TRY IT! Use `my_newton` to compute $\sqrt{2}$ to within a tolerance of 1e-6 starting at `x0 = 1.5`.

```
In [3]: estimate = my_newton(f, f_prime, 1.5, 1e-6)
        print("estimate =", estimate)
        print("sqrt(2) =", np.sqrt(2))

estimate = 1.4142135623746899
sqrt(2) = 1.4142135623730951
```

If x_0 is close to x_r, then it can be proven that, in general, the Newton–Raphson method converges to x_r much faster than the bisection method; however, since x_r is initially unknown, there is no way to know if the initial guess is close enough to the root to obtain this behavior unless some special information about the function is known *a priori* (e.g., the function has a root close to $x = 0$). In addition to this initialization problem, the Newton–Raphson method has other serious limitations. For example, if the derivative at a guess is close to zero, then the Newton step will be very large and probably lead far away from the root. Also, depending on the behavior of the function derivative between x_0 and x_r, the Newton–Raphson method may converge to a different root than x_r which may not be useful for the engineering application being considered.

TRY IT! Compute a single Newton step to get an improved approximation of the root of the function $f(x) = x^3 + 3x^2 - 2x - 5$ and initial guess $x_0 = 0.29$.

```
In [4]: x0 = 0.29
        x1 = x0-(x0**3+3*x0**2-2*x0-5)/(3*x0**2+6*x0-2)
        print("x1 =", x1)

x1 = -688.4516883116648
```

Note that $f'(x_0) = -0.0077$ (is close to zero), and the error at x_1 is approximately 324880000 (very large).

TRY IT! Consider the polynomial $f(x) = x^3 - 100x^2 - x + 100$. This polynomial has a root at $x = 1$ and $x = 100$. Use the Newton–Raphson method to find a root of f starting at $x_0 = 0$.

At $x_0 = 0$, $f(x_0) = 100$, and $f'(x) = -1$, the Newton step gives $x_1 = 0 - \frac{100}{-1} = 100$, which is a root of f. Note that this root is much farther from the initial guess than the other root at $x = 1$, and it may not be the root you wanted from an initial guess of zero.

19.5 ROOT FINDING IN PYTHON

Unsurprisingly, Python has root-finding functions. The function we will use to find roots is `f_solve` from the `scipy.optimize`.

The `f_solve` function takes in many arguments (study the documentation for addition information), but the most important two are: (1) the function that you want to find the root and (2) the initial guess.

TRY IT! Compute the root of the function $f(x) = x^3 - 100x^2 - x + 100$ using `f_solve`.

```
In [1]: from scipy.optimize import fsolve

In [2]: f = lambda x: x**3-100*x**2-x+100

        fsolve(f, [2, 80])
```

```
Out[2]: array([  1., 100.])
```

19.6 SUMMARY AND PROBLEMS

19.6.1 SUMMARY

1. Roots are an important property of functions.
2. The bisection method is a way of finding roots based on divide-and-conquer. Although stable, it might converge slowly compared to the Newton–Raphson method.
3. The Newton–Raphson method is a different way of finding roots based on an approximation of the function. Although the Newton–Raphson method converges quickly and stops near to the actual root, it can be unstable.

19.6.2 PROBLEMS

1. Write a function `my_nth_root(x,n,tol)` where `x` and `tol` are strictly positive scalars, and `n` is an integer strictly greater than 1. The output argument, `r`, should be an approximation $r = \sqrt[N]{x}$, the Nth root of `x`. This approximation should be computed by using the Newton–Raphson method to find the root of the function $f(y) = y^N - x$. The error metric should be $|f(y)|$.

2. Write a function `my_fixed_point(f,g,tol,max_iter)` where `f` and `g` are function objects, and `tol` and `max_iter` are strictly positive scalars. The input argument, `max_iter`, is also an integer. The output argument, `X`, should be a scalar satisfying $|f(X) - g(X)| < \text{tol}$, that is, `X` is a point that (almost) satisfies $f(X) = g(X)$. To find `X`, you should use the bisection method with the error metric, $|F(m)| < \text{tol}$. The function `my_fixed_point` should "give up" after `max_iter` number of iterations and return $X = []$ if this occurs.

3. Why does the bisection method fail for $f(x) = 1/x$ with an error given by $|b - a|$? Hint: How does $f(x)$ violate the intermediate value theorem?

4. Write a function `my_bisection(f,a,b,tol)` that returns `[R,E]`, where `f` is a function object, `a` and `b` are scalars such that `a < b`, and `tol` is a strictly positive scalar value. The function should return an array, `R`, where `R[i]` is the estimation of the root of f defined by $(a+b)/2$ for the ith iteration of the bisection method. Remember to include the initial estimate. The function should also return an array, `E`, where `E[i]` is the value of $|f(R[i])|$ for the ith iteration of the bisection method. The function should terminate when `E(i) < tol`. Assume that $\text{sign}(f(a)) \neq \text{sign}(f(b))$.
   Clarification: The input `a` and `b` constitute the first iteration of bisection; therefore, `R` and `E` should never be empty.
   Test cases:

```
In: f = lambda x: x**2 - 2
    [R, E] = my_bisection(f, 0, 2, 1e-1)
Out: R = [1, 1.5, 1.25, 1.375, 1.4375]
     E = [1, 0.25, 0.4375, 0.109375, 0.06640625]
```

```
In: f = lambda x: np.sin(x) - np.cos(x)
    [R, E] = my_bisection(f, 0, 2, 1e-2)
Out: R = [1, 0.5, 0.75, 0.875, 0.8125, 0.78125]
     E = [0.30116867893975674, 0.39815702328616975,
          0.05005010885048666, 0.12654664407270177,
          0.038323093040207645, 0.005866372111545948]
```

5. Write a function `my_newton(f,df,x0,tol)` that returns `[R,E]`, where `f` is a function object, `df` is a function object giving the derivative of `f`, `x0` is an initial estimation of the root, and `tol` is a strictly positive scalar. The function should return an array, `R`, where `R[i]` is the Newton–Raphson estimate of the root of `f` for the ith iteration. Remember to include the initial estimate. The function should also return an array, `E`, where `E[i]` is the value of $|f(R[i])|$ for the ith iteration of the Newton–Raphson method. The function should terminate when `E(i) < tol`. Assume that the derivative of `f` will not hit zero during any iteration for any of the test cases given.
Test cases:

```
In: f = lambda x: x**2 - 2
    df = lambda x: 2*x
    [R, E] = my_newton(f, df, 1, 1e-5)
Out: R = [1, 1.5, 1.4166666666666667, 1.4142156862745099]
     E = [1, 0.25, 0.006944444444444642, 6.007304882871267e-06]
```

```
In: f = lambda x: np.sin(x) - np.cos(x)
    df = lambda x: np.cos(x) + np.sin(x)
    [R, E] = my_newton(f, df, 1, 1e-5)
Out: R = [1, 0.782041901539138, 0.7853981759997019]
     E = [0.30116867893975674, 0.004746462127804163,
          1.7822277875723103e-08]
```

6. Consider the problem of building a pipeline from an offshore oil platform, a distance H miles from the shoreline, to an oil refinery station on land, a distance L miles along the shore. The cost of building the pipe is $C_{ocean/mile}$ while the pipe is under the ocean, and $C_{land/mile}$ while the pipe is on land. The pipe will be built in a straight line toward the shore where it will make contact at some point, x, between 0 and L. It will continue along the shore on land until it reaches the oil refinery. See the following figure for clarification.
Write a function `my_pipe_builder(C_ocean,C_land,L,H)` where the input arguments are as described earlier, and `x` is the x-value that minimizes the total cost of the pipeline. Use the bisection method to determine this value to within a tolerance of 1×10^{-6}, starting at an initial bound of $a = 0$ and $b = L$.
Test cases:

```
In: my_pipe_builder(20, 10, 100, 50)
Out: 28.867512941360474
```

```
In: my_pipe_builder(30, 10, 100, 50)
Out: 17.677670717239380
```

```
In: my_pipe_builder(30, 10, 100, 20)
Out: 7.071067392826080
```

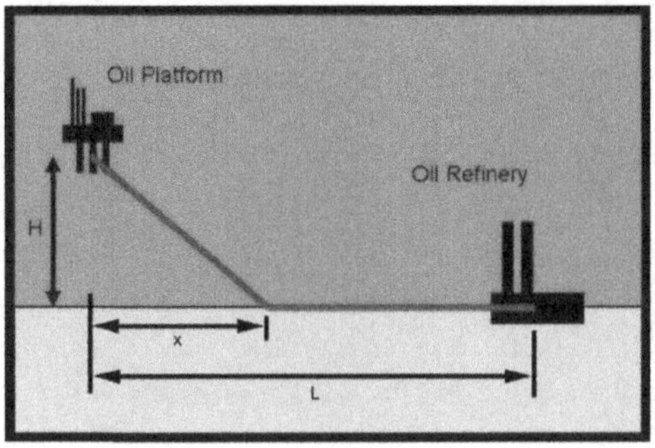

7. Find a function $f(x)$ and guess the root of f, namely x_0, such that the Newton–Raphson method will oscillate between x_0 and $-x_0$ indefinitely.

[illegible]

CHAPTER

NUMERICAL DIFFERENTIATION 20

CONTENTS

20.1 NUMERICAL DIFFERENTIATION PROBLEM STATEMENT

A **numerical grid** can be defined as an evenly spaced set of points over the domain of a function (i.e., the independent variable), over some interval. The **spacing** or **step size** of a numerical grid is the distance between adjacent points on the grid. For the purpose of this text, if x is a numerical grid, then x_j is the jth point in the numerical grid, and h is the spacing between x_{j-1} and x_j. Fig. 20.1 shows an example of a numerical grid.

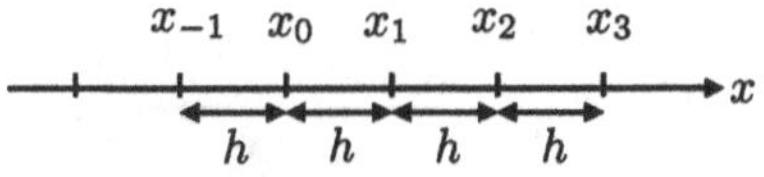

FIGURE 20.1

Numerical grid used to approximate functions.

There are several functions in Python that can be used to generate numerical grids. For numerical grids in one dimension, it is sufficient to use the `linspace` function, which you have already used for creating regularly spaced arrays.

In Python, a function $f(x)$ can be represented over an interval by computing its value on a grid. Although the function itself may be continuous, this **discrete** or **discretized** representation is useful for numerical calculations and corresponds to datasets that may be acquired in engineering and science practice. Specifically, the function value may only be known at discrete points. For example, a temperature sensor may deliver temperature versus time pairs at regular time intervals. Although temperature is a smooth function of time, the sensor only provides values at discrete time intervals; in this particular case, the underlying function would not even be known.

Python Programming and Numerical Methods. https://doi.org/10.1016/B978-0-12-819549-9.00030-0

Whether f is an analytic function or a discrete representation of one, we would like to derive methods of approximating the derivative of f over a numerical grid and determine its accuracy.

20.2 USING FINITE DIFFERENCE TO APPROXIMATE DERIVATIVES

The derivative $f'(x)$ of a function $f(x)$ at the point $x = a$ is defined as

$$f'(a) = \lim_{x\to a} \frac{f(x) - f(a)}{x - a}.$$

The derivative at $x = a$ is the slope at this point. In "finite difference" approximations of this slope, one uses values of the function in a neighborhood of the point $x = a$ to achieve the goal. There are various finite difference formulas used in different applications, and three of these, where the derivative is calculated using the values at two points, are presented below.

The **forward difference** estimates the slope of the function at x_j using the line that connects $(x_j, f(x_j))$ and $(x_{j+1}, f(x_{j+1}))$:

$$f'(x_j) = \frac{f(x_{j+1}) - f(x_j)}{x_{j+1} - x_j}.$$

The **backward difference** estimates the slope of the function at x_j using the line that connects $(x_{j-1}, f(x_{j-1}))$ and $(x_j, f(x_j))$:

$$f'(x_j) = \frac{f(x_j) - f(x_{j-1})}{x_j - x_{j-1}}.$$

The **central difference** estimates the slope of the function at x_j using the line that connects $(x_{j-1}, f(x_{j-1}))$ and $(x_{j+1}, f(x_{j+1}))$:

$$f'(x_j) = \frac{f(x_{j+1}) - f(x_{j-1})}{x_{j+1} - x_{j-1}}.$$

Fig. 20.2 illustrates the three different formulas required to estimate the slope.

20.2.1 USING FINITE DIFFERENCE TO APPROXIMATE DERIVATIVES WITH TAYLOR SERIES

To derive an approximation for the derivative of f, we return to the Taylor series. For an arbitrary function $f(x)$, the Taylor series of f around $a = x_j$ is

$$f(x) = \frac{f(x_j)(x - x_j)^0}{0!} + \frac{f'(x_j)(x - x_j)^1}{1!} + \frac{f''(x_j)(x - x_j)^2}{2!} + \frac{f'''(x_j)(x - x_j)^3}{3!} + \cdots.$$

If x is on a grid of points with spacing h, we can compute the Taylor series at $x = x_{j+1}$ to obtain

$$f(x_{j+1}) = \frac{f(x_j)(x_{j+1} - x_j)^0}{0!} + \frac{f'(x_j)(x_{j+1} - x_j)^1}{1!} + \frac{f''(x_j)(x_{j+1} - x_j)^2}{2!} + \frac{f'''(x_j)(x_{j+1} - x_j)^3}{3!} + \cdots.$$

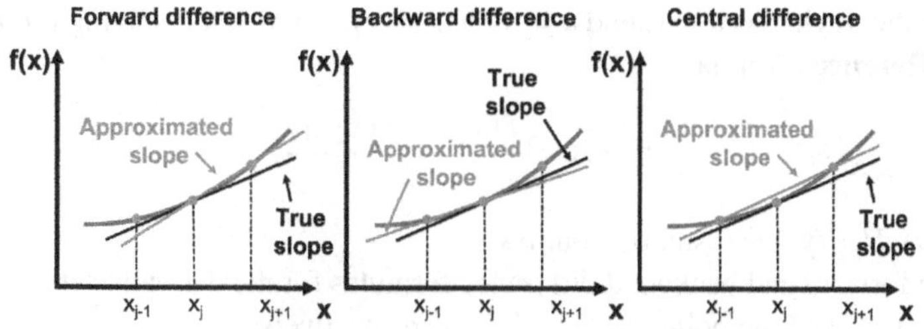

FIGURE 20.2

Finite difference approximation of the derivative.

Substituting $h = x_{j+1} - x_j$ and solving for $f'(x_j)$ gives the equation

$$f'(x_j) = \frac{f(x_{j+1}) - f(x_j)}{h} + \left(-\frac{f''(x_j)h}{2!} - \frac{f'''(x_j)h^2}{3!} - \cdots\right).$$

The terms that are in parentheses, $-\frac{f''(x_j)h}{2!} - \frac{f'''(x_j)h^2}{3!} - \cdots$, are called **higher order terms** of h. The higher order terms can be rewritten as

$$-\frac{f''(x_j)h}{2!} - \frac{f'''(x_j)h^2}{3!} - \cdots = h(\alpha + \epsilon(h)),$$

where α is some constant, and $\epsilon(h)$ is a function of h that goes to zero as h goes to zero. You can verify using algebra that this is true. We use the abbreviation "$O(h)$" for $h(\alpha + \epsilon(h))$, and in general, we use the abbreviation "$O(h^p)$" to denote $h^p(\alpha + \epsilon(h))$.

Substituting $O(h)$ into the previous equation gives

$$f'(x_j) = \frac{f(x_{j+1}) - f(x_j)}{h} + O(h).$$

This gives the **forward difference** formula for approximating derivatives as

$$f'(x_j) \approx \frac{f(x_{j+1}) - f(x_j)}{h},$$

and we denote this formula as $O(h)$.

Here, $O(h)$ describes the **accuracy** of the forward difference formula for approximating derivatives. For an approximation that is $O(h^p)$, we denote p as the **order** of the accuracy of the approximation. With few exceptions, higher-order accuracy is better than lower order. To illustrate this point, assume $q < p$. Then as the spacing, $h > 0$, goes to zero, h^p goes to zero faster than h^q. Therefore as h goes to zero, an approximation of a value that is $O(h^p)$ moves closer to the true value faster than one that is $O(h^q)$.

By computing the Taylor series around $a = x_j$ at $x = x_{j-1}$ and again solving for $f'(x_j)$, we obtain the **backward difference** formula

$$f'(x_j) \approx \frac{f(x_j) - f(x_{j-1})}{h},$$

which is also $O(h)$. Verify this result on your own.

Intuitively, the forward and backward difference formulas for the derivative at x_j are just the slopes between the point at x_j and the points x_{j+1} and x_{j-1}, respectively.

We can construct an improved approximation of the derivative by a clever manipulation of Taylor series terms taken at different points. To illustrate, we can compute the Taylor series around $a = x_j$ at both x_{j+1} and x_{j-1}. Written out, these equations are

$$f(x_{j+1}) = f(x_j) + f'(x_j)h + \frac{1}{2}f''(x_j)h^2 + \frac{1}{6}f'''(x_j)h^3 + \cdots$$

and

$$f(x_{j-1}) = f(x_j) - f'(x_j)h + \frac{1}{2}f''(x_j)h^2 - \frac{1}{6}f'''(x_j)h^3 + \cdots.$$

Subtracting the formulas above gives

$$f(x_{j+1}) - f(x_{j-1}) = 2f'(x_j)h + \frac{2}{3}f'''(x_j)h^3 + \cdots,$$

which, when solved for $f'(x_j)$, gives the **central difference** formula

$$f'(x_j) \approx \frac{f(x_{j+1}) - f(x_{j-1})}{2h}.$$

Because of how we subtracted the two equations, the h terms canceled out; therefore, the central difference formula is $O(h^2)$, even though it requires the same amount of computational effort as the forward and backward difference formulas! Thus the central difference formula gets an extra order of accuracy for free. In general, formulas that utilize symmetric points around x_j, for example, x_{j-1} and x_{j+1}, have better accuracy than asymmetric ones, such as the forward and background difference formulas.

Fig. 20.3 shows the forward difference (line joining (x_j, y_j) and (x_{j+1}, y_{j+1})), backward difference (line joining (x_j, y_j) and (x_{j-1}, y_{j-1})), and central difference (line joining (x_{j-1}, y_{j-1}) and (x_{j+1}, y_{j+1})) approximation of the derivative of a function f. As can be seen, the difference in the value of the slope can be significantly different based on the size of the step h and the nature of the function.

TRY IT! Take the Taylor series of f around $a = x_j$ and compute the series at $x = x_{j-2}, x_{j-1}, x_{j+1}, x_{j+2}$. Show that the resulting equations can be combined to form an approximation for $f'(x_j)$ which is $O(h^4)$.

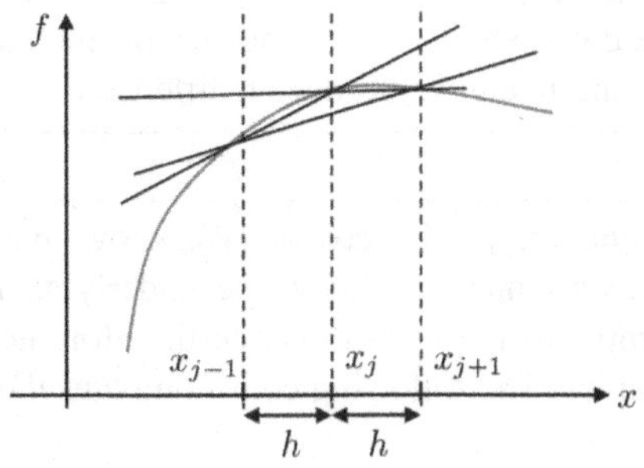

FIGURE 20.3

Illustration of the forward difference, the backward difference, and the central difference. Note the difference in slopes depending on the method used.

First, compute the Taylor series at the specified points.

$$f(x_{j-2}) = f(x_j) - 2hf'(x_j) + \frac{4h^2 f''(x_j)}{2} - \frac{8h^3 f'''(x_j)}{6} + \frac{16h^4 f''''(x_j)}{24} - \frac{32h^5 f'''''(x_j)}{120} + \cdots,$$
$$f(x_{j-1}) = f(x_j) - hf'(x_j) + \frac{h^2 f''(x_j)}{2} - \frac{h^3 f'''(x_j)}{6} + \frac{h^4 f''''(x_j)}{24} - \frac{h^5 f'''''(x_j)}{120} + \cdots,$$
$$f(x_{j+1}) = f(x_j) + hf'(x_j) + \frac{h^2 f''(x_j)}{2} + \frac{h^3 f'''(x_j)}{6} + \frac{h^4 f''''(x_j)}{24} + \frac{h^5 f'''''(x_j)}{120} + \cdots,$$
$$f(x_{j+2}) = f(x_j) + 2hf'(x_j) + \frac{4h^2 f''(x_j)}{2} + \frac{8h^3 f'''(x_j)}{6} + \frac{16h^4 f''''(x_j)}{24} + \frac{32h^5 f'''''(x_j)}{120} + \cdots.$$

To make sure the h^2, h^3, and h^4 terms cancel out, compute

$$f(x_{j-2}) - 8f(x_{j-1}) + 8f(x_{j-1}) - f(x_{j+2}) = 12hf'(x_j) - \frac{48h^5 f'''''(x_j)}{120},$$

which can be rearranged to

$$f'(x_j) = \frac{f(x_{j-2}) - 8f(x_{j-1}) + 8f(x_{j-1}) - f(x_{j+2})}{12h} + O(h^4).$$

This formula is a better approximation for the derivative at x_j than the central difference formula, but requires twice as many calculations.

TIP! Python has a command that can be used to compute finite differences directly: for a vector f, the command `d=np.diff(f)` produces an array `d` in which the entries are the differences of the adjacent elements in the initial array f. In other words, $d(i) = f(i+1) - f(i)$.

WARNING! When using the command `np.diff`, the size of the output is one less than the size of the input since it needs two arguments to produce a difference.

EXAMPLE: Consider the function $f(x) = \cos(x)$. We know that the derivative of $\cos(x)$ is $-\sin(x)$. Although in practice we may not know the underlying function we are finding the derivative for, we use the simple example to illustrate the aforementioned numerical differentiation methods and their accuracy. The following code computes the derivatives numerically.

```
In [1]: import numpy as np
        import matplotlib.pyplot as plt
        plt.style.use("seaborn-poster")
        %matplotlib inline

In [2]: # step size
        h = 0.1
        # define grid
        x = np.arange(0, 2*np.pi, h)
        # compute function
        y = np.cos(x)

        # compute vector of forward differences
        forward_diff = np.diff(y)/h
        # compute corresponding grid
        x_diff = x[:-1:]
        # compute exact solution
        exact_solution = -np.sin(x_diff)

        # Plot solution
        plt.figure(figsize = (12, 8))
        plt.plot(x_diff, forward_diff, "-", \
                 label = "Finite difference approximation")
        plt.plot(x_diff, exact_solution, label = "Exact solution")
        plt.legend()
        plt.show()

        # Compute max error between
        # numerical derivative and exact solution
        max_error = max(abs(exact_solution - forward_diff))
        print(max_error)
```

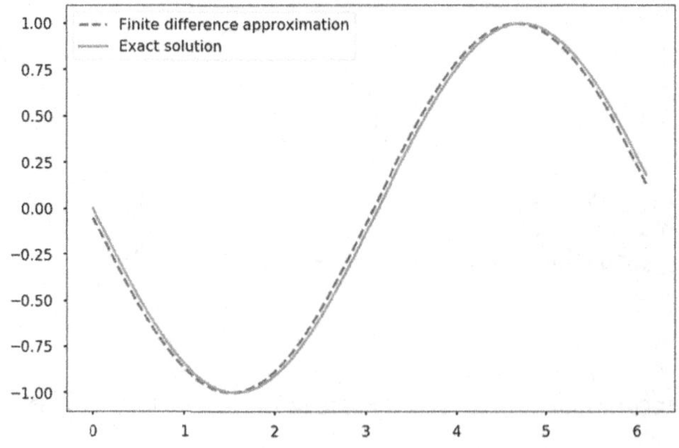


```
0.049984407218554114
```

As the above figure shows, there is a small offset between the two curves, which results from the numerical error in the evaluation of the numerical derivatives. The maximal error between the two numerical results is of the order 0.05 and expected to decrease with the size of the step.

As illustrated in the previous example, the finite difference scheme contains a numerical error due to the approximation of the derivative. This difference decreases with the size of the discretization step, which is illustrated in the following example.

EXAMPLE: The following code computes the numerical derivative of $f(x) = \cos(x)$ using the forward-difference formula for decreasing step size, h. It then plots the maximum error between the approximated derivative and the true derivative versus h as shown in the generated figure.

```
In [3]: # define step size
        h = 1
        # define number of iterations to perform
        iterations = 20
        # list to store our step sizes
        step_size = []
        # list to store max error for each step size
        max_error = []

        for i in range(iterations):
            # halve the step size
            h /= 2
            # store this step size
            step_size.append(h)
            # compute new grid
            x = np.arange(0, 2 * np.pi, h)
```

```
        # compute function value at grid
        y = np.cos(x)
        # compute vector of forward differences
        forward_diff = np.diff(y)/h
        # compute corresponding grid
        x_diff = x[:-1]
        # compute exact solution
        exact_solution = -np.sin(x_diff)

        # Compute max error between
        # numerical derivative and exact solution
        max_error.append(max(abs(exact_solution - forward_diff)))

    # produce log-log plot of max error versus step size
    plt.figure(figsize = (12, 8))
    plt.loglog(step_size, max_error, "v")
    plt.show()
```

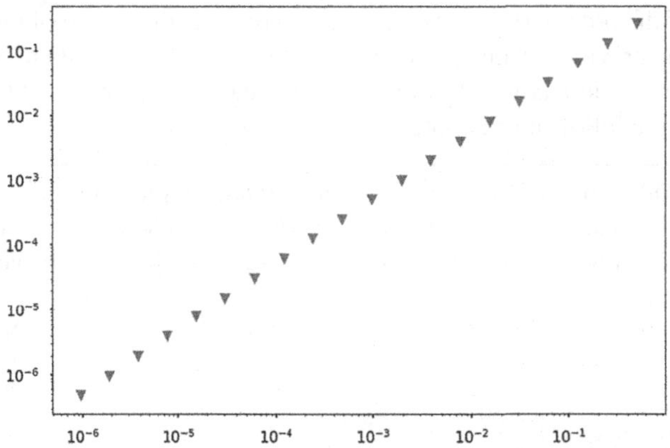

The slope of the line in log–log space is 1; therefore, the error is proportional to h^1, which means that, as expected, the forward-difference formula is $O(h)$.

20.3 APPROXIMATING OF HIGHER ORDER DERIVATIVES

It also possible to use the Taylor series to approximate higher-order derivatives (e.g., $f''(x_j)$, $f'''(x_j)$, etc.). For example, taking the Taylor series around $a = x_j$ and then computing it at $x = x_{j-1}$ and x_{j+1} gives

$$f(x_{j-1}) = f(x_j) - hf'(x_j) + \frac{h^2 f''(x_j)}{2} - \frac{h^3 f'''(x_j)}{6} + \cdots$$

and

$$f(x_{j+1}) = f(x_j) + hf'(x_j) + \frac{h^2 f''(x_j)}{2} + \frac{h^3 f'''(x_j)}{6} + \cdots.$$

If we add these two equations together, we get

$$f(x_{j-1}) + f(x_{j+1}) = 2f(x_j) + h^2 f''(x_j) + \frac{h^4 f''''(x_j)}{24} + \cdots,$$

and, with some rearrangement, this gives the approximation

$$f''(x_j) \approx \frac{f(x_{j+1}) - 2f(x_j) + f(x_{j-1})}{h^2},$$

which is $O(h^2)$.

20.4 NUMERICAL DIFFERENTIATION WITH NOISE

As stated earlier, sometimes f is given as a vector where f is the corresponding function value for independent data values in another vector x, which is gridded. Sometimes data can be contaminated with **noise**, meaning its value is off by a small amount from what it would be if it were computed from a pure mathematical function. This can often occur in engineering due to inaccuracies in measurement devices or the data itself can be slightly modified by perturbations outside the system of interest. For example, you may be trying to listen to your friend talk in a crowded room. The signal f might be the intensity and tonal values in your friend's speech; however, because the room is crowded, noise from other conversations is heard along with your friend's speech, and he becomes difficult to understand.

To illustrate this point, we numerically compute the derivative of a simple cosine wave corrupted by a small sin wave. Consider the following two functions:

$$f(x) = \cos(x)$$

and

$$f_{\epsilon,\omega}(x) = \cos(x) + \epsilon \sin(\omega x)$$

where $0 < \epsilon \ll 1$ is a very small number, and ω is a large number. When ϵ is small, it is clear that $f \simeq f_{\epsilon,\omega}$. To illustrate this point, we plot $f_{\epsilon,\omega}(x)$ for $\epsilon = 0.01$ and $\omega = 100$, and we can see it is very close to $f(x)$, as shown in the following figure.

```
In [1]: import numpy as np
        import matplotlib.pyplot as plt
        plt.style.use("seaborn-poster")
        %matplotlib inline
```

```
In [2]: x = np.arange(0, 2*np.pi, 0.01)
        # compute function
        omega = 100
        epsilon = 0.01

        y = np.cos(x)
        y_noise = y + epsilon*np.sin(omega*x)

        # Plot solution
        plt.figure(figsize = (12, 8))
        plt.plot(x, y_noise, "r-", label = "cos(x) + noise")
        plt.plot(x, y, "b-", label = "cos(x)")

        plt.xlabel("x")
        plt.ylabel("y")

        plt.legend()
        plt.show()
```

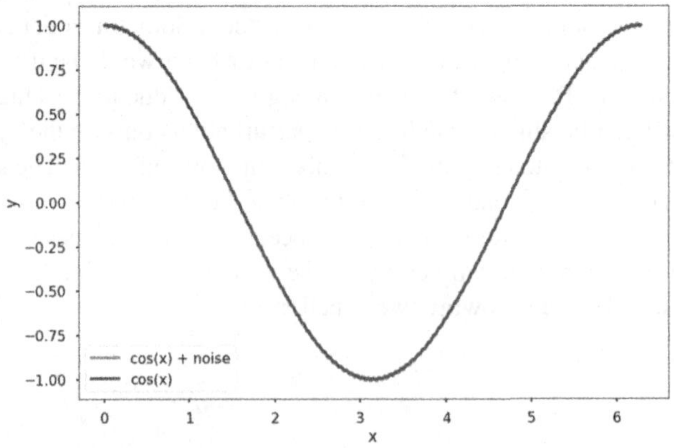

The derivatives of our two test functions are

$$f'(x) = -\sin(x)$$

and

$$f'_{\epsilon,\omega}(x) = -\sin(x) + \epsilon\omega\cos(\omega x).$$

Since $\epsilon\omega$ may not be small when ω is large, the contribution of the noise to the derivative may not be small. As a result, the derivative (analytic and numerical) may not be usable. For instance, the following figure shows $f'(x)$ and $f'_{\epsilon,\omega}(x)$ for $\epsilon = 0.01$ and $\omega = 100$.

```
In [3]: x = np.arange(0, 2*np.pi, 0.01)
        # compute function
        y = -np.sin(x)
        y_noise = y + epsilon*omega*np.cos(omega*x)

        # Plot solution
        plt.figure(figsize = (12, 8))
        plt.plot(x, y_noise, "r-", label = "Derivative cos(x) + noise")
        plt.plot(x, y, "b-", label = "Derivative of cos(x)")

        plt.xlabel("x")
        plt.ylabel("y")

        plt.legend()
        plt.show()
```

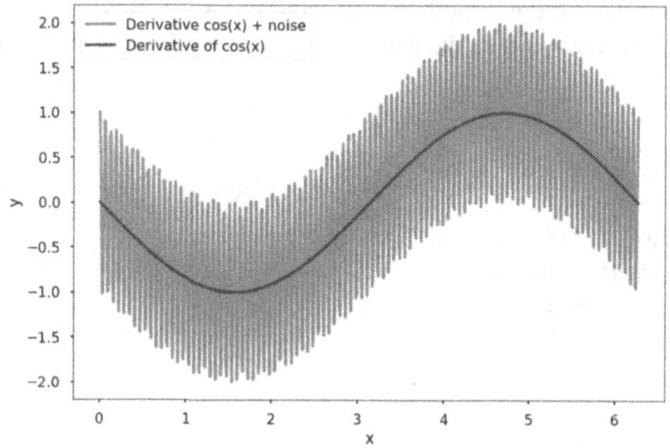

20.5 SUMMARY AND PROBLEMS

20.5.1 SUMMARY

1. Because explicit derivation of functions is sometimes cumbersome for engineering applications, numerical approaches are preferable.
2. Numerical approximation of derivatives can be done using a grid on which the derivative is approximated by finite differences.
3. Finite differences approximate the derivative by ratios of differences of the function values over small intervals.
4. Finite difference schemes have different approximation orders depending on the method used.
5. When the data is noisy, there are issues using finite differences for approximating derivatives.

20.5.2 PROBLEMS

1. Write a function `my_der_calc(f, a, b, N, option)` with the output as `[df,X]`, where `f` is a function object, `a` and `b` are scalars such that `a < b`, `N` is an integer bigger than 10, and `option` is the string `"forward"`, `"backward"`, or `"central"`. Let `x` be an array starting at `a`, ending at `b`, containing `N` evenly spaced elements, and let `y` be the array `f(x)`. The output argument, `df`, should be the numerical derivatives computed for `x` and `y` according to the method defined by the input argument, `option`. The output argument `X` should be an array the same size as `df`, containing the points in `x` for which `df` is valid. Specifically, the forward difference method "loses" the last point, the backward difference method loses the first point, and the central difference method loses the first and last points.

2. Write a function `my_num_diff(f,a,b,n,option)` with the output as `[df,X]`, where `f` is a function object. The function `my_num_diff` should compute the derivative of `f` numerically for `n` evenly spaced points starting at `a` and ending at `b`, according to the method defined by `option`. The input argument `option` is one of the following strings: `"forward"`, `"backward"`, and `"central"`. Note that for the forward and backward method, the output argument, `dy`, should be a 1D array of length $n-1$, and for the central difference method `dy` it should be a 1D array of length $n-2$. The function should also output a vector `X` that is the same size as `dy` and denotes the x-values for which `dy` is valid.
   Test cases:

```
x = np.linspace(0, 2*np.pi, 100)
f = lambda x: np.sin(x)
[dyf, Xf] = my_num_diff(f, 0, 2*np.pi, 10, "forward")
[dyb, Xb] = my_num_diff(f, 0, 2*np.pi, 10, "backward")
[dyc, Xc] = my_num_diff(f, 0, 2*np.pi, 10, "central")
plt.figure(figsize = (12, 8))
plt.plot(x, np.cos(x), label = "analytic")
plt.plot(Xf, dyf, label = "forward")
plt.plot(Xb, dyb, label = "backward")
plt.plot(Xc, dyc, label = "central")
plt.legend()
plt.title("Analytic and Numerical Derivatives of Sine")
plt.xlabel("x")
plt.ylabel("y")
plt.show()
```

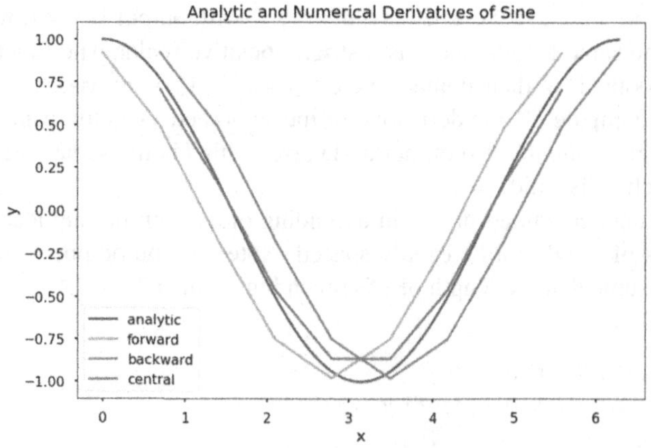


```
x = np.linspace(0, np.pi, 1000)
f = lambda x: np.sin(np.exp(x))
[dy10, X10] = my_num_diff(f, 0, np.pi, 10, "central")
[dy20, X20] = my_num_diff(f, 0, np.pi, 20, "central")
[dy100, X100] = my_num_diff(f, 0, np.pi, 100, "central")
plt.figure(figsize = (12, 8))
plt.plot(x, np.cos(np.exp(x)), label = "analytic")
plt.plot(X10, dy10, label = "10 points")
plt.plot(X20, dy20, label = "20 points")
plt.plot(X100, dy100, label = "100 points")
plt.legend()
plt.title("Analytic and Numerical Derivatives of Sine")
plt.xlabel("x")
plt.ylabel("y")
plt.show()
```

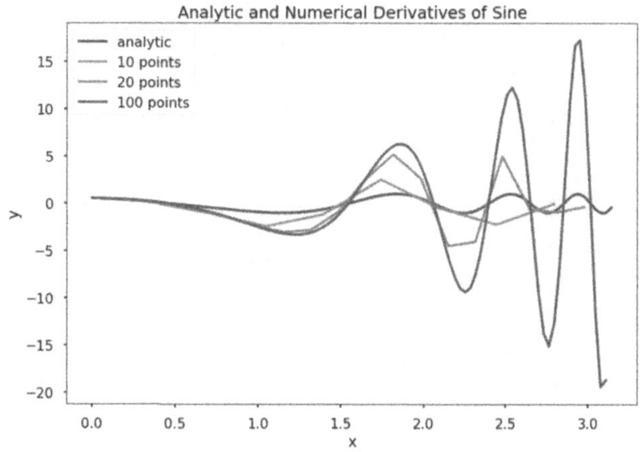

3. Write a function `my_num_diff_w_smoothing(x,y,n)` with output `[dy,X]`, where `x` and `y` are a 1D `NumPy` array of the same length, and `n` is a strictly positive scalar. The function should first create a vector of "smoothed" `y` data points where `y_smooth[i] = np.mean(y[i-n:i+n])`. The function should then compute `dy`, the derivative of the smoothed y-vector, using the central difference method. The function should also output a 1D array `X` that is the same size as `dy` and denotes the x-values for which `dy` is valid.
Assume that the data contained in `x` is in ascending order with no duplicate entries; it is possible that the elements of `x` will not be evenly spaced. Note that the output `dy` will have $2n+2$ fewer points than `y`. Assume that the length of `y` is much bigger than $2n+2$.
Test cases:

```
x = np.linspace(0, 2*np.pi, 100)
y = np.sin(x) + np.random.randn(len(x))/100
[dy, X] = my_num_diff_w_smoothing(x, y, 4)
plt.figure(figsize = (12, 12))
plt.subplot(211)
plt.plot(x, y)
plt.title("Noisy Sine function")
plt.xlabel("x")
plt.ylabel("y")
plt.subplot(212)
plt.plot(x, np.cos(x), "b", label = "cosine")
plt.plot(x[:-1], (y[1:] - y[:-1])/(x[1]-x[0]), "g", \
    label = "unsmoothed forward diff")
plt.plot(X, dy, "r", label = "smoothed")
plt.title("Analytic Derivative and Smoothed Derivative")
plt.xlabel("x")
plt.ylabel("y")
plt.legend()
plt.tight_layout()
plt.show()
```

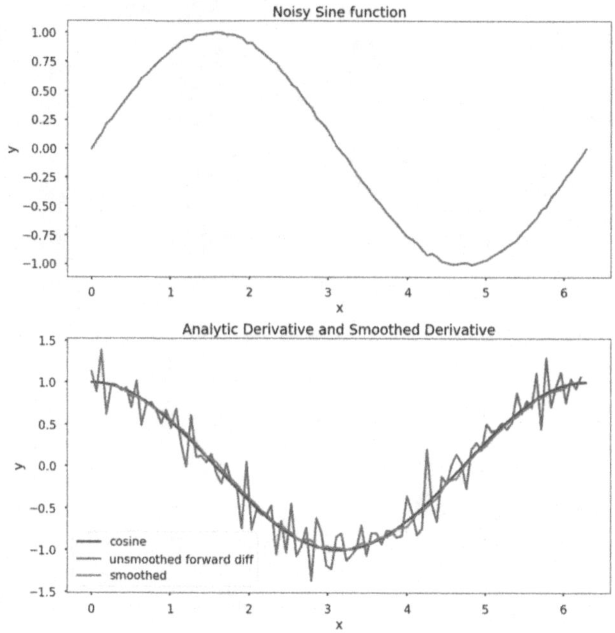

4. Use Taylor series to show the following approximations and their accuracy:

$$f''(x_j) = \frac{-f(x_{j+3}) + 4f(x_{j+2}) - 5f(x_{j+1}) + 2f(x_j)}{h^2} + O(h^2),$$

$$f'''(x_j) = \frac{f(x_{j+3}) - 3f(x_{j+2}) + 3f(x_{j+1}) - f(x_j)}{h^3} + O(h).$$

CHAPTER

NUMERICAL INTEGRATION 21

CONTENTS

21.1 NUMERICAL INTEGRATION PROBLEM STATEMENT

Given a function $f(x)$, approximate the integral of $f(x)$ over the total **interval**, $[a, b]$. Fig. 21.1 illustrates this area. To accomplish this goal, we assume that the interval has been discretized into a numeral grid, x, consisting of $n + 1$ points with spacing, $h = \frac{b-a}{n}$. Here, we denote each point in x by x_i, where $x_0 = a$ and $x_n = b$. Note that there are $n + 1$ grid points because the count starts at x_0. Assume that the function, $f(x)$, can be computed for any of the grid points, or that we have been given the function implicitly as $f(x_i)$. The interval $[x_i, x_{i+1}]$ is referred to as a **subinterval**.

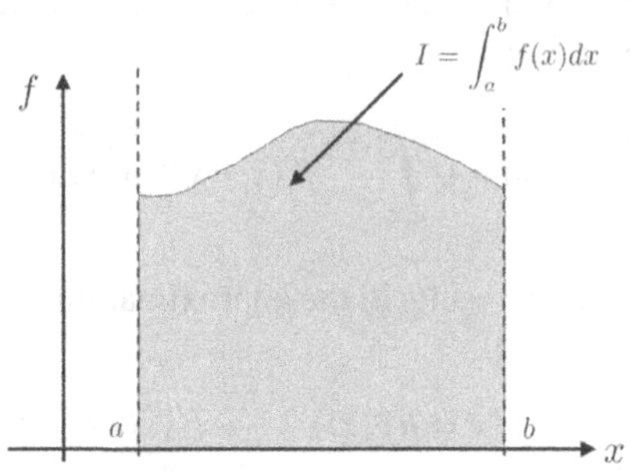

FIGURE 21.1

Illustration of the integral. The integral from a to b of the function f is the area below the curve (shaded in grey).

The following sections give some of the most common entry level methods of approximating $\int_a^b f(x)dx$. Each method approximates the area under $f(x)$ for each subinterval by a shape for which it is easy to compute the exact area, and then sums the area contributions of every subinterval.

Python Programming and Numerical Methods. https://doi.org/10.1016/B978-0-12-819549-9.00031-2

21.2 RIEMANN INTEGRAL

The simplest method for approximating integrals is by summing the area of rectangles that are defined for each subinterval. The width of the rectangle is $x_{i+1} - x_i = h$, and the height is defined by a function value $f(x)$ for some x in the subinterval. An obvious choice for the height is the function value at the left endpoint, x_i, or the right endpoint, x_{i+1}, because these values can be used even if the function itself is not known. This method gives the **Riemann integral** approximation, which is

$$\int_a^b f(x)dx \approx \sum_{i=0}^{n-1} hf(x_i),$$

or

$$\int_a^b f(x)dx \approx \sum_{i=1}^{n} hf(x_i),$$

depending on whether the left or right endpoint is chosen.

As with numerical differentiation, we want to characterize how the accuracy improves as h decreases. To determine this characterizing, we first rewrite the integral of $f(x)$ over an arbitrary subinterval in terms of the Taylor series. The Taylor series of $f(x)$ around $a = x_i$ is

$$f(x) = f(x_i) + f'(x_i)(x - x_i) + \cdots .$$

Thus

$$\int_{x_i}^{x_{i+1}} f(x)dx = \int_{x_i}^{x_{i+1}} (f(x_i) + f'(x_i)(x - x_i) + \cdots)\, dx$$

by substitution of the Taylor series for the function. Since the integral distributes, we can rearrange the right-hand side into the following form:

$$\int_{x_i}^{x_{i+1}} f(x_i)dx + \int_{x_i}^{x_{i+1}} f'(x_i)(x - x_i)dx + \cdots .$$

Computing each integral separately results in the approximation

$$\int_{x_i}^{x_{i+1}} f(x)dx = hf(x_i) + \frac{h^2}{2} f'(x_i) + O(h^3),$$

which is

$$\int_{x_i}^{x_{i+1}} f(x)dx = hf(x_i) + O(h^2).$$

Since the $hf(x_i)$ term is our Riemann integral approximation for a single subinterval, the Riemann integral approximation over a single interval is $O(h^2)$.

If we sum the $O(h^2)$ error over the entire Riemann sum, we obtain $nO(h^2)$. The relationship between n and h is

$$h = \frac{b-a}{n},$$

and so our total error becomes $\frac{b-a}{h}O(h^2) = O(h)$ over the whole interval. Thus the overall accuracy is $O(h)$.

The **Midpoint Rule** takes the rectangle height of the rectangle at each subinterval to be the function value at the midpoint between x_i and x_{i+1}, which for compactness we denote by $y_i = \frac{x_{i+1}+x_i}{2}$. The midpoint rule says

$$\int_a^b f(x)dx \approx \sum_{i=0}^{n-1} hf(y_i).$$

Similarly to the Riemann integral, we take the Taylor series of $f(x)$ around y_i, which is

$$f(x) = f(y_i) + f'(y_i)(x-y_i) + \frac{f''(y_i)(x-y_i)^2}{2!} + \cdots.$$

Then the integral over a subinterval is

$$\int_{x_i}^{x_{i+1}} f(x)dx = \int_{x_i}^{x_{i+1}} \left(f(y_i) + f'(y_i)(x-y_i) + \frac{f''(y_i)(x-y_i)^2}{2!} + \cdots \right) dx,$$

which distributes to

$$\int_{x_i}^{x_{i+1}} f(x)dx = \int_{x_i}^{x_{i+1}} f(y_i)dx + \int_{x_i}^{x_{i+1}} f'(y_i)(x-y_i)dx + \int_{x_i}^{x_{i+1}} \frac{f''(y_i)(x-y_i)^2}{2!}dx + \cdots.$$

Recognizing that x_i and x_{i+1} are symmetric around y_i, we get $\int_{x_i}^{x_{i+1}} f'(y_i)(x-y_i)dx = 0$. This is true for the integral of $(x-y_i)^p$ for any odd p. For the integral of $(x-y_i)^p$ and with p even, $\int_{x_i}^{x_{i+1}} (x-y_i)^p dx = \int_{-\frac{h}{2}}^{\frac{h}{2}} x^p dx$, which will result in some multiple of h^{p+1} with no lower order powers of h.

Utilizing these facts reduces the expression for the integral of $f(x)$ to

$$\int_{x_i}^{x_{i+1}} f(x)dx = hf(y_i) + O(h^3).$$

Since $hf(y_i)$ is the approximation of the integral over the subinterval, the midpoint rule is $O(h^3)$ for one subinterval; using similar arguments as those used for the Riemann integral, we get $O(h^2)$ over the whole interval. Since the midpoint rule requires the same number of calculations as the Riemann integral, we essentially get an extra order of accuracy for free; however, if $f(x_i)$ is given in the form of data points, then we will not be able to compute $f(y_i)$ for this integration scheme.

TRY IT! Use the left and right Riemann integral, as well as midpoint rule, to approximate $\int_0^{\pi} \sin(x)dx$ with 11 evenly spaced grid points over the whole interval. Compare this value to the exact value of 2.

```
In [1]: import numpy as np

        a = 0
        b = np.pi
        n = 11
        h = (b - a) / (n - 1)
        x = np.linspace(a, b, n)
        f = np.sin(x)

        I_riemannL = h * sum(f[:n-1])
        err_riemannL = 2 - I_riemannL

        I_riemannR = h * sum(f[1::])
        err_riemannR = 2 - I_riemannR

        I_mid = h * sum(np.sin((x[:n-1] + x[1:])/2))
        err_mid = 2 - I_mid

        print(I_riemannL)
        print(err_riemannL)

        print(I_riemannR)
        print(err_riemannR)

        print(I_mid)
        print(err_mid)

1.9835235375094546
0.01647646249054535
1.9835235375094546
0.01647646249054535
2.0082484079079745
-0.008248407907974542
```

21.3 TRAPEZOID RULE

The **Trapezoid Rule** fits a trapezoid into each subinterval and sums the areas of the trapezoids to approximate the total integral. This approximation for the integral to an arbitrary function is shown in Fig. 21.2. For each subinterval, the trapezoid rule computes the area of a trapezoid with corners at $(x_i, 0)$, $(x_{i+1}, 0)$, $(x_i, f(x_i))$, and $(x_{i+1}, f(x_{i+1}))$, which is $h\frac{f(x_i)+f(x_{i+1})}{2}$. Thus, the trapezoid rule approximates integrals according to the expression

$$\int_a^b f(x)dx \approx \sum_{i=0}^{n-1} h\frac{f(x_i) + f(x_{i+1})}{2}.$$

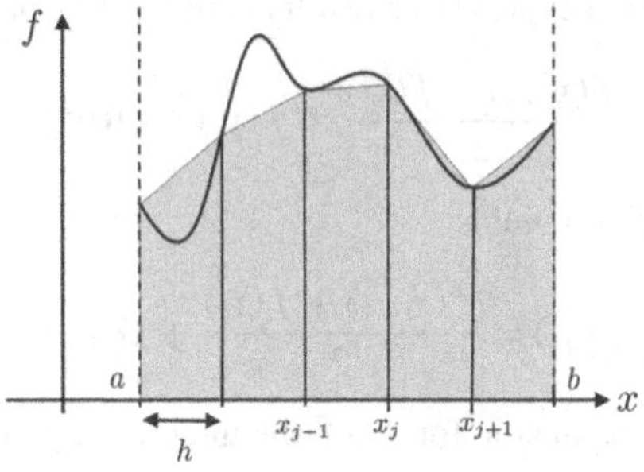

FIGURE 21.2

Illustration of the trapezoid integral procedure. The area below the curve is approximated by a sum of areas of trapezoids that approximate the function.

TRY IT! You may notice that the trapezoid rule "double-counts" most of the terms in the series. To illustrate this fact, consider the expansion of the trapezoid rule:

$$\begin{aligned}\sum_{i=0}^{n-1} h\frac{f(x_i)+f(x_{i+1})}{2} &= \frac{h}{2}\big[(f(x_0) + f(x_1)) + (f(x_1) + f(x_2)) + (f(x_2) \\ &\quad + f(x_3)) + \cdots + (f(x_{n-1}) + f(x_n))\big].\end{aligned}$$

Computationally, this is many extra additions and calls to $f(x)$ than are really necessary. We can be made more computationally efficient using the following expression:

$$\int_a^b f(x)dx \approx \frac{h}{2}\left(f(x_0) + 2\sum_{i=1}^{n-1} f(x_i) + f(x_n)\right).$$

To determine the accuracy of the trapezoid rule approximation, we first take Taylor series expansion of $f(x)$ around $y_i = \frac{x_{i+1}+x_i}{2}$, which is the midpoint between x_i and x_{i+1}. This Taylor series expansion is

$$f(x) = f(y_i) + f'(y_i)(x - y_i) + \frac{f''(y_i)(x - y_i)^2}{2!} + \cdots.$$

Computing the Taylor series at x_i and x_{i+1} and noting that $x_i - y_i = -\frac{h}{2}$ and $x_{i+1} - y_i = \frac{h}{2}$ results in the following expressions:

$$f(x_i) = f(y_i) - \frac{hf'(y_i)}{2} + \frac{h^2 f''(y_i)}{8} - \cdots$$

and

$$f(x_{i+1}) = f(y_i) + \frac{hf'(y_i)}{2} + \frac{h^2 f''(y_i)}{8} + \cdots.$$

Taking the average of these two expressions results in the new expression,

$$\frac{f(x_{i+1}) + f(x_i)}{2} = f(y_i) + O(h^2).$$

Solving this expression for $f(y_i)$ yields

$$f(y_i) = \frac{f(x_{i+1}) + f(x_i)}{2} + O(h^2).$$

Now returning to the Taylor expansion for $f(x)$, the integral of $f(x)$ over a subinterval is

$$\int_{x_i}^{x_{i+1}} f(x)dx = \int_{x_i}^{x_{i+1}} \left(f(y_i) + f'(y_i)(x - y_i) + \frac{f''(y_i)(x - y_i)^2}{2!} + \cdots \right) dx.$$

Distributing the integral results in the expression

$$\int_{x_i}^{x_{i+1}} f(x)dx = \int_{x_i}^{x_{i+1}} f(y_i)dx + \int_{x_i}^{x_{i+1}} f'(y_i)(x - y_i)dx + \int_{x_i}^{x_{i+1}} \frac{f''(y_i)(x - y_i)^2}{2!} dx + \cdots.$$

Since x_i and x_{i+1} are symmetric around y_i, the integrals of the odd powers of $(x - y_i)^p$ disappear, and the even powers resolve to a multiple h^{p+1}:

$$\int_{x_i}^{x_{i+1}} f(x)dx = hf(y_i) + O(h^3).$$

If we substitute $f(y_i)$ with the expression derived explicitly in terms of $f(x_i)$ and $f(x_{i+1})$, we obtain

$$\int_{x_i}^{x_{i+1}} f(x)dx = h\left(\frac{f(x_{i+1}) + f(x_i)}{2} + O(h^2)\right) + O(h^3),$$

which is equivalent to

$$h\left(\frac{f(x_{i+1}) + f(x_i)}{2}\right) + hO(h^2) + O(h^3)$$

and

$$\int_{x_i}^{x_{i+1}} f(x)dx = h\left(\frac{f(x_{i+1}) + f(x_i)}{2}\right) + O(h^3).$$

Since $\frac{h}{2}(f(x_{i+1}) + f(x_i))$ is the trapezoid rule approximation for the integral over the subinterval, it is $O(h^3)$ for a single subinterval and $O(h^2)$ over the whole interval.

TRY IT! Use the trapezoid rule to approximate $\int_0^{\pi} \sin(x)dx$ with 11 evenly spaced grid points over the whole interval. Compare this value to the exact value of 2.

```
In [1]: import numpy as np

        a = 0
        b = np.pi
        n = 11
        h = (b - a) / (n - 1)
        x = np.linspace(a, b, n)
        f = np.sin(x)

        I_trap = (h/2)*(f[0] + 2 * sum(f[1:n-1]) + f[n-1])
        err_trap = 2 - I_trap

        print(I_trap)
        print(err_trap)

1.9835235375094546
0.01647646249054535
```

21.4 SIMPSON'S RULE

Consider *two* consecutive subintervals, $[x_{i-1}, x_i]$ and $[x_i, x_{i+1}]$. **Simpson's Rule** approximates the area under $f(x)$ over these two subintervals by fitting a quadratic polynomial through the points $(x_{i-1}, f(x_{i-1}))$, $(x_i, f(x_i))$, and $(x_{i+1}, f(x_{i+1}))$, which is a unique polynomial, and then integrates the quadratic exactly. Fig. 21.3 shows this integral approximation for an arbitrary function.

First, we construct the quadratic polynomial approximation of the function over the two subintervals. The easiest way to do this is to use Lagrange polynomials, which was discussed in Chapter 17. By applying the formula for constructing Lagrange polynomials, we obtain

$$P_i(x) = f(x_{i-1})\frac{(x - x_i)(x - x_{i+1})}{(x_{i-1} - x_i)(x_{i-1} - x_{i+1})} + f(x_i)\frac{(x - x_{i-1})(x - x_{i+1})}{(x_i - x_{i-1})(x_i - x_{i+1})} + f(x_{i+1})\frac{(x - x_{i-1})(x - x_i)}{(x_{i+1} - x_{i-1})(x_{i+1} - x_i)},$$

and with substitutions for h results in

$$P_i(x) = \frac{f(x_{i-1})}{2h^2}(x - x_i)(x - x_{i+1}) - \frac{f(x_i)}{h^2}(x - x_{i-1})(x - x_{i+1}) + \frac{f(x_{i+1})}{2h^2}(x - x_{i-1})(x - x_i).$$

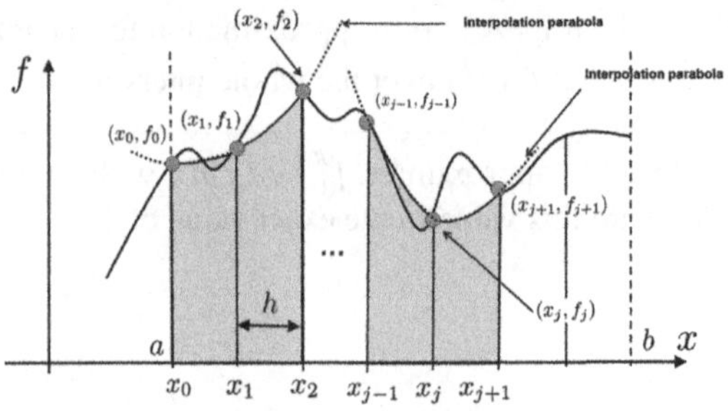

FIGURE 21.3

Illustration of the Simpson integral formula. Discretization points are grouped by three, and a parabola is fit between the three points. This can be done by a typical interpolation polynomial. The area under the curve is approximated by the area under the parabola.

You can confirm that the polynomial curve intersects the desired points. With some algebra and manipulation, the integral of $P_i(x)$ over the two subintervals is

$$\int_{x_{i-1}}^{x_{i+1}} P_i(x)dx = \frac{h}{3}(f(x_{i-1}) + 4f(x_i) + f(x_{i+1})).$$

To approximate the integral over (a, b), we must sum the integrals of $P_i(x)$ over all *pairs* of subintervals since $P_i(x)$ spans two subintervals. Substituting $\frac{h}{3}(f(x_{i-1}) + 4f(x_i) + f(x_{i+1}))$ for the integral of $P_i(x)$ and regrouping the terms for efficiency leads to the formula

$$\int_a^b f(x)dx \approx \frac{h}{3}\left[f(x_0) + 4\sum_{i=1,\, i \text{ odd}}^{n-1} f(x_i) + 2\sum_{i=2,\, i \text{ even}}^{n-2} f(x_i) + f(x_n)\right].$$

This regrouping is illustrated in Fig. 21.4:

> **WARNING!** Note that to use Simpson's rule, you **must** have an even number of intervals and, therefore, an odd number of grid points.

To compute the accuracy of the Simpson's rule, we take the Taylor series approximation of $f(x)$ as around x_i, which is

$$f(x) = f(x_i) + f'(x_i)(x - x_i) + \frac{f''(x_i)(x - x_i)^2}{2!} + \frac{f'''(x_i)(x - x_i)^3}{3!} + \frac{f''''(x_i)(x - x_i)^4}{4!} + \cdots.$$

Computing the Taylor series with x_{i-1} and x_{i+1}, and substituting for h where appropriate, gives the expressions

$$f(x_{i-1}) = f(x_i) - hf'(x_i) + \frac{h^2 f''(x_i)}{2!} - \frac{h^3 f'''(x_i)}{3!} + \frac{h^4 f''''(x_i)}{4!} - \cdots$$

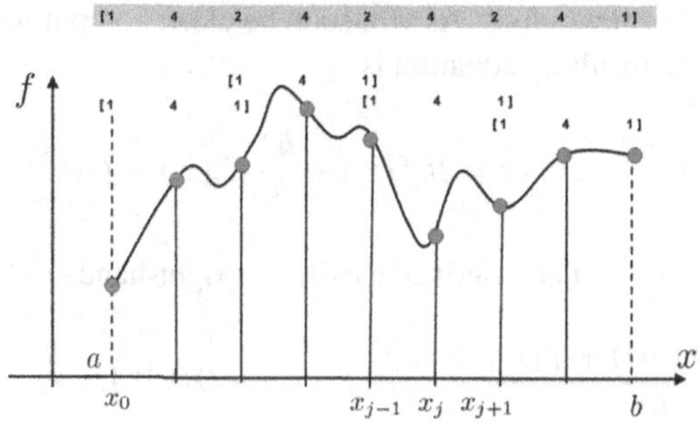

FIGURE 21.4

Illustration of the accounting procedure to approximate the function f by the Simpson rule for the entire interval $[a, b]$.

and

$$f(x_{i+1}) = f(x_i) + hf'(x_i) + \frac{h''(x_i)}{2!} + \frac{h^3 f'''(x_i)}{3!} + \frac{h^4 f''''(x_i)}{4!} + \cdots.$$

Consider the expression $\frac{f(x_{i-1})+4f(x_i)+f(x_{i+1})}{6}$. Substituting the Taylor series for the respective numerator values produces the equation

$$\frac{f(x_{i-1}) + 4f(x_i) + f(x_{i+1})}{6} = f(x_i) + \frac{h^2}{6} f''(x_i) + \frac{h^4}{72} f''''(x_i) + \cdots.$$

Note that the odd terms cancel out, implying

$$f(x_i) = \frac{f(x_{i-1}) + 4f(x_i) + f(x_{i+1})}{6} - \frac{h^2}{6} f''(x_i) + O(h^4).$$

By substitution of the Taylor series for $f(x)$, the integral of $f(x)$ over two subintervals is then

$$\int_{x_{i-1}}^{x_{i+1}} f(x)dx = \int_{x_{i-1}}^{x_{i+1}} \left(f(x_i) + f'(x_i)(x - x_i) + \frac{f''(x_i)(x - x_i)^2}{2!} + \frac{f'''(x_i)(x - x_i)^3}{3!} + \frac{f''''(x_i)(x - x_i)^4}{4!} + \cdots \right) dx.$$

Again, we distribute the integral and, without showing it, drop the integrals of terms with odd powers because they are zero to obtain

$$\int_{x_{i-1}}^{x_{i+1}} f(x)dx = \int_{x_{i-1}}^{x_{i+1}} f(x_i)dx + \int_{x_{i-1}}^{x_{i+1}} \frac{f''(x_i)(x - x_i)^2}{2!} dx + \int_{x_{i-1}}^{x_{i+1}} \frac{f''''(x_i)(x - x_i)^4}{4!} dx + \cdots,$$

at which point we perform the integrations. As will soon be clear, computing the integral of the second term exactly has benefits. The resulting equation is

$$\int_{x_{i-1}}^{x_{i+1}} f(x)dx = 2hf(x_i) + \frac{h^3}{3}f''(x_i) + O(h^5).$$

Substituting the expression for $f(x_i)$ derived earlier, the right-hand side becomes

$$2h\left(\frac{f(x_{i-1}) + 4f(x_i) + f(x_{i+1})}{6} - \frac{h^2}{6}f''(x_i) + O(h^4)\right) + \frac{h^3}{3}f''(x_i) + O(h^5),$$

which can be rearranged to

$$\left[\frac{h}{3}(f(x_{i-1}) + 4f(x_i) + f(x_{i+1})) - \frac{h^3}{3}f''(x_i) + O(h^5)\right] + \frac{h^3}{3}f''(x_i) + O(h^5).$$

Canceling and combining the appropriate terms results in the integral expression

$$\int_{x_{i-1}}^{x_{i+1}} f(x)dx = \frac{h}{3}(f(x_{i-1}) + 4f(x_i) + f(x_{i+1})) + O(h^5).$$

Recognizing that $\frac{h}{3}(f(x_{i-1}) + 4f(x_i) + f(x_{i+1}))$ is exactly the Simpson's rule approximation for the integral over this subinterval, this equation implies that Simpson's rule is $O(h^5)$ over a subinterval and $O(h^4)$ over the whole interval. Because the h^3 terms cancel out exactly, Simpson's rule gains another *two* orders of accuracy!

TRY IT! Use Simpson's rule to approximate $\int_0^{\pi} \sin(x)dx$ with 11 evenly spaced grid points over the whole interval. Compare this value to the exact value of 2.

```
In [1]: import numpy as np

        a = 0
        b = np.pi
        n = 11
        h = (b - a) / (n - 1)
        x = np.linspace(a, b, n)
        f = np.sin(x)

        I_simp = (h/3) * (f[0] + 2*sum(f[:n-2:2]) + 4*sum(f[1:n-1:2]) + f[n-1])
        err_simp = 2 - I_simp

        print(I_simp)
        print(err_simp)
```

```
2.0001095173150043
-0.00010951731500430384
```

21.5 COMPUTING INTEGRALS IN PYTHON

The `scipy.integrate` subpackage has several functions for computing integrals. The `trapz` takes as input arguments an array of function values f computed on a numerical grid x.

TRY IT! Use the `trapz` function to approximate $\int_0^{\pi} \sin(x)dx$ for 11 equally spaced points over the whole interval. Compare this value to the one computed in the earlier example using the trapezoid rule.

```
In [1]: import numpy as np
        from scipy.integrate import trapz

        a = 0
        b = np.pi
        n = 11
        h = (b - a) / (n - 1)
        x = np.linspace(a, b, n)
        f = np.sin(x)

        I_trapz = trapz(f,x)
        I_trap = (h/2)*(f[0] + 2 * sum(f[1:n-1]) + f[n-1])

        print(I_trapz)
        print(I_trap)

1.9835235375094542
1.9835235375094546
```

Sometimes we need to know the approximated cumulative integral. That is, $F(X) = \int_{x_0}^{X} f(x)dx$. For this purpose, it is useful to use the `cumtrapz` function, which takes the same input arguments as `trapz`.

TRY IT! Use the `cumtrapz` function to approximate the cumulative integral of $f(x) = \sin(x)$ from 0 to π, with a discretization step of 0.01. The exact solution of this integral is $F(x) = \sin(x)$. Plot the results.

```
In [2]: from scipy.integrate import cumtrapz
        import matplotlib.pyplot as plt
```

```
%matplotlib inline
plt.style.use("seaborn-poster")

x = np.arange(0, np.pi, 0.01)
F_exact = -np.cos(x)
F_approx = cumtrapz(np.sin(x), x)

plt.figure(figsize = (10,6))
plt.plot(x, F_exact)
plt.plot(x[1::], F_approx)
plt.grid()
plt.tight_layout()
plt.title("$F(x) = \int_0^{x} sin(y) dy$")
plt.xlabel("x")
plt.ylabel("f(x)")
plt.legend(["Exact with Offset", "Approx"])
plt.show()
```

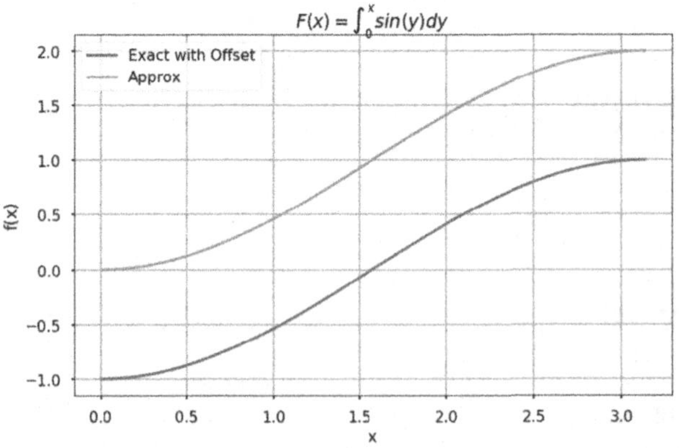

The `quad(f,a,b)` function uses a different numerical differentiation scheme to approximate integrals; `quad` integrates the function defined by the function object, `f`, from `a` to `b`.

TRY IT! Use the `integrate.quad` function to compute $\int_0^{\pi} \sin(x)dx$. Compare your answer with the correct answer of 2.

```
In [3]: from scipy.integrate import quad

        I_quad, est_err_quad = quad(np.sin, 0, np.pi)
        print(I_quad)
```

```
        err_quad = 2 - I_quad
        print(est_err_quad, err_quad)
```

```
2.0
2.220446049250313e-14 0.0
```

21.6 SUMMARY AND PROBLEMS

21.6.1 SUMMARY

1. Explicit integration of functions is often impossible or inconvenient, and numerical approaches must be used instead.
2. The Riemann integral, trapezoid rule, and Simpson's rule are common methods of approximating integrals.
3. Each method has an order of accuracy that depends on the approximation of the area below the function.

21.6.2 PROBLEMS

1. Write a function `my_int_calc(f,f0,a,b,N,option)` where `f` is a function object, `a` and `b` are scalars such that `a < b`, `N` is a positive integer, and `option` is the string `"rect"`, `"trap"`, or `"simp"`. Let `x` be an array starting at `a`, ending at `b`, and containing `N` evenly spaced elements. The output argument, `I`, should be an approximation to the integral of `f(x)`, with initial condition `f0` computed according to the input argument, `option`.

2. Write a function `my_poly_int(x,y)` where `x` and `y` are one-dimensional arrays of the same size, and the elements of `x` are unique and in ascending order. The function `my_poly_int` should (1) compute the Lagrange polynomial going through all the points defined by `x` and `y`; and (2) return an approximation to the area under the curve defined by `x` and `y`, `I`, defined as the analytic integral of the Lagrange interpolating polynomial.

3. When will `my_poly_int` work *worse* than the trapezoid method?

4. Write a function `my_num_calc(f,a,b,n,option)` where the output `I` is the numerical integral of `f`, a function object, computed on a grid of `n` evenly spaced points starting at `a` and ending at `b`. The integration method used should be one of the following strings defined by the option: `"rect"`, `"trap"`, `"simp"`. For the rectangle method, the function value should be taken from the right endpoint of the interval. Assume that `n` is odd.

   Warning: When programming your loops note that the `x` subscripts start at x_0 and not x_1. The odd–even indices will be reversed. Also the n term given in Simpson's rule denotes the number of subintervals, not the number of points as specified by the input argument, `n`.

Test cases:

```
In: f = lambda x: x**2
    my_num_int(f, 0, 1, 3, "rect")
Out: 0.625

In: my_num_int(f, 0, 1, 3, "trap")
Out: 0.375

In: my_num_int(f, 0, 1, 3, "simp")
Out: 0.3333333333333333

In: f = lambda x: np.exp(x**2)
    my_num_int(f, -1, 1, 101, "simp")
Out: 2.9253035883926493

In: my_num_int(f, -1, 1, 10001, "simp")
Out: 2.925303491814364

In: my_num_int(f, -1, 1, 100001, "simp")
Out: 2.9253034918143634
```

5. An earlier chapter demonstrated that some functions can be expressed as an infinite sum of polynomials (i.e., the Taylor series). Other functions, particularly periodic functions, can be written as an infinite sum of sine and cosine waves. For these functions,

$$f(x) = \frac{A_0}{2} + \sum_{n=1}^{\infty} A_n \cos(nx) + B_n \sin(nx).$$

It can be shown that the values of A_n and B_n can be computed using the following formulas:

$$A_n = \frac{1}{\pi}\int_{-\pi}^{\pi} f(x)\cos(nx)\,dx,$$

$$B_n = \frac{1}{\pi}\int_{-\pi}^{\pi} f(x)\sin(nx)\,dx.$$

Just like the Taylor series, functions can be approximated by truncating the Fourier series at some $n = N$. Fourier series can be used to approximate some particularly nasty functions, such as the step function, and they form the basis of many engineering applications, such as signal processing. Write a function `my_fourier_coef(f,n)` with output `[An,Bn]`, where `f` is an function object that is 2π-periodic. The function `my_fourier_coef` should compute the nth Fourier coefficients, `An` and `Bn`, in the Fourier series for `f` defined by the two formulas given earlier. Use the *quad* function to perform the integration.

Test cases:
Use the following plotting function to plot the analytic and approximation of functions using the Fourier series.

```
def plot_results(f, N):
    x = np.linspace(-np.pi, np.pi, 10000)
    [A0, B0] = my_fourier_coef(f, 0)
    y = A0*np.ones(len(x))/2
    for n in range(1, N):
        [An, Bn] = my_fourier_coef(f, n)
        y += An*np.cos(n*x)+Bn*np.sin(n*x)
    plt.figure(figsize = (10,6))
    plt.plot(x, f(x), label = "analytic")
    plt.plot(x, y, label = "approximate")
    plt.xlabel("x")
    plt.ylabel("y")
    plt.grid()
    plt.legend()
    plt.title(f"{N}th Order Fourier Approximation")
    plt.show()

f = lambda x: np.sin(np.exp(x))
N = 2
plot_results(f, N)
```

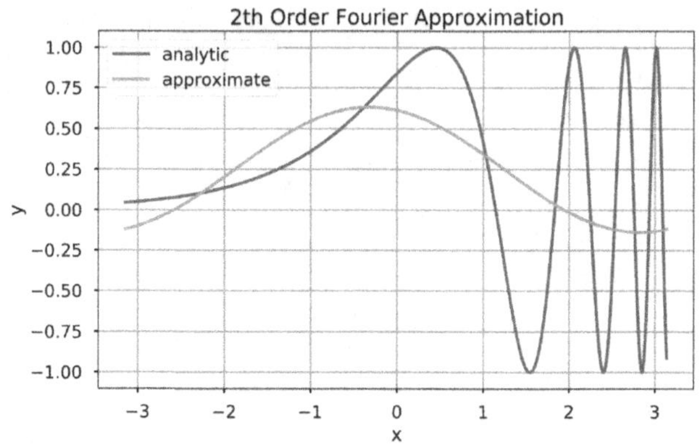


```
N = 2
plot_results(f, N)
```

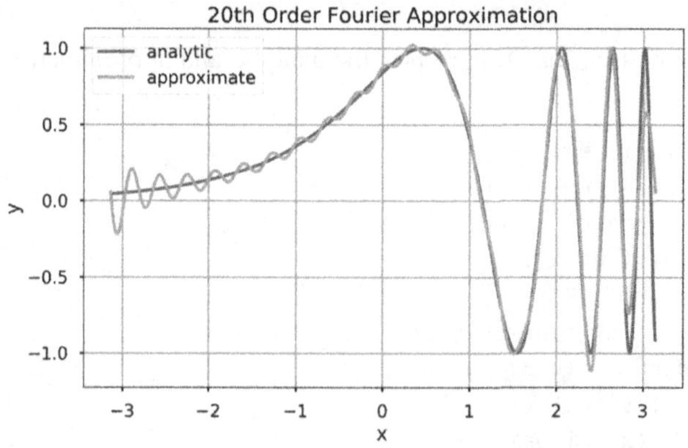


```
f = lambda x: np.mod(x, np.pi/2)
N = 5
plot_results(f, N)
```

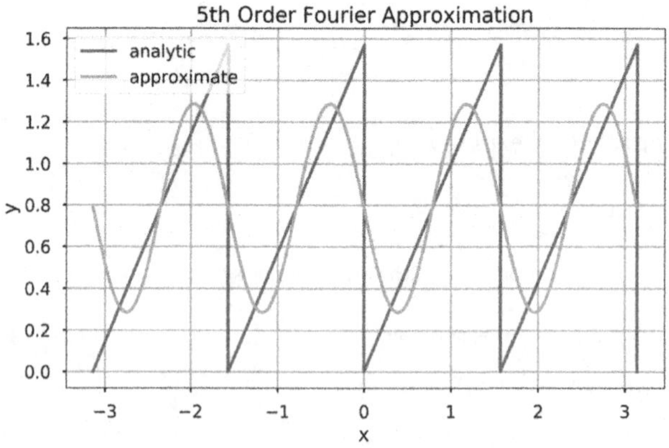


```
N = 20
plot_results(f, N)
```

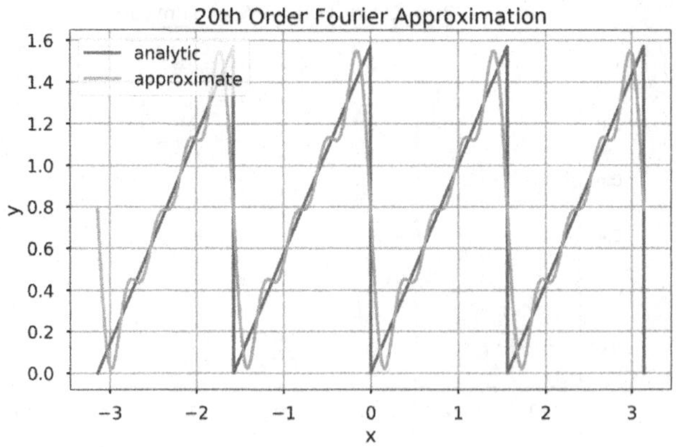


```
f = lambda x: (x > -np.pi/2) & (x < np.pi/2)
N = 2
plot_results(f, N)
```

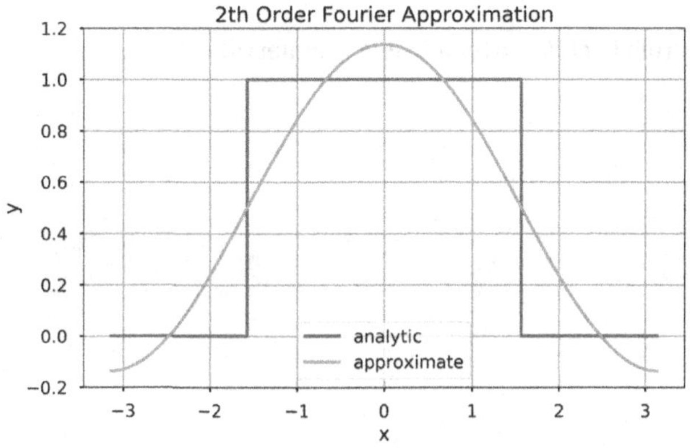


```
N = 20
plot_results(f, N)
```

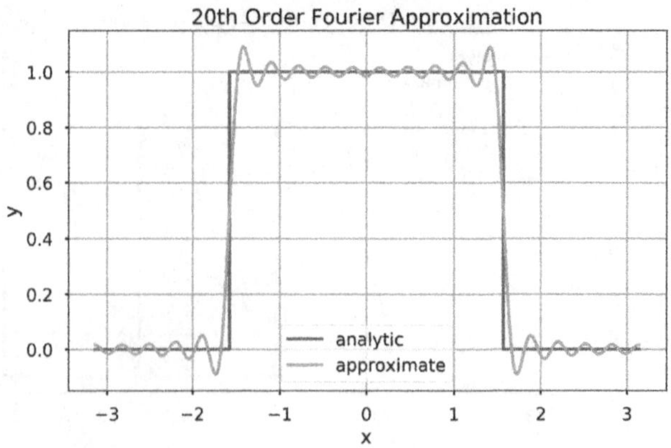

6. For a numerical grid with spacing h, Boole's rule for approximating integrals says that

$$\int_{x_i}^{x_{i+4}} f(x)dx \approx \frac{3h}{90}\left[7f(x_i) + 32f(x_{i+1}) + 12f(x_{i+2}) + 32f(x_{i+3}) + 7f(x_{i+4})\right].$$

Show that Boole's rule is $O(h^7)$ over a single subinterval.

CHAPTER

ORDINARY DIFFERENTIAL EQUATIONS (ODES) INITIAL-VALUE PROBLEMS

22

CONTENTS

22.1 ODE INITIAL VALUE PROBLEM STATEMENT

A **differential equation** is a relationship between a function, $f(x)$, its independent variable, x, and any number of its derivatives. An **ordinary differential equation** or **ODE** is a differential equation where the independent variable and its derivatives are in one dimension. For the purpose of this book, we assume that an ODE can be written as

$$F\left(x, f(x), \frac{df(x)}{dx}, \frac{d^2f(x)}{dx^2}, \frac{d^3f(x)}{dx^3}, \ldots, \frac{d^{n-1}f(x)}{dx^{n-1}}\right) = \frac{d^nf(x)}{dx^n},$$

where F is an arbitrary function that incorporates one or all of the input arguments, and n is the **order** of the differential equation. This equation is referred to as an **nth order ODE**.

To give an example of an ODE, consider a pendulum of length l with a mass, m, at its end; see Fig. 22.1. The angle the pendulum makes with the vertical axis over time, $\Theta(t)$, in the presence of vertical gravity, g, can be described by the pendulum equation, which is the ODE

$$ml\frac{d^2\Theta(t)}{dt^2} = -mg\sin(\Theta(t)).$$

Python Programming and Numerical Methods. https://doi.org/10.1016/B978-0-12-819549-9.00032-4

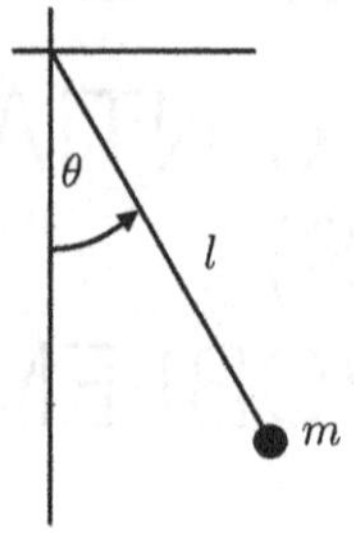

FIGURE 22.1

Pendulum system.

This equation can be derived by summing the forces in the x and y direction, and then changing them to polar coordinates.

In contrast, a **partial differential equation** or **PDE** is a general form differential equation where x is a vector containing the independent variables $x_1, x_2, x_3, \ldots, x_m$, and the partial derivatives can be of any order with respect to any combination of variables. An example of a PDE is the heat equation, which describes the evolution of temperature in space over time:

$$\frac{\partial u(t,x,y,z)}{\partial t} = \alpha\left(\frac{\partial u(t,x,y,z)}{\partial x} + \frac{\partial u(t,x,y,z)}{\partial y} + \frac{\partial u(t,x,y,z)}{\partial z}\right).$$

Here, $u(t,x,y,z)$ is the temperature at (x,y,z) at time t, and α is a thermal diffusion constant.

A **general solution** to a differential equation is a $g(x)$ that satisfies the differential equation. Although there are usually many solutions to a differential equation, they are still difficult to solve. For an ODE of order n, a **particular solution** is a $p(x)$ that satisfies the differential equation and has n explicitly **known values** of the solution, or its derivatives at certain points. Generally stated, $p(x)$ must satisfy the differential equation and $p^{(j)}(x_i) = p_i$, where $p^{(j)}$ is the jth derivative of p, for n triplets, (j, x_i, p_i). For the purpose of this text, we refer to the particular solution simply as the **solution**.

TRY IT! Returning to the pendulum example, if we assume the angles are very small (i.e., $\sin(\Theta(t)) \approx \Theta(t)$), then the pendulum equation reduces to

$$l\frac{d^2\Theta(t)}{dt^2} = -g\Theta(t).$$

Verify that $\Theta(t) = \cos\left(\sqrt{\frac{g}{l}}t\right)$ is a general solution to the pendulum equation. If the angle and angular velocities at $t = 0$ are the known values, Θ_0 and 0, respectively, verify that $\Theta(t) = \Theta_0 \cos\left(\sqrt{\frac{g}{l}}t\right)$ is a particular solution for these known values.

For the general solution, the derivatives of $\Theta(t)$ are

$$\frac{d\Theta(t)}{dt} = -\sqrt{\frac{g}{l}}\sin\left(\sqrt{\frac{g}{l}}t\right)$$

and

$$\frac{d^2\Theta(t)}{dt^2} = -\frac{g}{l}\cos\left(\sqrt{\frac{g}{l}}t\right).$$

By plugging the second derivative back into the differential equation on the left-hand side, it is easy to verify that $\Theta(t)$ satisfies the equation; thus, it is considered a general solution.

For the particular solution, the Θ_0 coefficient will carry through the derivatives, and it can be verified that the equation is satisfied: $\Theta(0) = \Theta_0\cos(0) = \Theta_0$, and $0 = -\Theta_0\sqrt{\frac{g}{l}}\sin(0) = 0$, therefore the particular solution also has the known values.

A pendulum swinging at small angles is a very uninteresting pendulum indeed. Unfortunately, there is no explicit solution for the pendulum equation with large angles that is as simple algebraically. Since this system is much simpler than most practical engineering systems and has no obvious analytical solution, the need for numerical solutions to ODEs is clear.

NOTE! An **analytical solution** of an ODE is a mathematical expression of the function $f(x)$ that satisfies the differential equation and has the initial value. But in many cases, an analytical solution is impossible in engineering and science. A numerical solution of an ODE is a set of discrete points (numerical grid) that approximate the function $f(x)$; we can obtain the solution using these grids.

A common set of known values for an ODE solution is the **initial value**. For an ODE of order n, the initial value is a known value for the 0th to $(n-1)$th derivatives at $x = 0$, namely $f(0), f^{(1)}(0), f^{(2)}(0), \ldots, f^{(n-1)}(0)$. For a certain class of ordinary differential equations, the initial value is sufficient to find a unique particular solution. Finding a solution to an ODE given an initial value is called the **initial value problem**. Although the name suggests we will only cover ODEs that evolve in time, initial value problems can also include systems that evolve in other dimensions such as space. Intuitively, the pendulum equation can be solved as an initial value problem because under only the force of gravity, an initial position and velocity should be sufficient to describe the motion of the pendulum for all time afterward.

The remainder of this chapter covers several methods of numerically approximating the solution to initial value problems on a numerical grid. Although initial value problems encompass more than just differential equations in time, we use time as the independent variable. We use several notations for the derivative of $f(t)$: $f'(t)$, $f^{(1)}(t)$, $\frac{df(t)}{dt}$, and $\dot{f}$, whichever is most convenient for the context.

22.2 REDUCTION OF ORDER

Many numerical methods for solving initial value problems are designed specifically to solve first-order differential equations. To make these solvers useful for solving higher-order differential equations, we must often **reduce the order** of the differential equation to a first order problem. To reduce the order of a differential equation, consider a vector, $S(t)$, which is the **state** of the system as a function of time. In general, the state of a system is a collection of all the dependent variables that are relevant to the behavior of the system. Recalling that the ODEs of interest in this book can be expressed as

$$f^{(n)}(t) = F\left(t, f(t), f^{(1)}(t), f^{(2)}(t), f^{(3)}(t), \ldots, f^{(n-1)}(t)\right),$$

for initial value problems, it is useful to take the state to be

$$S(t) = \begin{bmatrix} f(t) \\ f^{(1)}(t) \\ f^{(2)}(t) \\ f^{(3)}(t) \\ \vdots \\ f^{(n-1)}(t) \end{bmatrix}.$$

Then the derivative of the state is

$$\frac{dS(t)}{dt} = \begin{bmatrix} f^{(1)}(t) \\ f^{(2)}(t) \\ f^{(3)}(t) \\ f^{(4)}(t) \\ \vdots \\ f^{(n)}(t) \end{bmatrix} = \begin{bmatrix} f^{(1)}(t) \\ f^{(2)}(t) \\ f^{(3)}(t) \\ f^{(4)}(t) \\ \vdots \\ F\left(t, f(t), f^{(1)}(t), \ldots, f^{(n-1)}(t)\right) \end{bmatrix} = \begin{bmatrix} S_2(t) \\ S_3(t) \\ S_4(t) \\ S_5(t) \\ \vdots \\ F(t, S_1(t), S_2(t), \ldots, S_{n-1}(t)) \end{bmatrix},$$

where $S_i(t)$ is the ith element of $S(t)$. With the state written in this way, $\frac{dS(t)}{dt}$ can be written using only $S(t)$ (i.e., no $f(t)$) or its derivatives. In particular, $\frac{dS(t)}{dt} = \mathcal{F}(t, S(t))$, where $\mathcal{F}$ is a function that assembles the vector appropriately, describing the derivative of the state. This equation is in the form of a first-order differential equation in S. Essentially, what we have done is turn an nth order ODE into n first-order ODEs that are **coupled** together, meaning they share the same terms.

TRY IT! Reduce the second-order pendulum equation to a first-order equation, where

$$S(t) = \begin{bmatrix} \Theta(t) \\ \dot{\Theta}(t) \end{bmatrix}.$$

Taking the derivative of $S(t)$ and substituting gives the correct expression:

$$\frac{dS(t)}{dt} = \begin{bmatrix} S_2(t) \\ -\frac{g}{l}S_1(t) \end{bmatrix}.$$

This ODE can be written in matrix form:

$$\frac{dS(t)}{dt} = \begin{bmatrix} 0 & 1 \\ -\frac{g}{l} & 0 \end{bmatrix} S(t).$$

The ODEs that can be written in this way are said to be **linear ODEs**.

Although reducing the order of an ODE to first-order results in an ODE with multiple variables, all the derivatives are still taken with respect to the same independent variable, t; therefore, the ordinariness of the differential equation is retained.

Note that the state can hold multiple dependent variables and their derivatives as long as the derivatives are the same with respect to the independent variable.

TRY IT! A very simple model to describe the change in the population of rabbits, $r(t)$, due to wolves, $w(t)$, might be

$$\frac{dr(t)}{dt} = 4r(t) - 2w(t)$$

and

$$\frac{dw(t)}{dt} = r(t) + w(t).$$

The first ODE says that the rate of growth of the rabbit population is four times its value minus twice the size of the population of wolves (who eat the rabbits). The second ODE says that the growth rate of the wolf population is equal to the value of the wolf population plus the rabbit population. Write this system of differential equations as an equivalent differential equation in $S(t)$ where

$$S(t) = \begin{bmatrix} r(t) \\ w(t) \end{bmatrix}.$$

The following first-order ODE is equivalent to the pair of ODEs:

$$\frac{dS(t)}{dt} = \begin{bmatrix} 4 & -2 \\ 1 & 1 \end{bmatrix} S(t).$$

22.3 THE EULER METHOD

Let $\frac{dS(t)}{dt} = F(t, S(t))$ be an explicitly defined first order ODE, that is, F is a function that returns the derivative, or change, of a state given a time and state value. Also, let t be a numerical grid of the interval $[t_0, t_f]$ with spacing h. Without loss of generality, we assume that $t_0 = 0$ and that $t_f = Nh$ for some positive integer, N.

The linear approximation of $S(t)$ around t_j at t_{j+1} is

$$S(t_{j+1}) = S(t_j) + (t_{j+1} - t_j)\frac{dS(t_j)}{dt},$$

which can also be written

$$S(t_{j+1}) = S(t_j) + hF(t_j, S(t_j)).$$

This formula is called the **Explicit Euler Formula**. It allows us to compute an approximation for the state at $S(t_{j+1})$ given the state at $S(t_j)$. This is actually based on the Taylor series we discussed in Chapter 18, whereby we used only the first order item in Taylor series to linearly approximate the next solution. Later in this chapter, we will present a formula using higher terms to increase the accuracy. Starting from a given initial value of $S_0 = S(t_0)$, we can use this formula to integrate the states up to $S(t_f)$; these $S(t)$ values are then an approximation for the solution of the differential equation. The explicit Euler formula is the simplest and most intuitive method for solving initial value problems. At any state $(t_j, S(t_j))$ it uses F at that state to "point" linearly toward the next state and then moves in that direction a distance of h, as shown in Fig. 22.2.

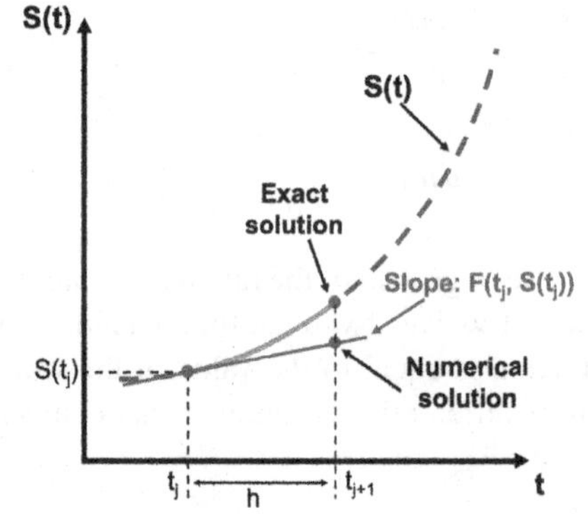

FIGURE 22.2

The illustration of the explicit Euler method.

Although there are more sophisticated and accurate methods for solving these problems, they all have the same fundamental structure. As such, we enumerate explicitly the steps for solving an initial value problem using the explicit Euler formula.

WHAT IS HAPPENING? Assume we are given a function $F(t, S(t))$ that computes $\frac{dS(t)}{dt}$, a numerical grid, t, of the interval, $[t_0, t_f]$, and an initial state value $S_0 = S(t_0)$. We can compute $S(t_j)$ for every t_j in t using the following steps:

1. Store $S_0 = S(t_0)$ in an array, S.
2. Compute $S(t_1) = S_0 + hF(t_0, S_0)$.
3. Store $S_1 = S(t_1)$ in S.
4. Compute $S(t_2) = S_1 + hF(t_1, S_1)$.
5. Store $S_2 = S(t_1)$ in S.
6. ...
7. Compute $S(t_f) = S_{f-1} + hF(t_{f-1}, S_{f-1})$.
8. Store $S_f = S(t_f)$ in S.
9. S is an approximation of the solution to the initial value problem.

When using a method with this structure, we say the method **integrates** the solution of the ODE.

TRY IT! The differential equation $\frac{df(t)}{dt} = e^{-t}$ with initial condition $f_0 = -1$ has the exact solution $f(t) = -e^{-t}$. Approximate the solution to this initial value problem between zero and 1 in increments of 0.1 using the explicit Euler formula. Plot the difference between the approximated solution and the exact solution.

```
In [1]: import numpy as np
        import matplotlib.pyplot as plt

        plt.style.use("seaborn-poster")
        %matplotlib inline

        # Define parameters
        f = lambda t, s: np.exp(-t) # ODE
        h = 0.1 # Step size
        t = np.arange(0, 1 + h, h) # Numerical grid
        s0 = -1 # Initial Condition

        # Explicit Euler Method
        s = np.zeros(len(t))
        s[0] = s0

        for i in range(0, len(t) - 1):
            s[i + 1] = s[i] + h*f(t[i], s[i])

        plt.figure(figsize = (12, 8))
        plt.plot(t, s, "b-", label="Approximate")
        plt.plot(t, -np.exp(-t), "g", label="Exact")
        plt.title("Approximate and Exact Solution for Simple ODE")
        plt.xlabel("t")
        plt.ylabel("f(t)")
```

```
        plt.grid()
        plt.legend(loc="lower right")
        plt.show()
```

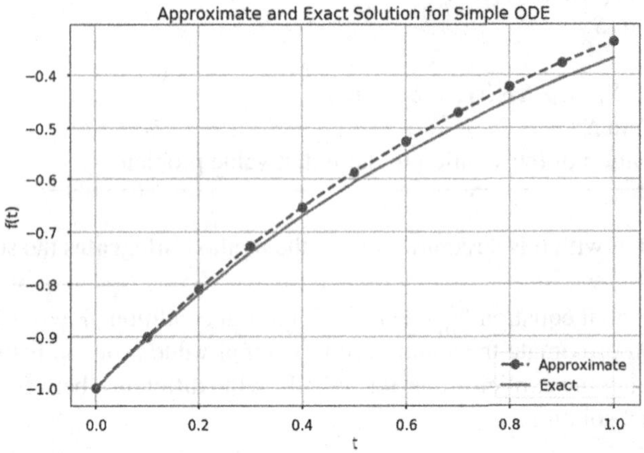

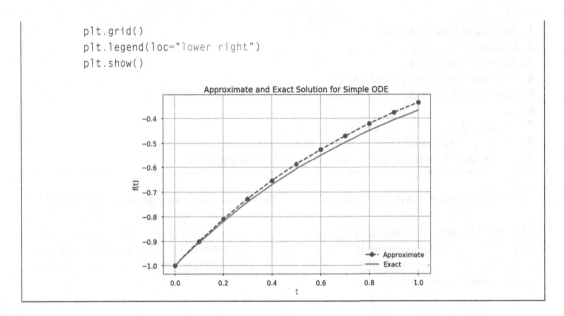

In the above figure, each dot is one approximation based on the previous dot in a linear fashion. From the initial value, we can eventually obtain an approximation of the solution on the numerical grid. If we repeat the process for $h = 0.01$, we obtain a better approximation for the solution:

```
In [2]: h = 0.01 # Step size
        t = np.arange(0, 1 + h, h) # Numerical grid
        s0 = -1 # Initial Condition

        # Explicit Euler Method
        s = np.zeros(len(t))
        s[0] = s0

        for i in range(0, len(t) - 1):
            s[i + 1] = s[i] + h*f(t[i], s[i])

        plt.figure(figsize = (12, 8))
        plt.plot(t, s, "b-", label="Approximate")
        plt.plot(t, -np.exp(-t), "g", label="Exact")
        plt.title("Approximate and Exact Solution for Simple ODE")
        plt.xlabel("t")
        plt.ylabel("f(t)")
        plt.grid()
        plt.legend(loc="lower right")
        plt.show()
```

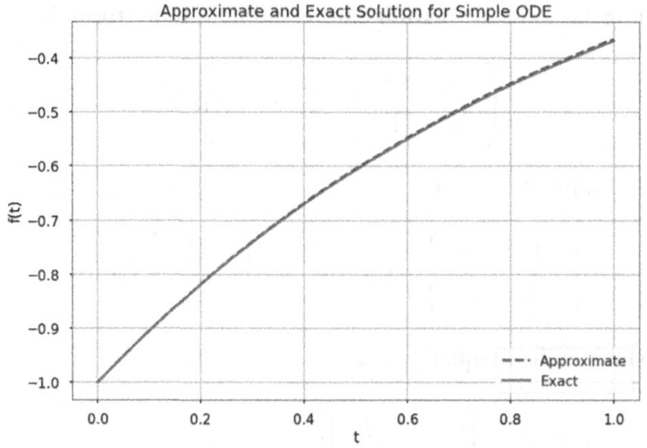

The explicit Euler formula is called "explicit" because it only requires information at t_j to compute the state at t_{j+1}. That is, $S(t_{j+1})$ can be written explicitly in terms of values we have (i.e., t_j and $S(t_j)$). The **Implicit Euler Formula** can be derived by taking the linear approximation of $S(t)$ around t_{j+1} and computing it at t_j:

$$S(t_{j+1}) = S(t_j) + hF(t_{j+1}, S(t_{j+1})).$$

This formula is peculiar because it requires that we know $S(t_{j+1})$ in order to compute $S(t_{j+1})$! However, it happens that sometimes we *can* use this formula to approximate the solution to initial value problems. Before we provide details on how to solve these problems using the implicit Euler formula, we introduce another implicit formula called the **Trapezoidal Formula**, which is the average of the explicit and implicit Euler formulas:

$$S(t_{j+1}) = S(t_j) + \frac{h}{2}(F(t_j, S(t_j)) + F(t_{j+1}, S(t_{j+1}))).$$

To illustrate how to solve these implicit schemes, consider again the pendulum equation, which has been reduced to a first-order equation:

$$\frac{dS(t)}{dt} = \begin{bmatrix} 0 & 1 \\ -\frac{g}{l} & 0 \end{bmatrix} S(t).$$

For this equation,

$$F(t_j, S(t_j)) = \begin{bmatrix} 0 & 1 \\ -\frac{g}{l} & 0 \end{bmatrix} S(t_j).$$

If we plug this expression into the explicit Euler formula, we obtain the following equation:

$$\begin{aligned} S(t_{j+1}) &= S(t_j) + h\begin{bmatrix} 0 & 1 \\ -\frac{g}{l} & 0 \end{bmatrix} S(t_j) \\ &= \begin{bmatrix} 1 & 0 \\ 0 & 1 \end{bmatrix} S(t_j) + h\begin{bmatrix} 0 & 1 \\ -\frac{g}{l} & 0 \end{bmatrix} S(t_j) = \begin{bmatrix} 1 & h \\ -\frac{gh}{l} & 1 \end{bmatrix} S(t_j). \end{aligned}$$

Similarly, we can plug the same expression into the implicit Euler formula to obtain

$$\begin{bmatrix} 1 & -h \\ \frac{gh}{l} & 1 \end{bmatrix} S(t_{j+1}) = S(t_j),$$

and into the trapezoidal formula to obtain

$$\begin{bmatrix} 1 & -\frac{h}{2} \\ \frac{gh}{2l} & 1 \end{bmatrix} S(t_{j+1}) = \begin{bmatrix} 1 & \frac{h}{2} \\ -\frac{gh}{2l} & 1 \end{bmatrix} S(t_j).$$

With some rearrangement, these equations become respectively

$$S(t_{j+1}) = \begin{bmatrix} 1 & -h \\ \frac{gh}{l} & 1 \end{bmatrix}^{-1} S(t_j),$$

$$S(t_{j+1}) = \begin{bmatrix} 1 & -\frac{h}{2} \\ \frac{gh}{2l} & 1 \end{bmatrix}^{-1} \begin{bmatrix} 1 & \frac{h}{2} \\ -\frac{gh}{2l} & 1 \end{bmatrix} S(t_j).$$

These equations allow us to solve the initial value problem since at each state, $S(t_j)$, we can compute the next state at $S(t_{j+1})$. In general, this is possible to do when an ODE is linear.

22.4 NUMERICAL ERROR AND INSTABILITY

There are two main issues to consider with regard to integration schemes for ODEs: **accuracy** and **stability**. Accuracy refers to a scheme's ability to get close to the exact solution, which is usually unknown, as a function of the step size h. Previous chapters have referred to accuracy using the notation $O(h^p)$. The same notation can be used to solve ODEs. The stability of an integration scheme is its ability to keep the error from growing as it integrates forward in time. If the error does not grow, then the scheme is stable; otherwise it is unstable. Some integration schemes are stable for certain choices of h and unstable for others; these integration schemes are also referred to as unstable.

To illustrate issues of stability, we numerically solve the pendulum equation using the explicit and implicit Euler, as well as trapezoidal, formulas.

TRY IT! Use the explicit and implicit Euler, as well as trapezoidal, formulas to solve the pendulum equation over the time interval $[0, 5]$ in increments of 0.1, and for an initial solution of $S_0 = \begin{bmatrix} 1 \\ 0 \end{bmatrix}$. For the model parameters using $\sqrt{\frac{g}{l}} = 4$, plot the approximate solution on a single graph.

```
In [1]: import numpy as np
        from numpy.linalg import inv
```

```
        import matplotlib.pyplot as plt

        plt.style.use("seaborn-poster")

        %matplotlib inline

In [2]: # define step size
        h = 0.1
        # define numerical grid
        t = np.arange(0, 5.1, h)
        # oscillation freq. of pendulum
        w = 4
        s0 = np.array([[1], [0]])

        m_e = np.array([[1, h],
                        [-w**2*h, 1]])
        m_i = inv(np.array([[1, -h],
                        [w**2*h, 1]]))
        m_t = np.dot(inv(np.array([[1, -h/2],
            [w**2*h/2,1]])), np.array(
              [[1,h/2], [-w**2*h/2, 1]]))

        s_e = np.zeros((len(t), 2))
        s_i = np.zeros((len(t), 2))
        s_t = np.zeros((len(t), 2))

        # do integrations
        s_e[0, :] = s0.T
        s_i[0, :] = s0.T
        s_t[0, :] = s0.T

        for j in range(0, len(t)-1):
            s_e[j+1, :] = np.dot(m_e,s_e[j, :])
            s_i[j+1, :] = np.dot(m_i,s_i[j, :])
            s_t[j+1, :] = np.dot(m_t,s_t[j, :])

        plt.figure(figsize = (12, 8))
        plt.plot(t,s_e[:,0],"b-")
        plt.plot(t,s_i[:,0],"g:")
        plt.plot(t,s_t[:,0],"r-")
        plt.plot(t, np.cos(w*t), "k")
        plt.ylim([-3, 3])
        plt.xlabel("t")
```

```
plt.ylabel("$\Theta (t)$")
plt.legend(["Explicit", "Implicit", "Trapezoidal", "Exact"])
plt.show()
```

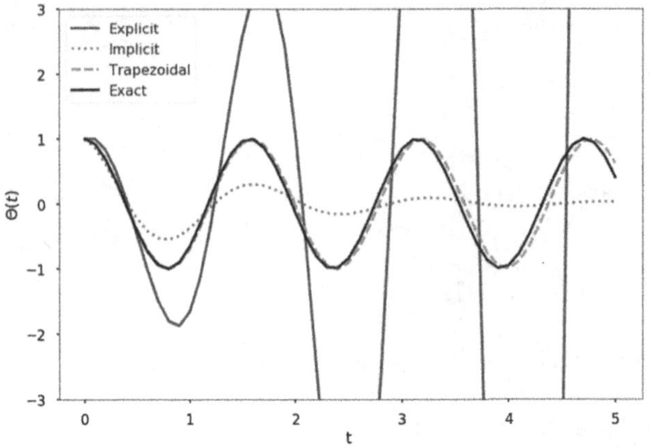

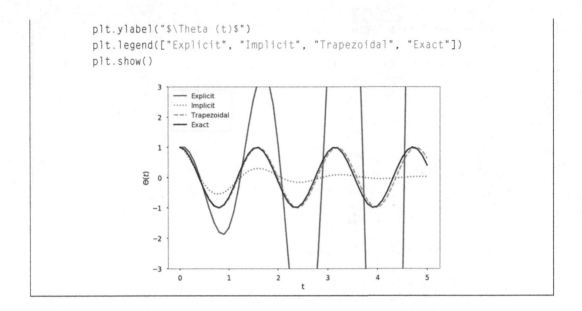

The generated figure above compares the numerical solution to the pendulum problem. The exact solution is a pure cosine wave. The explicit Euler scheme is clearly unstable. The implicit Euler scheme decays exponentially, which is not correct. The trapezoidal method captures the solution correctly, with a small phase shift as time increases.

22.5 PREDICTOR–CORRECTOR AND RUNGE–KUTTA METHODS

22.5.1 PREDICTOR–CORRECTOR METHODS

Given any time and state value, the function, $F(t, S(t))$, returns the change of state $\frac{dS(t)}{dt}$. The **predictor–corrector** methods of solving initial value problems improve the approximation accuracy of non-predictor–corrector methods by querying the F function several times at different locations (predictions). Then, using a weighted average of the results (corrections), updates the state. Essentially, it uses two formulas: **a predictor** and **a corrector**. The predictor is an explicit formula and estimates the solution at t_{j+1} first, i.e., we can use Euler method or some other methods to finish this step. After obtaining the solution $S(t_{j+1})$, we apply the corrector to improve the accuracy. Using the found $S(t_{j+1})$ on the right-hand side of an otherwise implicit formula, the corrector can calculate a new, more accurate solution.

The **midpoint method** has a predictor step:

$$S\left(t_j + \frac{h}{2}\right) = S(t_j) + \frac{h}{2}F(t_j, S(t_j)),$$

which is the prediction of the solution value halfway between t_j and t_{j+1}.

It then computes the corrector step:

$$S(t_{j+1}) = S(t_j) + hF\left(t_j + \frac{h}{2}, S\left(t_j + \frac{h}{2}\right)\right),$$

which computes the solution at $S(t_{j+1})$ from $S(t_j)$ but uses the derivative from $S\left(t_j + \frac{h}{2}\right)$.

22.5.2 RUNGE–KUTTA METHODS

Runge–Kutta (RK) methods are among the most widely used methods for solving ODEs. Recall that the Euler method uses the first two terms in Taylor series to approximate the numerical integration, which is linear: $S(t_{j+1}) = S(t_j + h) = S(t_j) + h \cdot S'(t_j)$.

We can greatly improve the accuracy of numerical integration if we keep more terms of the series as

$$S(t_{j+1}) = S(t_j + h) = S(t_j) + S'(t_j)h + \frac{1}{2!}S''(t_j)h^2 + \cdots + \frac{1}{n!}S^{(n)}(t_j)h^n.$$

In order to obtain this more accurate solution, we need to derive the expressions of $S''(t_j)$, $S'''(t_j)$, $\ldots$, $S^{(n)}(t_j)$. This extra work can be avoided using the RK methods, which are based on truncated Taylor series but do not require computation of these higher derivatives.

22.5.2.1 Second-Order Runge–Kutta Method

Let us first derive the second order RK method. Let $\frac{dS(t)}{dt} = F(t, S(t))$. Then we assume an integration formula the form of

$$S(t+h) = S(t) + c_1 F(t, S(t))h + c_2 F[t + ph, S(t) + qhF(t, S(t))]h. \tag{22.1}$$

We can attempt to find these parameters c_1, c_2, p, q by matching the above equation to the second-order Taylor series:

$$S(t+h) = S(t) + S'(t)h + \frac{1}{2!}S''(t)h^2 = S(t) + F(t, S(t))h + \frac{1}{2!}F'(t, S(t))h^2. \tag{22.2}$$

Note that

$$F'(t, s(t)) = \frac{\partial F}{\partial t} + \frac{\partial F}{\partial S}\frac{\partial S}{\partial t} = \frac{\partial F}{\partial t} + \frac{\partial F}{\partial S}F. \tag{22.3}$$

Therefore, Eq. (22.2) can be written as

$$S(t+h) = S + Fh + \frac{1}{2!}\left(\frac{\partial F}{\partial t} + \frac{\partial F}{\partial S}F\right)h^2. \tag{22.4}$$

In Eq. (22.1), we rewrite the last term by applying Taylor series in several variables:

$$F[t + ph, S + qhF)] = F + \frac{\partial F}{\partial t}ph + qh\frac{\partial F}{\partial S}F,$$

thus Eq. (22.1) becomes

$$S(t+h) = S + (c_1 + c_2)Fh + c_1\left[\frac{\partial F}{\partial t}p + q\frac{\partial F}{\partial S}F\right]h^2. \tag{22.5}$$

Comparing Eqs. (22.4) and (22.5), we can easily obtain

$$c_1 + c_2 = 1, c_2 p = \frac{1}{2}, c_2 q = \frac{1}{2}. \tag{22.6}$$

Because (22.6) has four unknowns and only three equations, we assign any value to one of the parameters and get the rest of the parameters. One popular choice is:

$$c_1 = \frac{1}{2}, \quad c_2 = \frac{1}{2}, \quad p = 1, \quad q = 1.$$

We can also define:

$$\begin{aligned} k_1 &= F(t_j, S(t_j)), \\ k_2 &= F\left(t_j + ph, S(t_j) + qhk_1\right), \end{aligned}$$

where we obtain

$$S(t_{j+1}) = S(t_j) + \frac{1}{2}(k_1 + k_2)h.$$

22.5.2.2 Fourth-Order Runge–Kutta Method

A classical method for integrating ODEs with a high order of accuracy is the **Fourth Order Runge–Kutta** (RK4) method. This method uses four points k_1, k_2, k_3, and k_4. A weighted average of these predictions is used to produce the approximation of the solution. The formula is as follows:

$$\begin{aligned} k_1 &= F(t_j, S(t_j)), \\ k_2 &= F\left(t_j + \tfrac{h}{2}, S(t_j) + \tfrac{1}{2}k_1 h\right), \\ k_3 &= F\left(t_j + \tfrac{h}{2}, S(t_j) + \tfrac{1}{2}k_2 h\right), \\ k_4 &= F(t_j + h, S(t_j) + k_3 h). \end{aligned}$$

Therefore, we will have

$$S(t_{j+1}) = S(t_j) + \frac{h}{6}\left(k_1 + 2k_2 + 2k_3 + k_4\right).$$

As indicated by its name, the RK4 method is fourth-order accurate, or $O(h^4)$.

22.6 PYTHON ODE SOLVERS

In `SciPy`, there are several built-in functions for solving initial value problems. The most common function is the `scipy.integrate.solve_ivp` function. The function construction is shown below:

CONSTRUCTION:

Let F be a function object to the function that computes

$$\frac{dS(t)}{dt} = F(t, S(t)),$$

$$S(t_0) = S_0.$$

The variable t is a one-dimensional independent variable (time), $S(t)$ is an n-dimensional vector-valued function (state), and $F(t, S(t))$ defines the differential equations; S_0 is an initial value for S. The function F *must* have the form $dS = F(t, S)$, although the name does not have to be F. The goal is to find the $S(t)$ that approximately satisfies the differential equations given the initial value $S(t_0) = S_0$.

Using the solver to solve the differential equation is as follows:

```
solve_ivp(fun,t_span,s0,method "RK45",t_eval=None)
```

where `fun` takes in the function in the right-hand side of the system; `t_span` is the interval of integration (t_0, t_f) where t_0 is the start and t_f is the end of the interval; s_0 is the initial state. There are a couple of methods to choose from: the default is "`RK45`", which is the explicit Runge–Kutta method of order 5(4). There are other methods you can use as well; see the end of this section for more information; `t_eval` takes in the times at which to store the computed solution, and must be sorted and lie within `t_span`.

Let us try one example below.

EXAMPLE: Consider the ODE

$$\frac{dS(t)}{dt} = \cos(t)$$

for an initial value of $S_0 = 0$. The exact solution to this problem is $S(t) = \sin(t)$. Use `solve_ivp` to approximate the solution to this initial value problem over the interval $[0, \pi]$. Plot the approximate solution versus the exact solution and the relative error over time.

```
In [1]: import matplotlib.pyplot as plt
        import numpy as np
        from scipy.integrate import solve_ivp

        plt.style.use("seaborn-poster")

        %matplotlib inline

        F = lambda t, s: np.cos(t)
```

```
t_eval = np.arange(0, np.pi, 0.1)
sol = solve_ivp(F, [0, np.pi], [0], t_eval=t_eval)

plt.figure(figsize = (12, 4))
plt.subplot(121)
plt.plot(sol.t, sol.y[0])
plt.xlabel("t")
plt.ylabel("S(t)")
plt.subplot(122)
plt.plot(sol.t, sol.y[0] - np.sin(sol.t))
plt.xlabel("t")
plt.ylabel("S(t) - sin(t)")
plt.tight_layout()
plt.show()
```

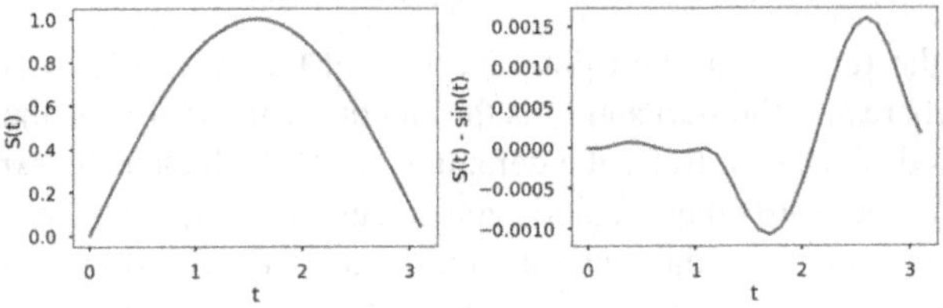

The above left figure shows the integration of $\frac{dS(t)}{dt} = \cos(t)$ with `solve_ivp`. The right figure computes the difference between the solution of the integration by `solve_ivp` and evaluates the analytical solution to this ODE. As can be seen from the figure, the difference between the approximate and exact solution to this ODE is small. Also, we can control the relative and absolute tolerances using the `rtol` and `atol` arguments; the solver keeps the local error estimates less than `atol + rtol*abs(S)`. The default values are 1e-3 for `rtol` and 1e-6 for `atol`.

TRY IT! Using the `rtol` and `atol` to make the difference between the approximate and exact solution less than 1e-7.

```
In [2]: sol = solve_ivp(F, [0, np.pi], [0], t_eval=t_eval, \
                        rtol = 1e-8, atol = 1e-8)

        plt.figure(figsize = (12, 4))
        plt.subplot(121)
        plt.plot(sol.t, sol.y[0])
        plt.xlabel("t")
        plt.ylabel("S(t)")
        plt.subplot(122)
        plt.plot(sol.t, sol.y[0] - np.sin(sol.t))
```

```
plt.xlabel("t")
plt.ylabel("S(t) - sin(t)")
plt.tight_layout()
plt.show()
```

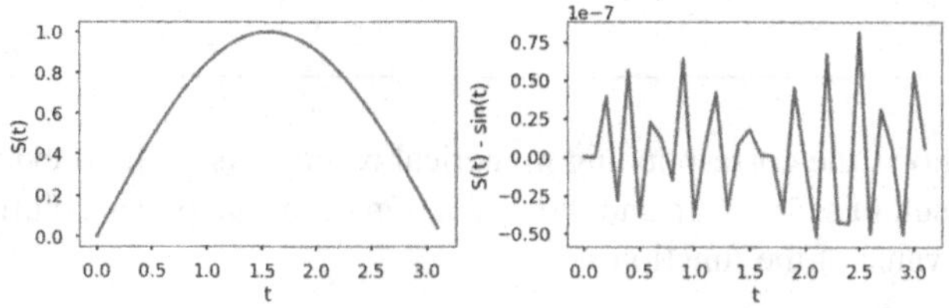

EXAMPLE: Consider the ODE

$$\frac{dS(t)}{dt} = -S(t),$$

with an initial value of $S_0 = 1$. The exact solution to this problem is $S(t) = e^{-t}$. Use `solve_ivp` to approximate the solution to this initial value problem over the interval [0, 1]. Plot the approximate solution versus the exact solution, and the relative error over time.

```
In [3]: F = lambda t, s: -s

        t_eval = np.arange(0, 1.01, 0.01)
        sol = solve_ivp(F, [0, 1], [1], t_eval=t_eval)

        plt.figure(figsize = (12, 4))
        plt.subplot(121)
        plt.plot(sol.t, sol.y[0])
        plt.xlabel("t")
        plt.ylabel("S(t)")
        plt.subplot(122)
        plt.plot(sol.t, sol.y[0] - np.exp(-sol.t))
        plt.xlabel("t")
        plt.ylabel("S(t) - exp(-t)")
        plt.tight_layout()
        plt.show()
```

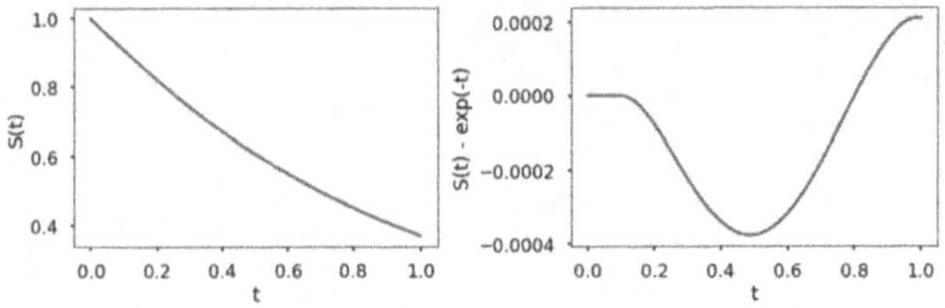

The above figure shows the corresponding numerical results. As in the previous example, the difference between the result of solve_ivp and the evaluation of the analytical solution by Python is very small compared to the value of the function.

EXAMPLE: Let the state of a system be defined by $S(t) = \begin{bmatrix} x(t) \\ y(t) \end{bmatrix}$, and let the evolution of the system be defined by the ODE

$$\frac{dS(t)}{dt} = \begin{bmatrix} 0 & t^2 \\ -t & 0 \end{bmatrix} S(t).$$

Use solve_ivp to solve this ODE for the time interval [0, 10] with an initial value of $S_0 = \begin{bmatrix} 1 \\ 1 \end{bmatrix}$. Plot the solution in $(x(t), y(t))$.

```
In [4]: F=lambda t, s: np.dot(np.array([[0, t**2], [-t,0]]),s)

        t_eval = np.arange(0, 10.01, 0.01)
        sol = solve_ivp(F, [0, 10], [1, 1], t_eval=t_eval)

        plt.figure(figsize = (12, 8))
        plt.plot(sol.y.T[:, 0], sol.y.T[:, 1])
        plt.xlabel("x")
        plt.ylabel("y")
        plt.show()
```

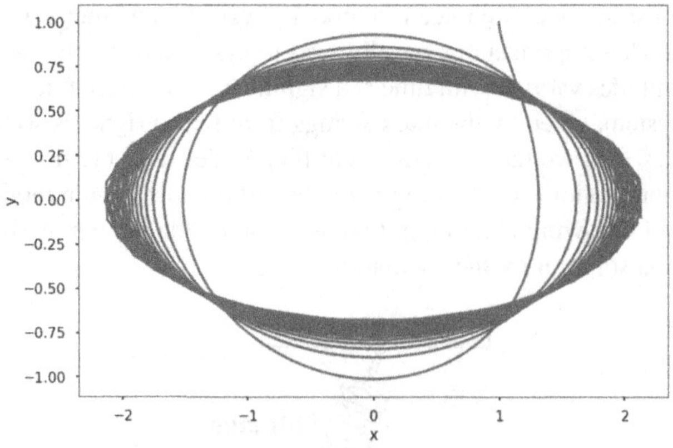

22.7 ADVANCED TOPICS

This section discusses briefly some more advanced topics in IVP ODE. Although we will not go into great detail, we suggest you check out some great books such as Ordinary Differential Equations by Morris Tenenbaum and Harry Pollard, Numerical Methods for Engineers and Scientists by Amos Gilat and Vish Subramaniam, as well as Numerical Methods for Ordinary Differential Equations by J.C. Butcher, if you are interested.

22.7.1 MULTISTEP METHODS

So far, most of the methods we discussed are called "one-step methods" because the approximation for the next point t_{j+1} is obtained by using information only from $S(t_j)$ and t_j at the previous point. Although some of the methods, such as RK methods, might use function-evaluation information at points between t_j and t_{j+1}, they do not retain the information for direct use in future approximations. In an attempt to gain more efficiency, the **multistep methods** use two or more previous points to approximate the solution at the next point t_{j+1}. For linear multistep methods, a linear combination of the previous points and derivative values are used to approximate the next point. The coefficients can be determined using polynomial interpolation, as discussed in Chapter 17.

There are three families of linear multistep methods commonly used: Adams–Bashforth methods, Adams–Moulton methods, and the backward differentiation formulas (BDFs).

22.7.2 STIFFNESS ODE

Stiffness is a difficult and important concept in the numerical solution of ODEs. A stiff ODE equation will make the solution being sought vary slowly and will not be stable, i.e., if there are nearby solutions, the solution will change dramatically. This will force us into taking small steps to obtain reasonable results. Therefore, stiffness is usually an efficiency issue: if we do not care about computation cost, we would not be concerned about stiffness.

In science and engineering, we often need to model physical phenomena with very different time scales or spatial scales. These applications usually lead to systems of ODEs, whose solution include several terms with magnitudes varying with time at a significantly different rate. For example, Fig. 22.3 shows a spring–mass system, whereby the mass swings from left to right, as well as oscillates up and down due to the spring. Thus, we have two different time scales, i.e., the time scale of the swinging motion and the oscillation motion. If the spring is really stiff, the oscillation motion time scale will be much smaller than that of the swinging motion. In order to study the system, we have to use a very tiny time step to obtain a good solution for the oscillation.

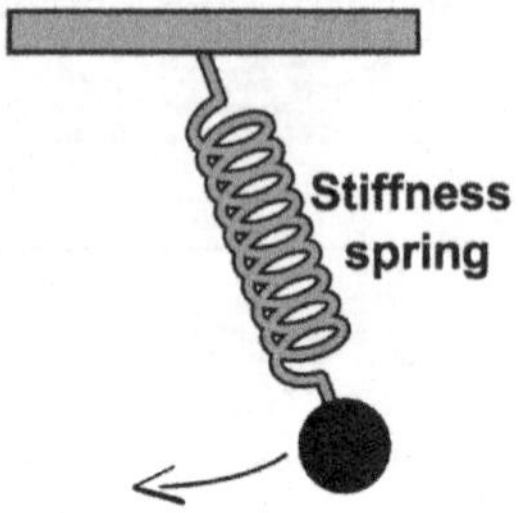

FIGURE 22.3

The illustration of the stiffness equation.

Depending on the properties of the ODE and the desired level of accuracy, you might need to use different methods for `solve_ivp`.

There are many methods to choose from for the `method` argument in `solve_ivp`; browse through the documentation for additional information. As suggested by the documentation, use the `"RK45"` or `"RK23"` method for non-stiff problems and `"Radau"` or `"BDF"` for stiff problems. If not sure, first try to run `"RK45"`. Should this solution experience an unusually high number of iterations, diverge, or fail, this problem is likely to be stiff, and you should use `"Radau"` or `"BDF"`. `"LSODA"` can also be a good universal choice, but it might be somewhat less convenient to work with as it wraps old Fortran code.

22.8 SUMMARY AND PROBLEMS

22.8.1 SUMMARY

1. Ordinary differential equations (ODEs) are equations that relate a function to its derivatives, and initial value problems are a specific kind of ODE-solving problem.
2. Because most initial value problems cannot be integrated explicitly, they require numerical solutions.
3. There are explicit, implicit, and predictor–corrector methods for numerically solving initial value problems.
4. The accuracy of the scheme used depends on its order of approximation of the ODE.
5. The stability of the scheme used depends on the ODE, scheme, and choice of the integration parameters.

22.8.2 PROBLEMS

1. The logistic equation is a simple differential equation model that can be used to relate the change in population $\frac{dP}{dt}$ to the current population, P, given a growth rate, r, and a carrying capacity, K. The logistic equation can be expressed by

$$\frac{dP}{dt} = rP\left(1 - \frac{P}{K}\right).$$

Write a function `my_logistic_eq(t, P, r, K)` that represents the logistic equation with a return of `dP`. Note that this format allows `my_logistic_eq` to be used as an input argument to `solve_ivp`. Assume that the arguments `dP`, `t`, `P`, `r`, and `K` are all scalars, and `dP` is the value $\frac{dP}{dt}$ given r, P, and K. Note that the input argument, `t`, is obligatory if `my_logistic_eq` is to be used as an input argument to `solve_ivp`, even though it is part of the differential equation.
Note that the logistic equation has an analytic solution defined by

$$P(t) = \frac{KP_0e^{rt}}{K + P_0(e^{rt} - 1)}$$

where P_0 is the initial population. Verify that this equation is a solution to the logistic equation.
Test cases:

```
In [1]: import numpy as np
        from scipy.integrate import solve_ivp
        import matplotlib.pyplot as plt
        from functools import partial
        plt.style.use("seaborn-poster")

        %matplotlib inline

In [2]: def my_logistic_eq(t, P, r, K):
            # put your code here

            return dP

        dP = my_logistic_eq(0, 10, 1.1, 15)
        dP

Out[2]: 3.666666666666667

In [3]: from functools import partial

        t0 = 0
        tf = 20
        P0 = 10
        r = 1.1
        K = 20
        t = np.linspace(0, 20, 2001)
```

```
f = partial(my_logistic_eq, r=r, K=K)
sol=solve_ivp(f,[t0,tf],[P0],t_eval=t)

plt.figure(figsize = (10, 8))
plt.plot(sol.t, sol.y[0])
plt.plot(t, K*P0*np.exp(r*t)/(K+P0*(np.exp(r*t)-1)),"r:")
plt.xlabel("time")
plt.ylabel("population")

plt.legend(["Numerical Solution", "Exact Solution"])
plt.grid(True)
plt.show()
```

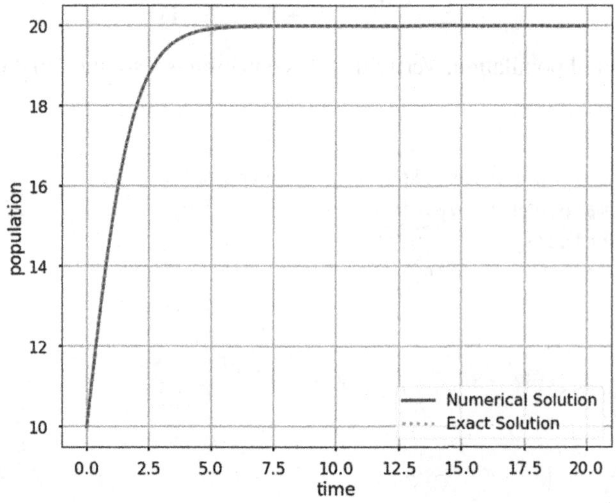

2. The Lorenz attractor is a system of ordinary differential equations that was originally developed to model convection currents in the atmosphere. The Lorenz equations can be written as follows:

$$\frac{dx}{dt} = \sigma(y - x),$$

$$\frac{dy}{dt} = x(\rho - z) - y,$$

$$\frac{dz}{dt} = xy - \beta z,$$

where x, y, and z represent the position in three dimensions, and σ, ρ, and β are scalar parameters of the system. Read more about the Lorenz attractor on Wikipedia[1] or more details in the book "Viability Theory – New Directions." Write a function `my_lorenz(t,S,sigma,rho,beta)` where `t` is a scalar denoting time, `S` is a 3D array denoting the position (x, y, z), and `sigma`, `rho`, and `beta` are strictly positive scalars representing σ, ρ, and β. The output argument dS should be the same size as S.
Test cases:

```
In [4]: def my_lorenz(t, S, sigma, rho, beta):
            # put your code here

            return dS

        s = np.array([1, 2, 3])
        dS = my_lorenz(0, s, 10, 28, 8/3)
        dS

Out[4]: array([10., 23., -6.])
```

3. Write a function `my_lorenz_solver(t_span,s0,sigma,rho,beta)` solves the Lorenz equations using `solve_ivp`, the function returns `[T,X,Y,Z]`. The input argument `t_span` should be a list of the form $[t_0, t_f]$, where t_0 is the initial time, and t_f is the final time of consideration. The input argument `s0` should be a 3D array of the form $[x_0, y_0, z_0]$, where (x_0, y_0, z_0) represents an initial position. Finally, the input arguments `sigma`, `rho`, and β are the scalar parameters σ, ρ, and β of the Lorenz system. The output argument `T` should be an array of times given as the output argument of `solve_ivp`. The output arguments, `X`, `Y`, and `Z` should be the numerically integrated solution produced from `my_lorenz` in the previous problem and `solve_ivp`.
Test cases:

```
In [5]: def my_lorenz_solver(t_span, s0, sigma, rho, beta):
            # put your code here

            return [T, X, Y, Z]

        sigma = 10
        rho = 28
        beta = 8/3
        t0 = 0
        tf = 50
        s0 = np.array([0, 1, 1.05])

        [T, X, Y, Z] = my_lorenz_solver([t0, tf], s0, sigma, rho, beta)
```

[1] https://en.wikipedia.org/wiki/Lorenz_system.

```
from mpl_toolkits import mplot3d

fig = plt.figure(figsize = (10,10))
ax = plt.axes(projection="3d")
ax.grid()

ax.plot3D(X, Y, Z)

# Set axes label
ax.set_xlabel("x", labelpad=20)
ax.set_ylabel("y", labelpad=20)
ax.set_zlabel("z", labelpad=20)

plt.show()
```

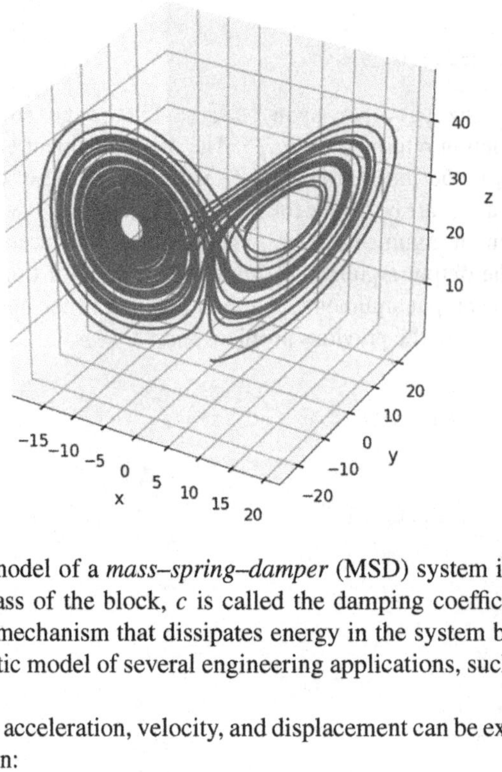

4. Consider the following model of a *mass–spring–damper* (MSD) system in one dimension. In this figure, m denotes the mass of the block, c is called the damping coefficient, and k is the spring stiffness. A damper is a mechanism that dissipates energy in the system by resisting velocity. The MSD system is a simplistic model of several engineering applications, such as shock observers and structural systems.
The relationship between acceleration, velocity, and displacement can be expressed by the following MSD differential equation:

$$m\ddot{x} + c\dot{x} + kx = 0,$$

which can be rewritten as

$$\ddot{x} = \frac{-(c\dot{x} + kx)}{m}.$$

Let the state of the system be denoted by the vector $S = [x; v]$ where x is the displacement of the mass from its resting configuration, and v is its velocity. Rewrite the MSD equation as a first-order differential equation in terms of the state, S. In other words, rewrite the MSD equation as $dS/dt = f(t, S)$.

Write a function `my_msd(t,S,m,c,k)` where `t` is a scalar denoting time, `S` is a 2D vector denoting the state of the MSD system, and `m`, `c`, and `k` are the mass, damping, and stiffness coefficients of the MSD equation, respectively.

Test cases:

```
In [6]: def my_msd(t, S, m, c, k):
            # put your code here
            return ds

        my_msd(0, [1, -1], 10, 1, 100)

Out[6]: array([-1. , -9.9])

In [7]: m = 1
        k = 10
        f = partial(my_msd, m=m, c=0, k=k)
        t_e = np.arange(0, 20, 0.1)
        sol_1=solve_ivp(f,[0,20],[1,0],t_eval=t_e)

        f = partial(my_msd, m=m, c=1, k=k)
        sol_2=solve_ivp(f,[0,20],[1,0],t_eval=t_e)

        f = partial(my_msd, m=m, c=10, k=k)
        sol_3=solve_ivp(f,[0,20],[1,0],t_eval=t_e)

        plt.figure(figsize = (10, 8))
        plt.plot(sol_1.t, sol_1.y[0])
        plt.plot(sol_2.t, sol_2.y[0])
        plt.plot(sol_3.t, sol_3.y[0])
        plt.title("Numerical Solution of MSD System with Varying Dampling")
        plt.xlabel("time")
        plt.ylabel("displacement")
        plt.legend(["no dampling", "c=1", ">critically damped"], loc=1)
```

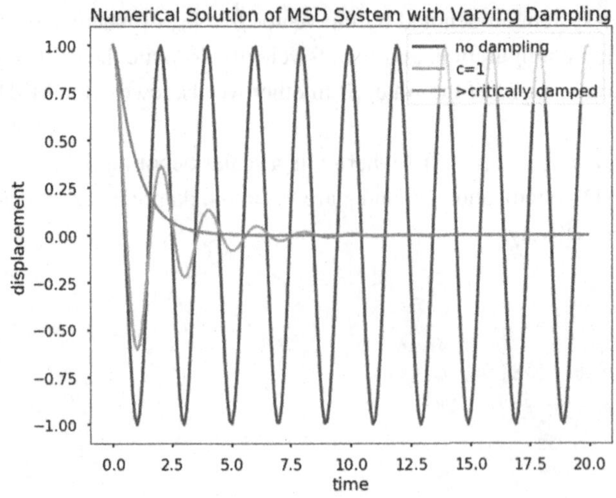

5. Write a function `my_forward_euler(ds,t_span,s0)` where `ds` is a function object, $f(t, s)$, describing a first-order differential equation, `t_span` is an array of times for which numerical solutions of the differential equation are desired, and `s0` is the initial condition of the system. Assume that the size of the state is one. The output argument should be a list of `[t,s]`, such that `t[i] = t_span[i]` for all `i`, and `s` should be the integrated values of `ds` at times `t`. Perform the integration using the forward Euler method, $s[t_i] = s[t_{i-1}] + (t_i - t_{i-1})ds(t_{i-1}, s[t_{i-1}])$. Note that $s[0]$ should equal s_0.
Test cases:

```
In [8]: def my_forward_euler(ds, t_span, s0):
            # put your code here

            return [t, s]

        t_span = np.linspace(0, 1, 10)
        s0 = 1

        # Define parameters
        f = lambda t, s: t*np.exp(-s)

        t_eul, s_eul = my_forward_euler(f, t_span, s0)

        print(t_eul)
        print(s_eul)

[0.         0.11111111 0.22222222 0.33333333 0.44444444 0.55555556
 0.66666667 0.77777778 0.88888889 1.        ]
[1.         1.         1.00454172 1.013584   1.02702534 1.04470783
 1.06642355 1.09192262 1.12092255 1.153118  ]
```

```
In [9]: plt.figure(figsize = (10, 8))

        # Exact solution
        t = np.linspace(0, 1, 1000)
        s = np.log(np.exp(s0) + (t**2-t[0])/2)
        plt.plot(t, s, "r", label="Exact")

        # Forward Euler
        plt.plot(t_eul, s_eul, "g", label="Euler")

        # Python solver
        sol = solve_ivp(f, [0, 1], [s0], t_eval=t)
        plt.plot(sol.t, sol.y[0], "b-", label="Python Solver")

        plt.xlabel("t")
        plt.ylabel("f(t)")
        plt.grid()
        plt.legend(loc=2)
        plt.show()
```

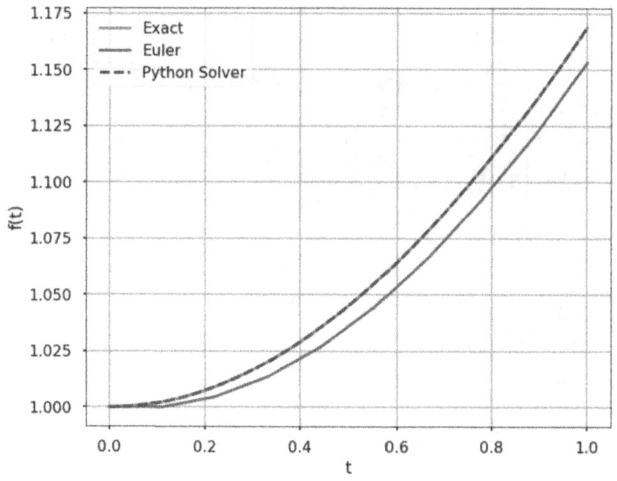

6. Write a function `myRK4(ds,t_span,s0)`, where the input and output arguments are the same as in Problem 5. The function `myRK4` should numerically integrate `ds` using the fourth-order Runge–Kutta method.
Test cases:

```
In [10]: def myRK4(ds, t_span, s0):
             # put your code here

             return [t, s]
```

```
f = lambda t, s: np.sin(np.exp(s))/(t+1)
t_span = np.linspace(0, 2*np.pi, 10)
s0 = 0

plt.figure(figsize = (10, 8))

# Runge-Kutta method
t, s = myRK4(f, t_span, s0)
plt.plot(t, s, "r", label="RK4")

# Python solver
sol = solve_ivp(f, [0, 2*np.pi], [s0], t_eval=t)
plt.plot(sol.t, sol.y[0], "b-", label="Python Solver")

plt.xlabel("t")
plt.ylabel("f(t)")
plt.grid()
plt.legend(loc=2)
plt.show()
```

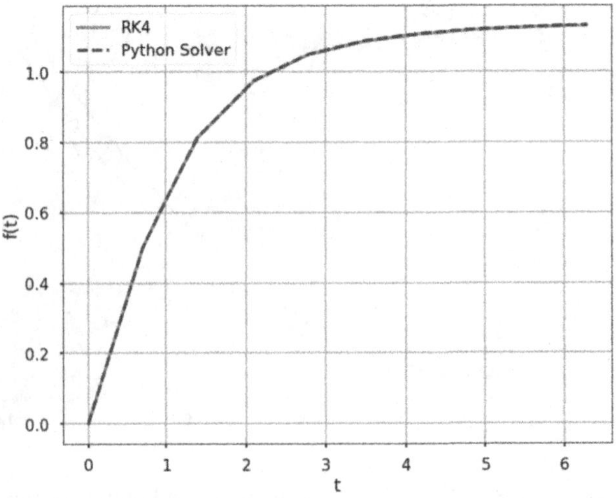

CHAPTER

BOUNDARY-VALUE PROBLEMS FOR ORDINARY DIFFERENTIAL EQUATIONS (ODES) 23

CONTENTS

23.1 ODE BOUNDARY VALUE PROBLEM STATEMENT

The previous chapter introduced ordinary differential equation initial-value problems. For initial value problems, all the known values are specified at the same value of the independent variable, usually at the lower boundary of the interval; this is where the term "initial" comes from. This chapter introduces another type of problem, namely the **boundary-value problem.** As the name suggests, the known values are specified at the extremes of the independent variable, i.e., the boundaries of the interval.

For example, if we have a simple second-order ordinary differential equation,

$$\frac{d^2 f(x)}{dx^2} = \frac{df(x)}{dx} + 3$$

and if the independent variable varies over the domain of $[0, 20]$, the initial value problem will have the two conditions at the value of zero: that is, we know the value of $f(0)$ and $f'(0)$. In contrast, boundary-value problems will specify the values at $x = 0$ and $x = 20$. Note that to solve a first-order ODE to obtain a particular solution requires one constraint, while an nth-order ODE requires n constraints.

The boundary-value problem for an nth-order ordinary differential equation,

$$F\left(x, f(x), \frac{df(x)}{dx}, \frac{d^2 f(x)}{dx^2}, \frac{d^3 f(x)}{dx^3}, \ldots, \frac{d^{n-1} f(x)}{dx^{n-1}}\right) = \frac{d^n f(x)}{dx^n},$$

specifies n known boundary conditions at a and b, to solve this equation on an interval of $x \in [a, b]$. For the second-order case, since the boundary condition can be either be a value of $f(x)$ or a value of the derivative $f'(x)$, we can have several different cases for the specified values. For example, we can have the boundary condition values specified as:

Python Programming and Numerical Methods. https://doi.org/10.1016/B978-0-12-819549-9.00033-6

1. Two values of $f(x)$, that is, $f(a)$ and $f(b)$ are known.
2. Two derivatives of $f'(x)$, that is, $f'(a)$ and $f'(b)$ are known.
3. Mixed conditions from the above two cases are known: that is, either $f(a)$ and $f'(b)$ are known, or $f'(a)$ and $f(b)$ are known.

To get the particular solution, we need two boundary conditions. The second-order ODE boundary-value problem is also called the "Two-Point Boundary-Value Problem." The higher-order ODE problems need additional boundary conditions, which are usually the values of higher derivatives of the independent variables. This chapter focuses on the two-point boundary-value problems.

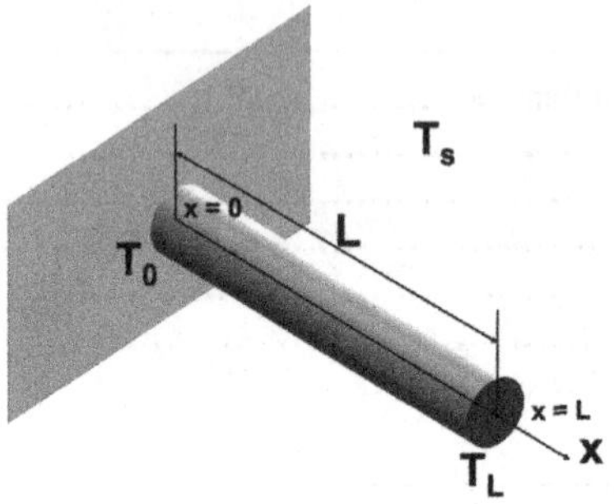

FIGURE 23.1

Heat flow in a pin fin. The variable L is the length of the pin fin, which starts at $x = 0$ and finishes at $x = L$. The temperatures at two ends are T_0 and T_L, with T_s being the surrounding environment temperature.

Below is an example of a boundary-value problem and its solution.

Fins are used in many applications to increase the heat transfer from surfaces. Usually, the design of cooling pin fins is encountered in many applications, e.g., the pin fin used as a heat sink for cooling an object. We can model the temperature distribution in a pin fin as shown in Fig. 23.1, where the length of the fin is L, and the start and end of the fin are $x = 0$ and $x = L$, respectively. The temperatures at the two ends are T_0 and T_L, while T_s is the temperature of the surrounding environment. If we consider both convection and radiation, the steady-state temperature distribution of the pin fin $T(x)$ can be modeled as follows:

$$\frac{d^2T}{dx^2} - \alpha_1(T - T_s) - \alpha_2(T^4 - T_s^4) = 0$$

with the boundary conditions, $T(0) = T_0$ and $T(L) = T_L$; α_1 and α_2 are coefficients. This is a second-order ODE with two boundary conditions; therefore, we can solve it to get particular solutions.

The remainder of this chapter covers two methods of numerically approximating the solution to boundary-value problems on a numerical grid. We will cover both the shooting and finite difference methods to solve ODE boundary-value problems.

23.2 THE SHOOTING METHOD

The **shooting method** was developed with the goal of transforming an ODE boundary value problem into an equivalent initial value problem so that we can solve it using the methods we learned from the previous chapter. With initial value problems, we start at the initial value and march forward to get the solution. This method does not work for boundary value problems because there are not enough initial value conditions to solve the ODE to get a unique solution. The shooting method was developed to overcome this difficulty.

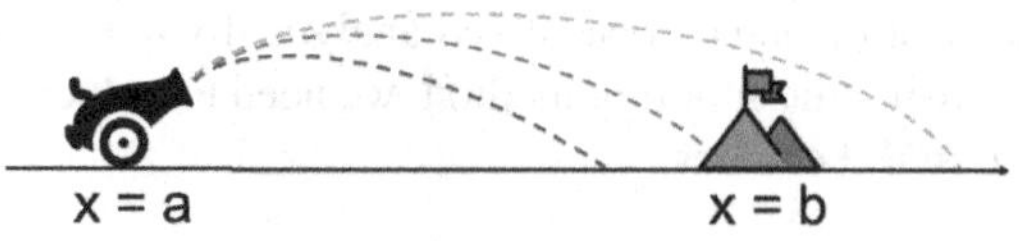

FIGURE 23.2

Target shooting analogy to the shooting method.

The name "shooting method" is analogous with the target shooting: as shown in Fig. 23.2, we shoot at the target and observe where we hit the target. Based on the errors, we can adjust our aim and shoot again hoping that we will hit closer to the target. We can see from the analogy that the shooting method is an iterative optimization method.

Let us see how the shooting method works using the second-order ODE given $f(a) = f_a$ and $f(b) = f_b$, as well as

$$F\left(x, f(x), \frac{df(x)}{dx}\right) = \frac{d^2 f(x)}{dx^2}.$$

Step 1. We start the whole process by guessing $f'(a) = \alpha$, then, together with $f(a) = f_a$, we turn the above problem into an initial value problem with two conditions all at the value $x = a$. This is the **aim** step.

Step 2. Using what we learned from the previous chapter, i.e., we can use a Runge–Kutta method, to integrate to the other boundary b to find $f(b) = f_\beta$. This is called the **shooting** step.

Step 3. Now we compare the value of f_β with f_b. Usually, our initial guess is not good, and $f_\beta \neq f_b$, but what we want is $f_\beta - f_b = 0$; therefore, we adjust our initial guesses and repeat the process until the error is acceptable, at which time we can stop. This is the iterative step.

Although the ideas behind the shooting method are very simple, comparing and finding the best guesses is not easy; this procedure can be very tedious. Finding the best guess to obtain $f_\beta - f_b = 0$ is a root-finding problem and can be tedious, but it does offer a systematic way to search for the best guess. Since f_β is a function of α, the problem becomes finding the root of $g(\alpha) - f_b = 0$. We can use any methods from Chapter 19 to solve the problem.

TRY IT! Say, we want to launch a rocket, and let $y(t)$ be the altitude (in meters from the surface) of the rocket at time t. We know the gravity $g = 9.8\ \text{m/s}^2$. If we want to have the rocket at 50 m off the ground after 5 s after launch, what should be the velocity at launch? (Assuming we ignore the drag of the air resistance.)

To answer this question, we can frame the problem as a boundary-value problem for a second-order ODE. The ODE is

$$\frac{d^2y}{dt^2} = -g,$$

and the two boundary conditions are $y(0) = 0$ and $y(5) = 50$. We want to answer the question: What's $y'(0)$ at launch?

This is a quite simple question and can be solved analytically quite easily; the correct answer $y'(0) = 34.5$. If we solve it using the shooting method, we need to reduce the order of the function first, and the second-order ODE becomes:

$$\frac{dy}{dt} = v,$$

$$\frac{dv}{dt} = -g.$$

Therefore, we have $S(t) = \begin{bmatrix} y(t) \\ v(t) \end{bmatrix}$ which satisfies

$$\frac{dS(t)}{dt} = \begin{bmatrix} 0 & 1 \\ 0 & -g/v \end{bmatrix} S(t).$$

```
In [1]: import numpy as np
        import matplotlib.pyplot as plt
        from scipy.integrate import solve_ivp
        plt.style.use("seaborn-poster")
        %matplotlib inline
```

For our first guess, we take the velocity at launch as 25 m/s.

```
In [2]: F = lambda t, s: np.dot(np.array([[0,1],[0,-9.8/s[1]]]),s)

        t_span = np.linspace(0, 5, 100)
        y0 = 0
        v0 = 25
        t_eval = np.linspace(0, 5, 10)
        sol = solve_ivp(F, [0, 5], [y0, v0], t_eval = t_eval)

        plt.figure(figsize = (10, 8))
        plt.plot(sol.t, sol.y[0])
        plt.plot(5, 50, "ro")
        plt.xlabel("time (s)")
```

```
    plt.ylabel("altitude (m)")
    plt.title(f"first guess v={v0} m/s")
    plt.show()
```

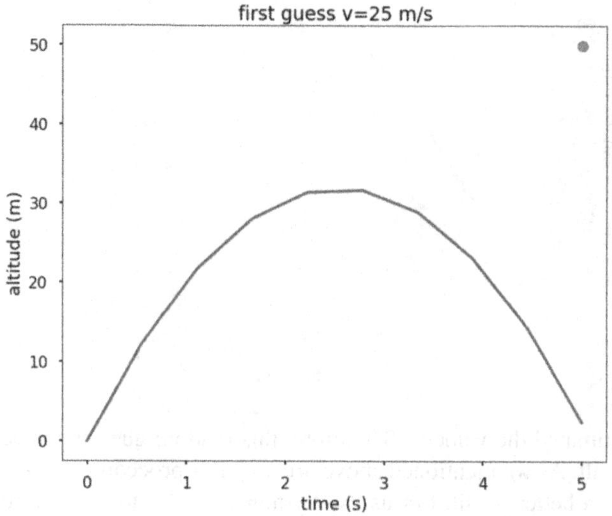

The figure shows that the first guess is a little too small, since after 5 s for the chosen initial velocity, the altitude of the rocket is less than 10 m. The red dot in the figure is the target we want to hit. If we adjust our guess and increase the velocity to 40 m/s, then we obtain:

```
In [3]: v0 = 40
        sol = solve_ivp(F, [0, 5], [y0, v0], t_eval = t_eval)

        plt.figure(figsize = (10, 8))
        plt.plot(sol.t, sol.y[0])
        plt.plot(5, 50, "ro")
        plt.xlabel("time (s)")
        plt.ylabel("altitude (m)")
        plt.title(f"second guess v={v0} m/s")
        plt.show()
```

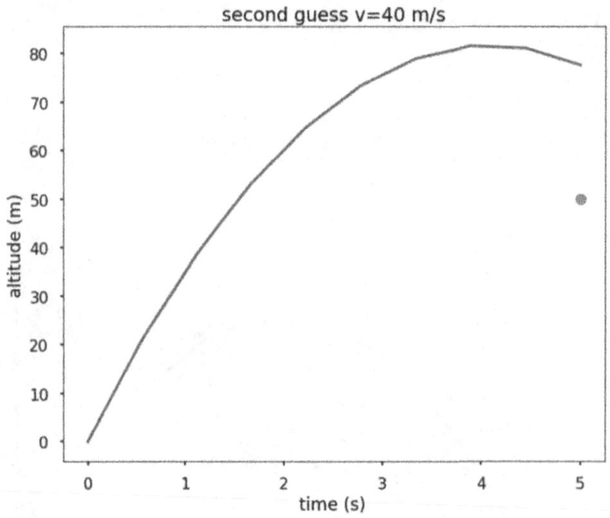

Here, we overestimated the velocity. Therefore, this random guessing is perhaps not the best way to obtain the result. As we mentioned above, treating this procedure as a root-finding problem will provide us with a better result. Let us use Python's `fsolve` to find the root. The following example will demonstrate how to find the correct answer directly.

```
In [4]: from scipy.optimize import fsolve

        def objective(v0):
            sol = solve_ivp(F, [0, 5], [y0, v0], t_eval = t_eval)
            y = sol.y[0]
            return y[-1] - 50

        v0, = fsolve(objective, 10)
        print(v0)

34.499999999999986
```

```
In [5]: sol = solve_ivp(F, [0, 5], [y0, v0], t_eval = t_eval)

        plt.figure(figsize = (10, 8))
        plt.plot(sol.t, sol.y[0])
        plt.plot(5, 50, "ro")
        plt.xlabel("time (s)")
        plt.ylabel("altitude (m)")
```

```
        plt.title(f"root finding v={v0} m/s")
        plt.show()
```

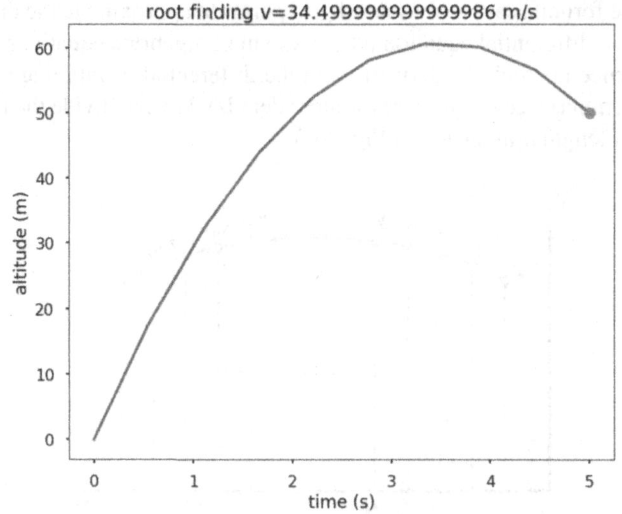

TRY IT! Let us change the initial guess and see if this changes the result.

```
In [6]: for v0_guess in range(1, 100, 10):
            v0, = fsolve(objective, v0_guess)
            print("Init: %d, Result: %.1f" %(v0_guess, v0))

Init: 1, Result: 34.5
Init: 11, Result: 34.5
Init: 21, Result: 34.5
Init: 31, Result: 34.5
Init: 41, Result: 34.5
Init: 51, Result: 34.5
Init: 61, Result: 34.5
Init: 71, Result: 34.5
Init: 81, Result: 34.5
Init: 91, Result: 34.5
```

Note that changing the initial guesses does not change the result, which means that this method is stable; see below regarding the problem of stability.

23.3 THE FINITE DIFFERENCE METHOD

Another way to solve the ODE boundary value problems is to use the **finite difference method**, where we use finite difference formulas at evenly spaced grid points to approximate the differential equations, which then transforms a differential equation into a system of algebraic equations to solve.

In the finite difference method, the derivatives in the differential equation are approximated using the finite difference formulas (see Chapter 20 for more details). We can divide the interval of $[a, b]$ into n equal subintervals of length h as shown in Fig. 23.3.

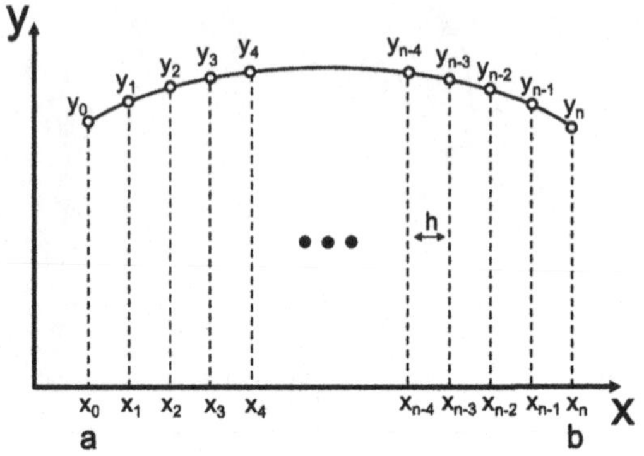

FIGURE 23.3

Illustration of the finite difference method.

Usually, we use the central difference formulas in the finite difference methods because they yield better accuracy. The differential equation is enforced only at the grid points, and the first and second derivatives are:

$$\frac{dy}{dx} = \frac{y_{i+1} - y_{i-1}}{2h},$$
$$\frac{d^2y}{dx^2} = \frac{y_{i-1} - 2y_i + y_{i+1}}{h^2}.$$

These finite difference expressions are used to replace the derivatives of y in the differential equation, leading to a system of $n+1$ linear algebraic equations if the differential equation is linear. If the differential equation is nonlinear, the algebraic equations will also be nonlinear.

EXAMPLE: Solve the rocket problem in the previous section using the finite difference method; plot the altitude of the rocket after launch. The ODE is

$$\frac{d^2y}{dt^2} = -g$$

with the boundary conditions $y(0) = 0$ and $y(5) = 50$. Let us take $n = 10$.

Since the time interval is $[0, 5]$ and we have $n = 10$, $h = 0.5$, using the finite difference approximated derivatives, we obtain

$$y_0 = 0, \quad y_{i-1} - 2y_i + y_{i+1} = -gh^2, \; i = 1, 2, \ldots, n-1, \quad y_{10} = 50.$$

If we use matrix notation, we obtain

$$\begin{bmatrix} 1 & 0 & & & \\ 1 & -2 & 1 & & \\ & \ddots & \ddots & \ddots & \\ & & 1 & -2 & 1 \\ & & & & 1 \end{bmatrix} \begin{bmatrix} y_0 \\ y_1 \\ \vdots \\ y_{n-1} \\ y_n \end{bmatrix} = \begin{bmatrix} 0 \\ -gh^2 \\ \vdots \\ -gh^2 \\ 50 \end{bmatrix}.$$

With 11 equations in the system, we can solve it using the method we presented in Chapter 14.

```
In [1]: import numpy as np
        import matplotlib.pyplot as plt
        plt.style.use("seaborn-poster")
        %matplotlib inline

        n = 10
        h = (5-0) / n

        # Get A
        A = np.zeros((n+1, n+1))
        A[0, 0] = 1
        A[n, n] = 1
        for i in range(1, n):
            A[i, i-1] = 1
            A[i, i] = -2
            A[i, i+1] = 1

        print(A)

        # Get b
        b = np.zeros(n+1)
        b[1:-1] = -9.8*h**2
        b[-1] = 50
        print(b)

        # solve the linear equations
        y = np.linalg.solve(A, b)
```

```
    t = np.linspace(0, 5, 11)

    plt.figure(figsize=(10,8))
    plt.plot(t, y)
    plt.plot(5, 50, "ro")
    plt.xlabel("time (s)")
    plt.ylabel("altitude (m)")
    plt.show()
```

```
[[ 1.  0.  0.  0.  0.  0.  0.  0.  0.  0.  0.]
 [ 1. -2.  1.  0.  0.  0.  0.  0.  0.  0.  0.]
 [ 0.  1. -2.  1.  0.  0.  0.  0.  0.  0.  0.]
 [ 0.  0.  1. -2.  1.  0.  0.  0.  0.  0.  0.]
 [ 0.  0.  0.  1. -2.  1.  0.  0.  0.  0.  0.]
 [ 0.  0.  0.  0.  1. -2.  1.  0.  0.  0.  0.]
 [ 0.  0.  0.  0.  0.  1. -2.  1.  0.  0.  0.]
 [ 0.  0.  0.  0.  0.  0.  1. -2.  1.  0.  0.]
 [ 0.  0.  0.  0.  0.  0.  0.  1. -2.  1.  0.]
 [ 0.  0.  0.  0.  0.  0.  0.  0.  1. -2.  1.]
 [ 0.  0.  0.  0.  0.  0.  0.  0.  0.  0.  1.]]
[0.  -2.45 -2.45 -2.45 -2.45 -2.45 -2.45 -2.45 -2.45 -2.45 50.]
```

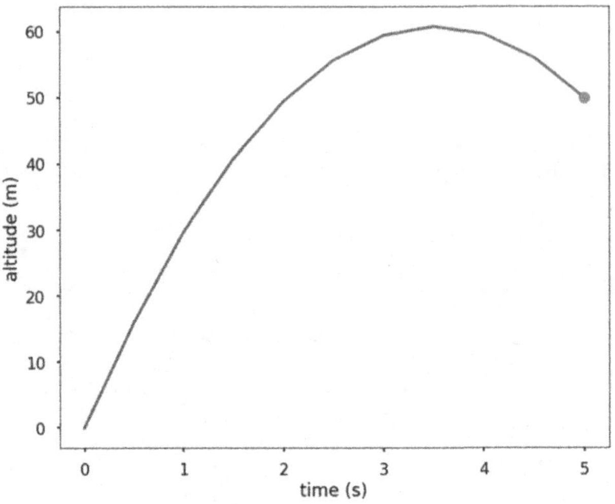

Let us solve for $y'(0)$. From the finite difference formula, we know that $\frac{dy}{dx} = \frac{y_{i+1}-y_{i-1}}{2h}$, which means that $y'(0) = \frac{y_1-y_{-1}}{2h}$; but we do not know what is y_{-1}. We can calculate y_{-1} since we know the y values on each grid point. From the second derivative finite difference formula, we know that $\frac{y_{-1}-2y_0+y_1}{h^2} = -g$; therefore, we can solve for y_{-1} to obtain the launching velocity. See the calculation below.

```
In [2]: y_n1 = -9.8*h**2 + 2*y[0] - y[1]
        (y[1] - y_n1) / (2*h)

Out[2]: 34.5
```

We obtain the correct launching velocity using the finite difference method. To given you additional exposure to this concept, let us see another example.

TRY IT! Use the finite difference method to solve the following linear boundary value problem:

$$\frac{d''y(t)}{dt^2} = -4y + 4x$$

with the boundary conditions as $y(0) = 0$ and $y'(\pi/2) = 0$. The exact solution of the problem is $y = x - \sin 2x$, plot the errors against the n grid points (n from 3 to 100) for the boundary point $y(\pi/2)$.

Using the finite difference approximated derivatives, we have

$$y_0 = 0, \quad y_{i-1} - 2y_i + y_{i+1} - h^2(-4y_i + 4x_i) = 0, \ i = 1, 2, ..., n-1,$$

$$2y_{n-1} - 2y_n - h^2(-4y_n + 4x_n) = 0.$$

The last equation is derived from $\frac{y_{n+1}-y_{n-1}}{2h} = 0$ (the boundary condition $y'(\pi/2) = 0$); therefore, $y_{n+1} = y_{n-1}$.

If we use matrix notation, we will obtain:

$$\begin{bmatrix} 1 & 0 & & & \\ 1 & -2+4h^2 & 1 & & \\ & \ddots & \ddots & \ddots & \\ & & 1 & -2+4h^2 & 1 \\ & & & 2 & -2+4h^2 \end{bmatrix} \begin{bmatrix} y_0 \\ y_1 \\ \vdots \\ y_{n-1} \\ y_n \end{bmatrix} = \begin{bmatrix} 0 \\ 4h^2x_1 \\ \vdots \\ 4h^2x_{n-1} \\ 4h^2x_n \end{bmatrix}.$$

```
In [3]: def get_a_b(n):
            h = (np.pi/2-0) / n
            x = np.linspace(0, np.pi/2, n+1)
            # Get A
            A = np.zeros((n+1, n+1))
            A[0, 0] = 1
            A[n, n] = -2+4*h**2
```

```
    A[n, n-1] = 2
    for i in range(1, n):
        A[i, i-1] = 1
        A[i, i] = -2+4*h**2
        A[i, i+1] = 1

    # Get b
    b = np.zeros(n+1)
    for i in range(1, n+1):
        b[i] = 4*h**2*x[i]

    return x, A, b

x = np.pi/2
v = x - np.sin(2*x)

n_s = []
errors = []

for n in range(3, 100, 5):
    x, A, b = get_a_b(n)
    y = np.linalg.solve(A, b)
    n_s.append(n)
    e = v - y[-1]
    errors.append(e)

plt.figure(figsize = (10,8))
plt.plot(n_s, errors)
plt.yscale("log")
plt.xlabel("n gird points")
plt.ylabel("errors at x = $\pi/2$")
plt.show()
```

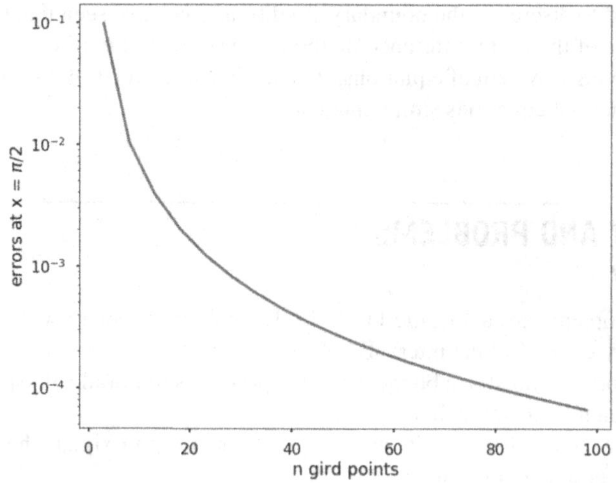

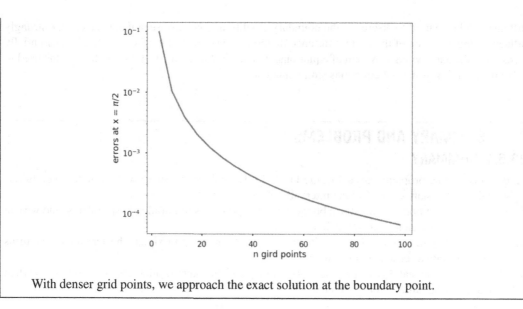

With denser grid points, we approach the exact solution at the boundary point.

The finite difference method can be also applied to higher-order ODEs, but it needs approximation of the higher-order derivatives using a finite difference formula. For example, to solve a fourth-order ODE requires performing the following:

$$\frac{d^4y}{dx^4} = \frac{y_{i-2} - 4y_{i-1} + 6y_i - 4y_{i+1} + y_{i+2}}{h^4}.$$

We will not discuss higher-order ODEs, since the idea behind its solution is similar to the second-order ODE we presented above.

23.4 NUMERICAL ERROR AND INSTABILITY

Boundary-value problems also exhibit two main issues discussed in Chapter 22: the **numerical error (accuracy)** and **stability**. Depending on the different methods used, keep in mind that the shooting and finite difference methods are different in terms of the errors they present.

For the shooting method, the *numerical error* is similar to that described for initial-value problems as the shooting method essentially transforms the boundary-value problem into a series of initial value problems. In terms of the stability of the method, as shown in the example in Section 23.2, in terms of stability, even though our initial guesses were not close to the true answer, the method returned an accurate numerical solution. This is due to the adding of the right-most constraint, which keeps the errors from increasing unboundedly.

In the case of finite difference method, the numerical error is determined by the order of accuracy of the numerical scheme used. The accuracy of the different scheme used for derivative approximations are discussed in Section 20.2. The accuracy of the finite difference method is determined by the larger of the two truncation errors: the difference scheme used for the differential equation or that of the difference scheme used to discretize the boundary conditions (we have seen that the step size strongly influences the accuracy of the finite difference method). Since the finite difference method essentially turns the BVP into solving a system of equations, it is dependent on the stability of the scheme used to solve the resulting system of equations simultaneously.

23.5 SUMMARY AND PROBLEMS

23.5.1 SUMMARY

1. Boundary-value problems are a specific kind of ODE-solving problem with boundary conditions specified at the start and end of the interval.
2. The shooting method can transform boundary-value problems to initial value problems, and we can use a root-finding method to solve them.
3. The finite difference method uses a finite difference scheme to approximate the derivatives and turns the problem into a set of equations to solve.
4. The accuracy and stability of the boundary-value problems have similarities and differences when compared to the initial-value problems.

23.5.2 PROBLEMS

1. Describe the difference between boundary-value problems and the initial-value problems in ODEs.
2. Try to describe the intuition behind the shooting method and its links to initial value problems.
3. What is the finite difference method for boundary value problems? How do we apply it?
4. Solve the following boundary value problem with $y(0) = 0$ and $y(\pi/2) = 1$:

$$y'' + (1 - 0.2x)y^2 = 0.$$

5. Solve the following ODE with $y(0) = 0$ and $y(\pi) = 0$:

$$y'' + \sin y + 1 = 0.$$

6. Given the ODE with the boundary conditions $y(0) = 0$ and $y(12) = 0$,

$$y'' + 0.5x^2 - 6x = 0,$$

what is the value of $y'(0)$?

7. Solve the following ODE with boundary conditions $y(1) = 0$, $y''(1) = 0$ and $y(2) = 1$:

$$y''' + \frac{1}{x}y'' - \frac{1}{x^2}y' - 0.1(y')^3 = 0.$$

8. A flexible cable is suspended between two points, as shown in the following figure. The density of the cable is uniform. The shape of the cable $y(x)$ is governed by the differential equation:

$$\frac{d^2y}{d^2x} = C\sqrt{1+\left(\frac{dy}{dx}\right)^2}$$

where C is a constant that equal to the ratio of the weight per unit length of the cable to the magnitude of the horizontal component of tension in the cable at its lowest point. The cable hangs between two points specified by $y(0) = 8$ m and $y(10) = 10$ m, and $C = 0.039$ m^{-1}. Can you determine and plot the shape of the cable between $x = 0$ and $x = 10$?

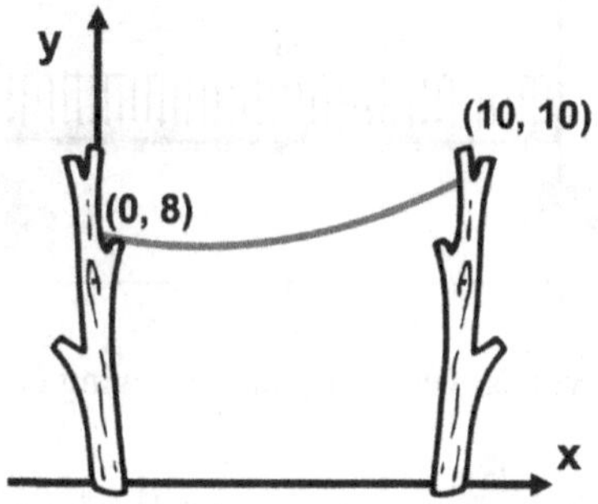

9. Fins are used in many applications to increase the heat transfer from surfaces. The design of cooling pin fins is used for many applications, e.g., as a heat sink for cooling an object. We model the temperature distribution in a pin fin as shown in the following figure, where the length of the fin is L, and the start and end of the fin are $x = 0$ and $x = L$, respectively. The temperatures at the two ends are T_0 and T_L, while T_s is the temperature of the surrounding environment. If we consider both convection and radiation, the steady-state temperature distribution of the pin fin $T(x)$ between $x = 0$ and $x = L$ can be modeled with the following equation:

$$\frac{d^2T}{dx^2} - \alpha_1(T - T_s) - \alpha_2(T^4 - T_s^4) = 0$$

with the boundary conditions $T(0) = T_0$ and $T(L) = T_L$, and α_1 and α_2 being coefficients. They are defined as $\alpha_1 = \frac{h_c P}{kA_c}$ and $\alpha_2 = \frac{\epsilon\sigma_{SB}P}{kA_c}$, where h_c is the convective heat transfer coefficient, P is the perimeter bounding the cross-section of the fin, ϵ is the radiative emissivity of the surface of the fin, k is the thermal conductivity of the fin material, A_c is the cross-sectional area of the fin, and $\sigma_{SB} = 5.67 \times 10^{-8}$ W/(m^{2}K^2) is the Stefan–Boltzmann constant.

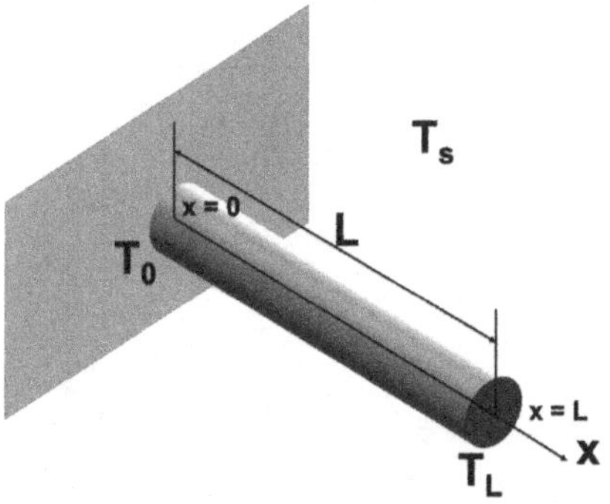

Determine the temperature distribution if $L = 0.2$ m, $T(0) = 475$ K, $T(0.1) = 290$ K, and $T_s = 290$ K. Use the following values for the parameters: $h_c = 40$ W/m^2/K, $P = 0.015$ m, $\epsilon = 0.4$, $k = 240$W/m/K, and $A_c = 1.55 \times 10^{-5}$ m^2.

10. A simply supported beam carries a uniform load of intensity ω_0 as shown in the following figure.

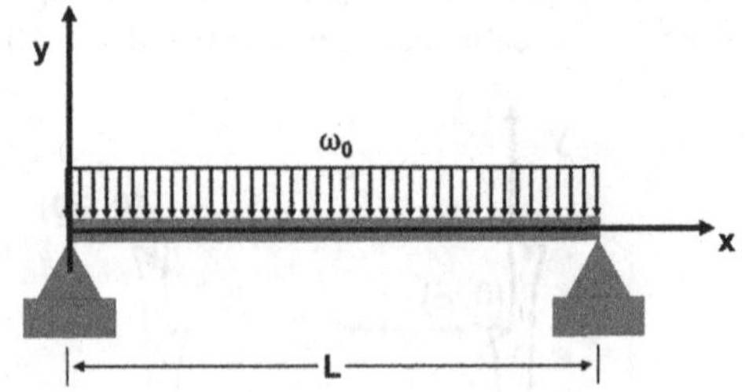

The deflection of the beam, y, is governed by the following ODE:

$$EI\frac{d^2y}{dx^2} = \frac{1}{2}\omega_0(Lx - x^2)\left[1 + \left(\frac{dy}{dx}^2\right)\right]^{\frac{3}{2}}$$

where EI is the flexural rigidity.
If $L = 5$ m, and the two boundary conditions are $y(0) = 0$ and $y(L) = 0$, $EI = 1.8 \times 10^7\,\text{N}\cdot\text{m}^2$, and $\omega_0 = 15 \times 10^3$ N/m, determine and plot the deflection of the beam as a function of x.

CHAPTER

FOURIER TRANSFORM

24

CONTENTS

24.1 THE BASICS OF WAVES

There are many types of waves in our environment. For example, if you throw a rock into a pond, you can see the waves form and travel in the water. There are many types of waves. Some of them are difficult to see, such as sound waves, earthquake waves, and microwaves (that we use to cook our food in the kitchen). But in physics, a wave is a disturbance that travels through space and matter, with a transferring energy from one place to another. It is important to study waves in our life to understand how they form, travel, etc. This chapter will cover a basic tool that helps us understand and study waves – the **Fourier Transform**. But before we proceed, let us first familiarize ourselves with how to model waves mathematically.

24.1.1 MODELING A WAVE USING MATHEMATICAL TOOLS

We can model a single wave as a field with a function $F(x,t)$, where x is the location of a point in space, while t is the time. One simplest case is the shape of a sine wave change over x.

Python Programming and Numerical Methods. https://doi.org/10.1016/B978-0-12-819549-9.00034-8

```
In [1]: import matplotlib.pyplot as plt
        import numpy as np

        plt.style.use("seaborn-poster")
        %matplotlib inline

In [2]: x = np.linspace(0, 20, 201)
        y = np.sin(x)

        plt.figure(figsize = (8, 6))
        plt.plot(x, y, "b")
        plt.ylabel("Amplitude")
        plt.xlabel("Location (x)")
        plt.show()
```

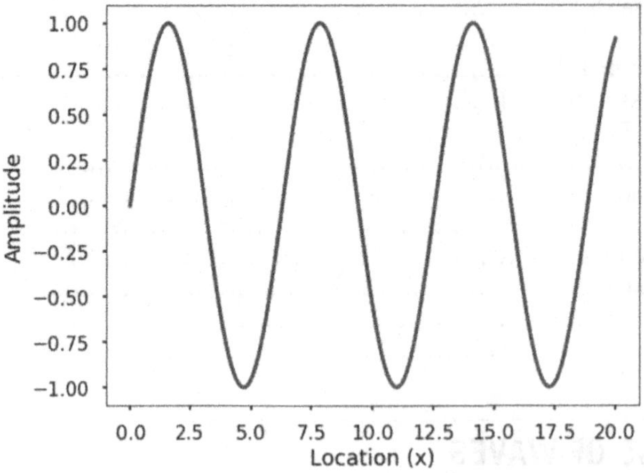

A sine wave can change both in time and space. If we plot the changes at various locations, each time snapshot will be a sine wave changing with location. See the following figure with a fixed point at $x = 2.5$, showing as a red dot. Of course, you can see the changes over time at specific location as well; plot this by yourself.

```
In [3]: fig = plt.figure(figsize = (8,8))

        times = np.arange(5)

        n = len(times)

        for t in times:
            plt.subplot(n, 1, t+1)
```

```
        y = np.sin(x + t)
        plt.plot(x, y, "b")
        plt.plot(x[25], y [25], "ro")
        plt.ylim(-1.1, 1.1)
        plt.ylabel("y")
        plt.title(f"t = {t}")

    plt.xlabel("location (x)")
    plt.tight_layout()
    plt.show()
```

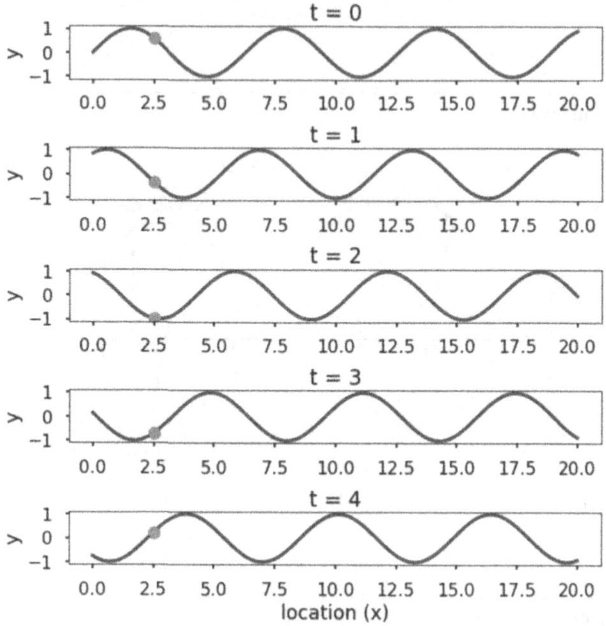

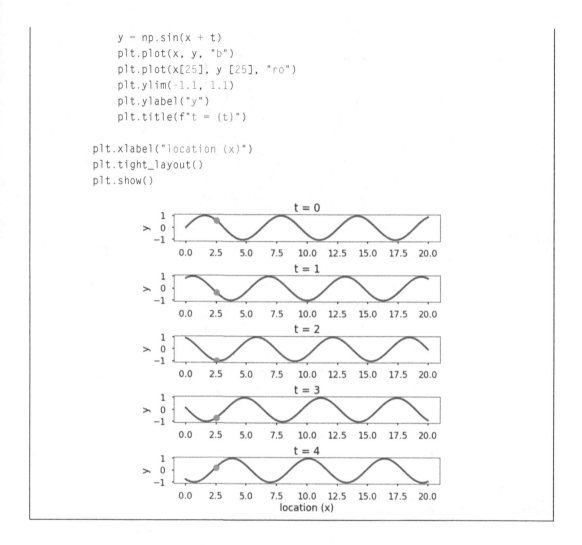

24.1.2 CHARACTERISTICS OF A WAVE

Waves can be a continuous entity both in time and space. For numerical purposes, one needs to digitize time and space at various points. For example, in the context of geophysics, one can use sensors such as accelerometers (which measure the acceleration of a movement) at different locations on the Earth to monitor earthquakes; this is called spatial discretization. Similarly, these sensors usually record the data at certain times, which is called temporal discretization. For a single wave, it has different characteristics. See Figs. 24.1 and 24.2.

Amplitude is used to describe the difference between the maximum values to the baseline value (see the above figures). A sine wave is a periodic signal, which means it repeats itself after a certain time and can be measured by **period**. The period of a wave is the time it takes to finish the complete cycle. In the figure, the period can be measured from the two adjacent peaks. The **wavelength** measures

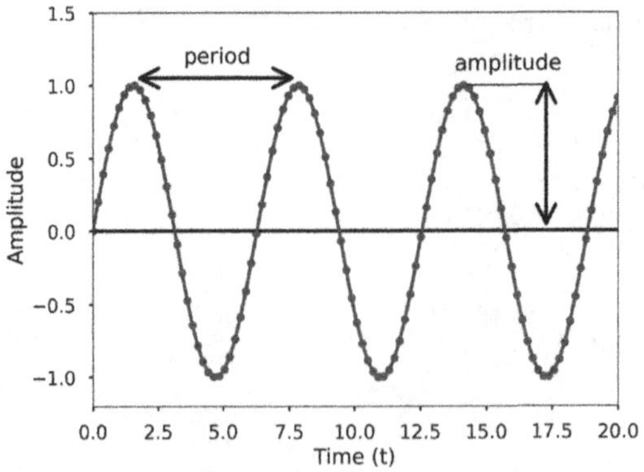

FIGURE 24.1

Period and amplitude of a sine wave.

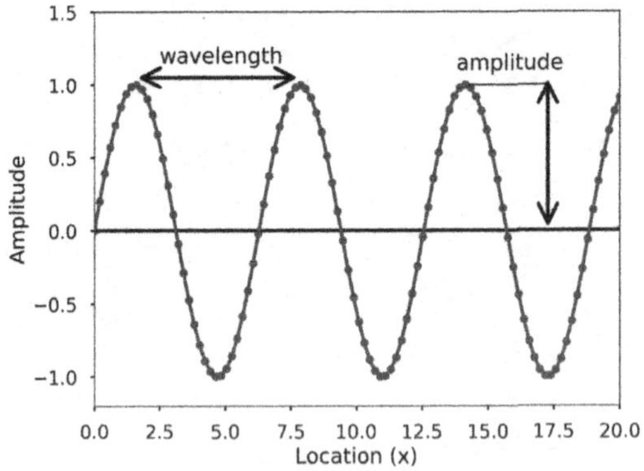

FIGURE 24.2

Wavelength and amplitude of a sine wave.

the distance between two successive crests or troughs of a wave. the **frequency** describes the number of waves that pass a fixed location in a given amount of time. Frequency can be measured by how many cycles pass within 1 s. Therefore, the unit of frequency is cycles/second, or more commonly used **Hertz** (abbreviated **Hz**). Frequency is different from period, but they are related to each other. Frequency refers to how often something happens while period refers to the time it takes to complete something, mathematically,

$$\text{period} = \frac{1}{\text{frequency}}.$$

Note in the two figures the blue dots on the sine waves; these are the discretization points we plotted both in time and space; only at these dots did we sample the value of the wave. Usually when we record a wave, we need to specify how often we sample the wave in time; this is called **sampling**. This rate is called the **sampling rate**, in Hz. For example, if we sample a wave at 2 Hz, it means that at every second we sample two data points. Now that we understand more about the basics of a wave, let us study a sine wave, which can be represented by the following equation:

$$y(t) = A \sin(\omega t + \phi)$$

where A is the amplitude of the wave, and ω is the **angular frequency**, which specifies how many cycles occur in a second, in radians per second; ϕ is the **phase** of the signal. If T is the period of the wave and f is the frequency of the wave, then ω has the following relationship to them:

$$\omega = \frac{2\pi}{T} = 2\pi f.$$

TRY IT! Generate two sine waves with time between zero and 1 s whose frequency is 5 and 10 Hz, respectively, sampled at 100 Hz. Plot the two waves and see the difference. Count how many cycles there are in 1 s.

```
In [4]: # sampling rate
        sr = 100.0
        # sampling interval
        ts = 1.0/sr
        t = np.arange(0,1,ts)

        # frequency of the signal
        freq = 5
        y = np.sin(2*np.pi*freq*t)

        plt.figure(figsize = (8, 8))
        plt.subplot(211)
        plt.plot(t, y, "b")
        plt.ylabel("Amplitude")

        freq = 10
        y = np.sin(2*np.pi*freq*t)

        plt.subplot(212)
        plt.plot(t, y, "b")
        plt.ylabel("Amplitude")

        plt.xlabel("Time (s)")
        plt.show()
```

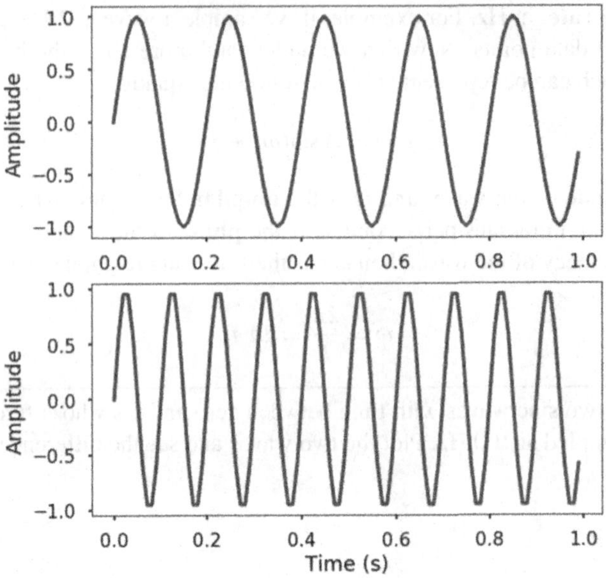

TRY IT! Generate two sine waves with time between zero and 1 s. Both waves have a frequency of 5 Hz, sampled at 100 Hz, but their phase are zero and 10, respectively. The amplitudes of the two waves are 5 and 10, respectively. Plot the two waves and see the difference.

```
In [5]: # frequency of the signal
        freq = 5
        y = 5*np.sin(2*np.pi*freq*t)

        plt.figure(figsize = (8, 8))
        plt.subplot(211)
        plt.plot(t, y, "b")
        plt.ylabel("Amplitude")

        y = 10*np.sin(2*np.pi*freq*t + 10)

        plt.subplot(212)
        plt.plot(t, y, "b")
        plt.ylabel("Amplitude")

        plt.xlabel("Time (s)")
        plt.show()
```

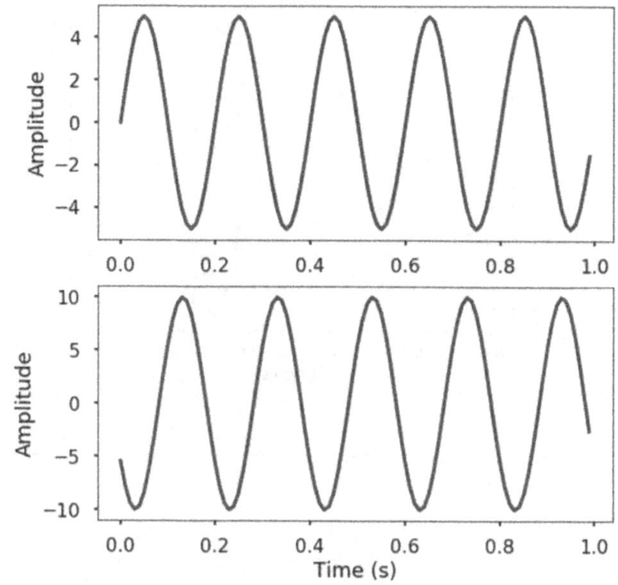

24.2 DISCRETE FOURIER TRANSFORM (DFT)

The previous section demonstrated how easy it is to characterize a wave with period/frequency, amplitude, and phase. This only applies to simple periodic signals, such as sine or cosine waves. For complicated waves, such characterization is not easy anymore. For example, Fig. 24.3 is a relatively more complicated wave, and it is hard to say what is the frequency, and amplitude of the wave.

There are more complicated cases in the real world. The **Fourier Transform** can be used for the purpose to study these waves. It decomposes any signal into a sum of fundamental sine and cosine waves that easily measures the frequency, amplitude, and phase. The Fourier transform can be applied to continuous or discrete waves; this chapter will only discuss the Discrete Fourier Transform (DFT).

Using the DFT, we can compose the above signal to a series of sinusoids, with each of them having a different frequency. The following 3D figure (Fig. 24.4) shows the idea behind the DFT: the above signal is actually the result of the sum of three different sine waves. The time domain signal, which is the above signal that can be transformed into a figure in the frequency domain, is called DFT amplitude spectrum, where the signal frequencies are shown as vertical bars. The height of the bar after normalization is the amplitude of the signal in the time domain. Note that the three vertical bars correspond to the three frequencies of the sine wave, which are also plotted in the figure.

This section will discuss how to use the DFT to compute and plot the DFT amplitude spectrum.

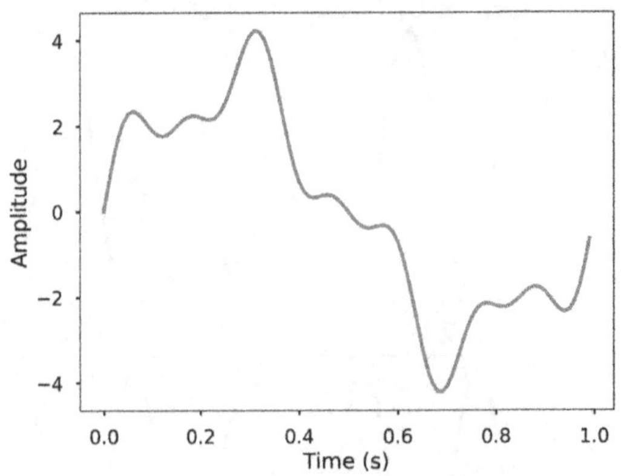

FIGURE 24.3

More general wave form.

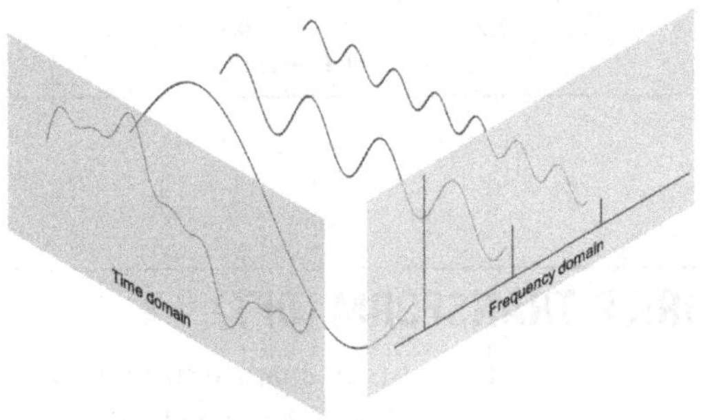

FIGURE 24.4

Illustration of Fourier transform with time and frequency domain signal.

24.2.1 DFT

The DFT can transform a sequence of evenly spaced signals to the information about the frequency of all the sine waves needed to sum to obtain the time-domain signal. It is defined as

$$X_k = \sum_{n=0}^{N-1} x_n \cdot e^{-i2\pi kn/N} = \sum_{n=0}^{N-1} x_n[\cos(2\pi kn/N) - i \cdot \sin(2\pi kn/N)]$$

- N = number of samples
- n = current sample
- k = current frequency, where $k \in [0, N-1]$
- x_n = the sine value at sample n
- X_k = the DFT that includes information of both amplitude and phase

Also, the last expression in the above equation is derived from the *Euler's formula*, which links the trigonometric functions to the complex exponential function: $e^{i \cdot x} = \cos x + i \cdot \sin x$.

Due to the nature of the transform, $X_0 = \sum_{n=0}^{N-1} x_n$. If N is an odd number, the elements $X_1, X_2, \ldots, X_{(N-1)/2}$ contain the positive frequency terms, and the elements $X_{(N+1)/2}, \ldots, X_{N-1}$ contain the negative frequency terms, in order of decreasingly negative frequency. If N is even, the elements $X_1, X_2, \ldots, X_{N/2-1}$ contain the positive frequency terms, and the elements $X_{N/2}, \ldots, X_{N-1}$ contain the negative frequency terms, in order of decreasingly negative frequency. In the case here, our input signal x is a real-valued sequence; therefore, the DFT output X_n for positive frequencies is the conjugate of the values X_n for negative frequencies, and the spectrum will be symmetric. Usually, we only plot the DFT corresponding to the positive frequencies.

Note that X_k is a complex number that encodes both the amplitude and phase information of a complex sinusoidal component $e^{i \cdot 2\pi kn/N}$ of function x_n. The amplitude and phase of the signal can be calculated as

$$\text{amp} = \frac{|X_k|}{N} = \frac{\sqrt{\text{Re}(X_k)^2 + \text{Im}(X_k)^2}}{N}$$

$$\text{phase} = \text{atan2}(\text{Im}(X_k), \text{Re}(X_k))$$

where $\text{Im}(X_k)$ and $\text{Re}(X_k)$ are the imaginary and real parts of the complex number, and atan2 is the two-argument form of the arctan function.

The amplitudes returned by DFT are equal to the amplitudes of the signals fed into the DFT if we normalize them by the number of sample points. We mentioned previously that for a real-valued signal, the output of the DFT will be mirrored about the half of the sampling rate (as shown in the following examples). This half of the sampling rate is called the **Nyquist frequency**. This usually means that we only need to look at one side of the DFT result and drop the duplicate information in the other half; therefore, to obtain the amplitude corresponding to that of the time domain signal, we divide by $N/2$ instead of N.

Now that we have the basic knowledge of DFT, let us see how we can use it.

TRY IT! Generate three sine waves with frequencies 1, 4, and 7 Hz, amplitudes 3, 1, and 0.5, and phases being all zeros. Add these three sine waves together with a sampling rate of 100 Hz; it is the same signal shown at the beginning of the section.

```
In [1]: import matplotlib.pyplot as plt
        import numpy as np

        plt.style.use("seaborn-poster")
        %matplotlib inline

In [2]: # sampling rate
        sr = 100
        # sampling interval
        ts = 1.0/sr
```

```
t = np.arange(0,1,ts)

freq = 1.
x = 3*np.sin(2*np.pi*freq*t)

freq = 4
x += np.sin(2*np.pi*freq*t)

freq = 7
x += 0.5* np.sin(2*np.pi*freq*t)

plt.figure(figsize = (8, 6))
plt.plot(t, x, "r")
plt.ylabel("Amplitude")

plt.show()
```

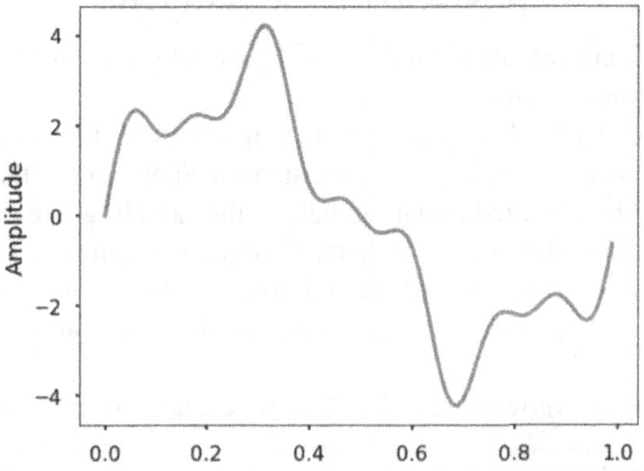

TRY IT! Write a function DFT(x) which takes in one argument, x, with a 1D real-valued signal. The function will calculate the DFT of the signal and return the DFT values. Apply this function to the signal we generated above and plot the result.

```
In [3]: def DFT(x):
            """
            Function to calculate the
            discrete Fourier Transform
            of a 1D real-valued signal x
            """

            N = len(x)
```

```
        n = np.arange(N)
        k = n.reshape((N, 1))
        e = np.exp(-2j * np.pi * k * n / N)

        X = np.dot(e, x)

        return X

In [4]: X = DFT(x)

    # calculate the frequency
    N = len(X)
    n = np.arange(N)
    T = N/sr
    freq = n/T

    plt.figure(figsize = (8, 6))
    plt.stem(freq, abs(X), "b", markerfmt=" ", basefmt="-b")
    plt.xlabel("Freq (Hz)")
    plt.ylabel("DFT Amplitude |X(freq)|")
    plt.show()
```

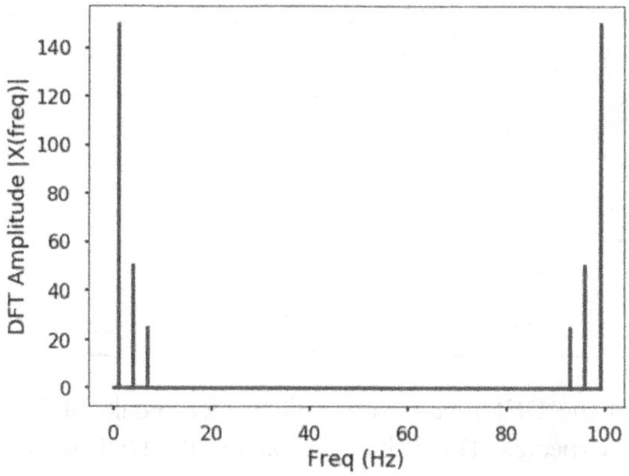

Note that the output of the DFT is symmetric at half of the sampling rate (try different sampling rates for fun). As mentioned earlier, this half of the sampling rate is called the **Nyquist frequency** or the folding frequency. It is named after the electrical engineer Harry Nyquist. He and Claude Shannon formulated the Nyquist–Shannon sampling theorem, which states that a signal sampled at a rate can be fully reconstructed if it contains only frequency components below half that sampling frequency; thus, the highest frequency output from the DFT is half the sampling rate.

```
In [5]: n_oneside = N//2
        # get the one side frequency
        f_oneside = freq[:n_oneside]

        # normalize the amplitude
        X_oneside =X[:n_oneside]/n_oneside

        plt.figure(figsize = (12, 6))
        plt.subplot(121)
        plt.stem(f_oneside, abs(X_oneside), "b", markerfmt=" ", basefmt="-b")
        plt.xlabel("Freq (Hz)")
        plt.ylabel("DFT Amplitude |X(freq)|")

        plt.subplot(122)
        plt.stem(f_oneside, abs(X_oneside), "b", markerfmt=" ", basefmt="-b")
        plt.xlabel("Freq (Hz)")
        plt.xlim(0, 10)
        plt.tight_layout()
        plt.show()
```

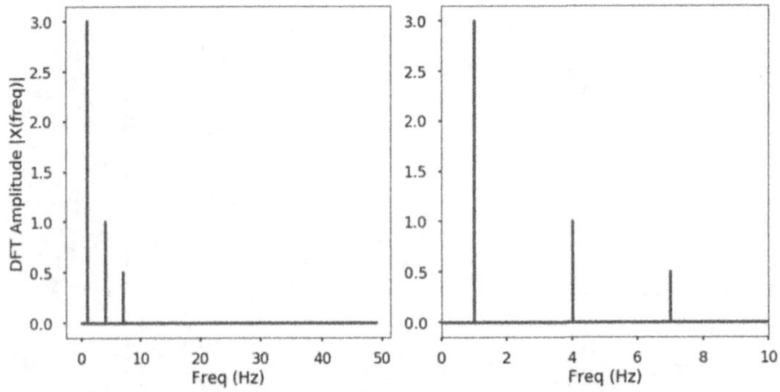

Plotting the first half of the DFT results shows three clear peaks at frequency of 1, 4, and 7 Hz, with amplitude 3, 1, 0.5, as expected. This is how we can use the DFT to analyze an arbitrary signal by decomposing it into simple sine waves.

24.2.2 THE INVERSE DFT

We can easily compute the inverse transform of the DFT:

$$x_n = \frac{1}{N}\sum_{k=0}^{N-1} X_k \cdot e^{i\cdot 2\pi kn/N}.$$

We will leave this as an exercise for you to write a function.

24.2.3 THE LIMIT OF DFT

The main issue with the above DFT implementation is that it is not efficient if we have a signal with many data points. It may take a long time to compute the DFT if the signal is large.

TRY IT Write a function to generate a simple signal with a different sampling rate, and see the difference in computing time by varying the sampling rate.

```
In [6]: def gen_sig(sr):
            """
            function to generate
            a simple 1D signal with
            different sampling rate
            """
            ts = 1.0/sr
            t = np.arange(0,1,ts)

            freq = 1.
            x = 3*np.sin(2*np.pi*freq*t)
            return x
```

```
In [7]: # sampling rate =2000
        sr = 2000
        %timeit DFT(gen_sig(sr))
```

120 ms ± 8.27 ms per loop (mean ± std. dev. of 7 runs, 10 loops each)

```
In [8]: # sampling rate 20000
        sr = 20000
        %timeit DFT(gen_sig(sr))
```

15.9 s ± 1.51 s per loop (mean ±std. dev. of 7 runs, 1 loop each)

The increasing number of data points will require a lot of computation time using this DFT. Luckily, the *Fast Fourier Transform* (FFT), popularized by Cooley and Tukey in their 1965 paper,[1] can solve this problem efficiently. This is the topic for the next section.

[1] http://www.ams.org/journals/mcom/1965-19-090/S0025-5718-1965-0178586-1/.

24.3 FAST FOURIER TRANSFORM (FFT)

The **Fast Fourier Transform (FFT)** is an efficient algorithm used to calculate the DFT of a sequence. First described in Cooley and Tukey's classic paper in 1965, the idea can be traced back to Gauss's unpublished work in 1805. It is a divide-and-conquer algorithm that recursively breaks the DFT into smaller DFTs to reduce the number of computations. As a result, it successfully reduces the complexity of the DFT from $O(n^2)$ to $O(n \log n)$, where n is the size of the data. This reduction in computation time is significant, especially for data with large N, and FFT is widely used in engineering, science, and mathematics for this reason.

This section will explore how using FFT reduces computation time. The content of this section is heavily based on this great tutorial[2] put together by Jake VanderPlas.[3]

24.3.1 SYMMETRIES IN THE DFT

The answer to how does the FFT speedup the computing of DFT lies in the exploitation of the symmetries in the DFT. Let us study the symmetries in the DFT. From the definition of the DFT equation

$$X_k = \sum_{n=0}^{N-1} x_n \cdot e^{-i2\pi kn/N}.$$

We can calculate

$$X_{k+N} = \sum_{n=0}^{N-1} x_n \cdot e^{-i2\pi(k+N)n/N} = \sum_{n=0}^{N-1} x_n \cdot e^{-i2\pi n} \cdot e^{-i2\pi kn/N}.$$

Note that $e^{-i2\pi n} = 1$; therefore, we have

$$X_{k+N} = \sum_{n=0}^{N-1} x_n \cdot e^{-i2\pi kn/N} = X_k,$$

with a little extension, we can have

$$X_{k+i\cdot N} = X_k, \text{ for any integer } i.$$

Thus, within the DFT, clearly there are some symmetries that we can use to reduce the computation.

24.3.2 TRICKS IN FFT

Given that there are symmetries in the DFT, we can consider using it to reduce the computation, because if we need to calculate both X_k and X_{k+N}, we only need to do this once. This is exactly the idea behind the FFT. Cooley and Tukey showed that we can calculate DFT more efficiently if we continue to divide

[2] https://jakevdp.github.io/blog/2013/08/28/understanding-the-fft/.

[3] http://vanderplas.com.

the problem into smaller ones. Let us first divide the whole series into two parts, i.e., the even and odd number parts:

$$
\begin{aligned}
X_k &= \sum_{n=0}^{N-1} x_n \cdot e^{-i2\pi kn/N} \\
&= \sum_{m=0}^{N/2-1} x_{2m} \cdot e^{-i2\pi k(2m)/N} + \sum_{m=0}^{N/2-1} x_{2m+1} \cdot e^{-i2\pi k(2m+1)/N} \\
&= \sum_{m=0}^{N/2-1} x_{2m} \cdot e^{-i2\pi km/(N/2)} + e^{-i2\pi k/N} \sum_{m=0}^{N/2-1} x_{2m+1} \cdot e^{-i2\pi km/(N/2)}.
\end{aligned}
$$

As shown, the two smaller terms, which only have half of the size ($\frac{N}{2}$) in the above equation, are two smaller DFTs. For each term, the $0 \leq m \leq \frac{N}{2}$, but $0 \leq k \leq N$; therefore, half of the values will be the same due to the symmetry properties described above. Thus, we only need to calculate half of the fields in each term. Of course, we can continue to divide each term into half with the even and odd values until it reaches the last two numbers, at which point the calculation will be really simple.

This is how FFT works using this recursive approach. Let us perform a quick and dirty implementation of the FFT. Note that the input signal to FFT should have a length of power of 2. If it is not, then we need to fill up zeros to the next power of 2 size.

```
In [1]: import matplotlib.pyplot as plt
        import numpy as np

        plt.style.use("seaborn-poster")
        %matplotlib inline

In [2]: def FFT(x):
            """
            A recursive implementation of
            the 1D Cooley-Tukey FFT, the
            input should have a length of
            power of 2.
            """
            N = len(x)

            if N == 1:
                return x
            else:
                X_even = FFT(x[::2])
                X_odd = FFT(x[1::2])
                factor = np.exp(-2j*np.pi*np.arange(N)/ N)
```

```
            X = np.concatenate(\
                [X_even+factor[:int(N/2)]*X_odd,
                 X_even+factor[int(N/2):]*X_odd])
            return X

In [3]: # sampling rate
        sr = 128
        # sampling interval
        ts = 1.0/sr
        t = np.arange(0,1,ts)

        freq = 1.
        x = 3*np.sin(2*np.pi*freq*t)

        freq = 4
        x += np.sin(2*np.pi*freq*t)

        freq = 7
        x += 0.5* np.sin(2*np.pi*freq*t)

        plt.figure(figsize = (8, 6))
        plt.plot(t, x, "r")
        plt.ylabel("Amplitude")

        plt.show()
```

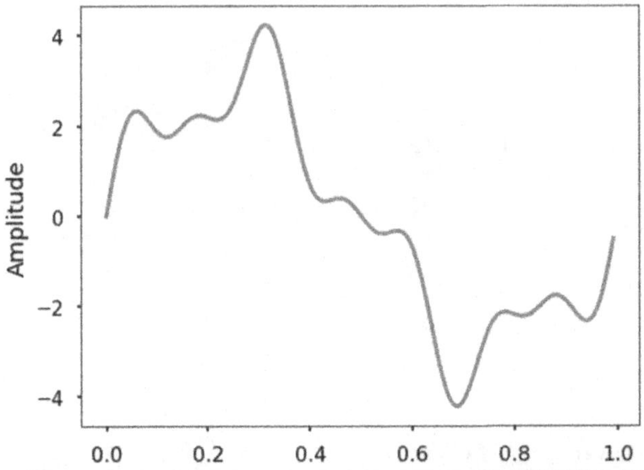

TRY IT! Use the FFT function to calculate the Fourier transform of the above signal. Plot the amplitude spectrum for both the two- and one-sided frequencies.

```
In [4]: X=FFT(x)

        # calculate the frequency
        N = len(X)
        n = np.arange(N)
        T = N/sr
        freq = n/T

        plt.figure(figsize = (12, 6))
        plt.subplot(121)
        plt.stem(freq, abs(X), "b", markerfmt=" ", basefmt="-b")
        plt.xlabel("Freq (Hz)")
        plt.ylabel("FFT Amplitude |X(freq)|")

        # Get the one-sided spectrum
        n_oneside = N//2
        # get the one side frequency
        f_oneside = freq[:n_oneside]

        # normalize the amplitude
        X_oneside =X[:n_oneside]/n_oneside

        plt.subplot(122)
        plt.stem(f_oneside, abs(X_oneside), "b", markerfmt=" ", basefmt="-b")
        plt.xlabel("Freq (Hz)")
        plt.ylabel("Normalized FFT Amplitude |X(freq)|")
        plt.tight_layout()
        plt.show()
```

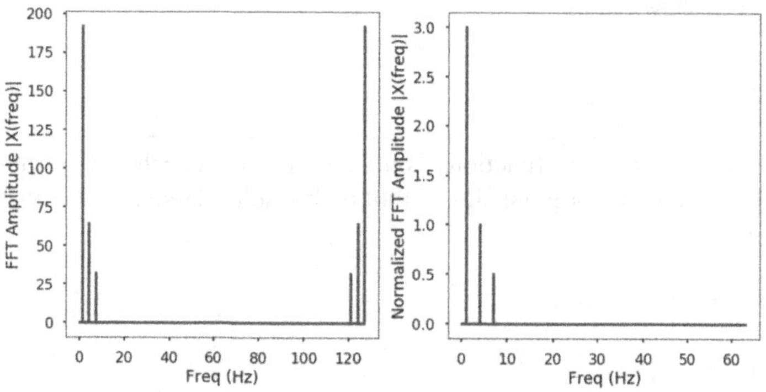

TRY IT! Generate a simple signal of length 2048, and record the time it will take to run the FFT; compare the speed with the DFT.

```
In [5]: def gen_sig(sr):
            """
            function to generate
            a simple 1D signal with
            different sampling rate
            """
            ts = 1.0/sr
            t = np.arange(0,1,ts)

            freq = 1.
            x = 3*np.sin(2*np.pi*freq*t)
            return x

In [6]: # sampling rate =2048
        sr = 2048
        %timeit FFT(gen_sig(sr))
```

16.9 ms ± 1.3 ms per loop (mean ± std. dev. of 7 runs, 100 loops each)

Thus, for a signal with length 2048 (about 2000), this implementation of FFT uses 16.9 ms instead of 120 ms using DFT. Note that there are lots of ways to optimize the FFT implementation to make it faster. The next section will introduce the Python built-in FFT functions, which will run much faster.

24.4 FFT IN PYTHON

Python has very mature FFT functions both in `NumPy` and `SciPy`. This section will take a look at both packages and see how easily they can be incorporated in our work. Let us use the signal generated above to run the test as shown in Fig. 24.5.

24.4.1 FFT IN NUMPY

EXAMPLE: Us the `fft` and `ifft` functions from `NumPy` to calculate the FFT amplitude spectrum and inverse FFT to obtain the original signal. Plot both results. Time the `fft` function using this 2000-length signal.

```
In [3]: from numpy.fft import fft, ifft

        X = fft(x)
```

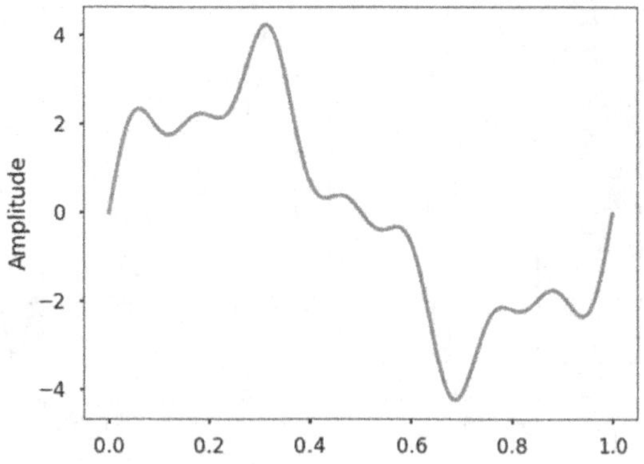

FIGURE 24.5

Signal generated before with 3 frequencies.

```
N = len(X)
n = np.arange(N)
T = N/sr
freq = n/T

plt.figure(figsize = (12, 6))
plt.subplot(121)

plt.stem(freq, np.abs(X), "b", markerfmt=" ", basefmt="-b")
plt.xlabel("Freq (Hz)")
plt.ylabel("FFT Amplitude |X(freq)|")
plt.xlim(0, 10)

plt.subplot(122)
plt.plot(t, ifft(X), "r")
plt.xlabel("Time (s)")
plt.ylabel("Amplitude")
plt.tight_layout()
plt.show()
```

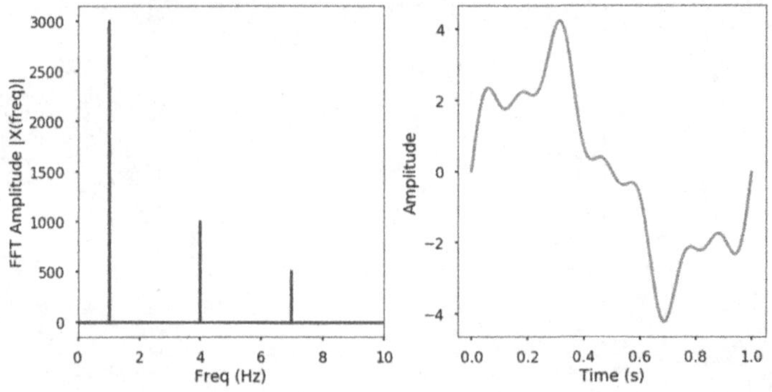


```
In [4]: %timeit fft(x)
```

42.3 µs ± 5.03 µs per loop (mean ± std. dev. of 7 runs, 10000 loops each)

24.4.2 FFT IN SCIPY

EXAMPLE: Use `fft` and `ifft` functions from `SciPy` to calculate the FFT amplitude spectrum and inverse FFT to obtain the original signal. Plot both results. Time the `fft` function using this 2000-length signal.

```
In [5]: from scipy.fftpack import fft, ifft

        X = fft(x)

        plt.figure(figsize = (12, 6))
        plt.subplot(121)

        plt.stem(freq, np.abs(X), "b", markerfmt=" ", basefmt="-b")
        plt.xlabel("Freq (Hz)")
        plt.ylabel("FFT Amplitude |X(freq)|")
        plt.xlim(0, 10)

        plt.subplot(122)
        plt.plot(t, ifft(X), "r")
        plt.xlabel("Time (s)")
        plt.ylabel("Amplitude")
        plt.tight_layout()
        plt.show()
```

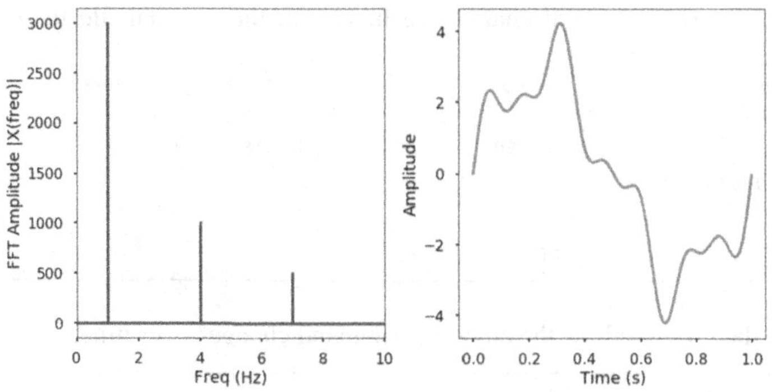


```
In [6]: %timeit fft(x)
```

12.6 μs ± 222 ns per loop (mean ± std. dev. of 7 runs, 100000 loops each)

Note that the built-in FFT functions are much faster and easier to use, especially when using the `SciPy` version. Here are the results for comparison:

- Implemented DFT: 120 ms
- Implemented FFT: 16.9 ms
- `NumPy` FFT: 42.3 μs
- `SciPy` FFT: 12.6 μs

24.4.3 MORE EXAMPLES

More examples how to use FFT in real-world applications are shown below.

24.4.3.1 Electricity Demand in California

First, we explore the electricity demand from California from 2019-11-30 to 2019-12-30. You can download data from US Energy Information Administration.[4] Herein, the data has been downloaded for you.

The electricity demand data from California is stored in "930-data-export.csv" in 3 columns. Remember that we learned how to read CSV file using `NumPy`. Here, we will use another package, `pandas`, which is a very popular package that deals with time series data. This package will not be discussed herein. We highly recommend learning how to use it by yourself.

First, let us study the data.

```
In [7]: import pandas as pd
```

The `read_csv` function will read the CSV file. Pay attention to the `parse_dates` parameter, which will find the date and time in column one. The data will be read into a panda `DataFrame`, and

4 https://www.eia.gov/beta/electricity/gridmonitor/dashboard/electric_overview/US48/US48.

we use `df` to store it. Then we will change the header in the original file to something easier to use.

```
In [8]: df = pd.read_csv("./data/930-data-export.csv",
                             delimiter=",", parse_dates=[1])
        df.rename(columns={"Timestamp (Hour Ending)":"hour",
                              "Total CAL Demand (MWh)":"demand"},
                              inplace=True)
```

By plotting the data, we see how the electricity demand changes over time.

```
In [9]: plt.figure(figsize = (12, 6))
        plt.plot(df["hour"], df["demand"])
        plt.xlabel("Datetime")
        plt.ylabel("California electricity demand (MWh)")
        plt.xticks(rotation=25)
        plt.show()
```

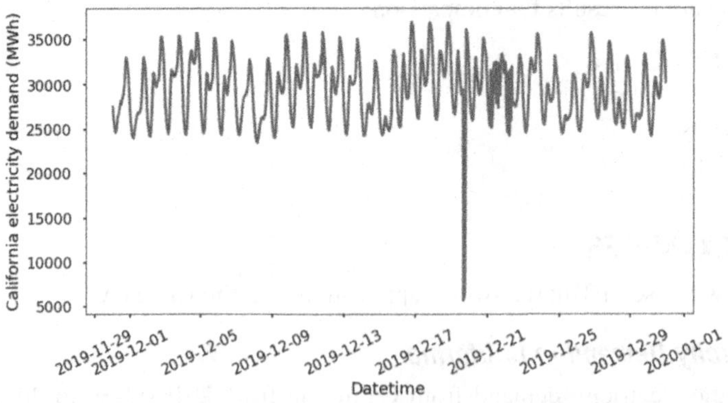

From the plotted time series, it is hard to discern if there are any patterns behind the data. Let us transform the data into frequency domain and see if there is anything interesting.

```
In [10]: X = fft(df["demand"])
         N = len(X)
         n = np.arange(N)
         # get the sampling rate
         sr = 1 / (60*60)
         T = N/sr
         freq = n/T

         # Get the one-sided spectrum
         n_oneside = N//2
```

```
# get the one side frequency
f_oneside = freq[:n_oneside]

plt.figure(figsize = (12, 6))
plt.plot(f_oneside, np.abs(X[:n_oneside]), "b")
plt.xlabel("Freq (Hz)")
plt.ylabel("FFT Amplitude |X(freq)|")
plt.show()
```

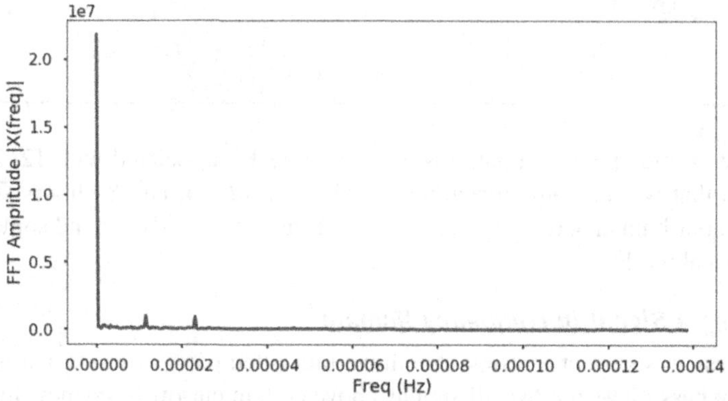

Note the clear peaks in the FFT amplitude figure, but it is hard to tell what they are in terms of frequency. Let us plot the results using hours and highlight some of the hours associated with the peaks.

```
In [11]: # convert frequency to hour
         t_h = 1/f_oneside / (60 * 60)

         plt.figure(figsize=(12,6))
         plt.plot(t_h, np.abs(X[:n_oneside])/n_oneside)
         plt.xticks([12, 24, 84, 168])
         plt.xlim(0, 200)
         plt.xlabel("Period ($hour$)")
         plt.show()
```

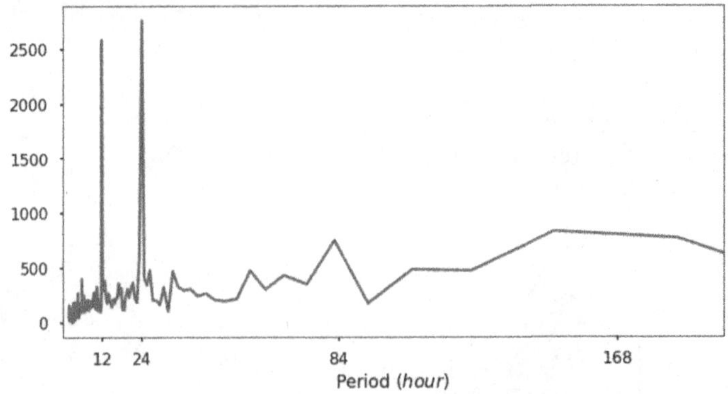

We can now see some interesting patterns, i.e., three peaks associated with 12, 24, and 84 hours. These peaks mean that we see some repeating signal every 12, 24, and 84 hours. This makes sense and corresponds to our human activity pattern. The FFT can help us understand some of the repeating signals in our physical world.

24.4.3.2 Filtering a Signal in Frequency Domain

Filtering is a process in signal processing to remove unwanted parts of the signal within certain frequency range. Low-pass filters remove all signals above certain cut-off frequency; high-pass filters do the opposite. Combining low- and high-pass filters allows constructing a band-pass filter, which means we only keep the signals within a pair of frequencies. Using FFT, we can easily do this. Let us play with the following example to illustrate the basics of a band-pass filter. Note that we just want to show the idea of filtering using very basic operations, in reality, filtering processes are much more sophisticated.

EXAMPLE: Use the signal generated at the beginning of this section (the mixed sine waves with 1, 4, and 7 Hz) and high-pass filter this signal at 6 Hz. Plot the filtered signal and the FFT amplitude before and after filtering.

```
In [12]: from scipy.fftpack import fftfreq
```

```
In [13]: plt.figure(figsize = (8, 6))
         plt.plot(t, x, "r")
         plt.ylabel("Amplitude")
         plt.title("Original signal")
         plt.show()
```

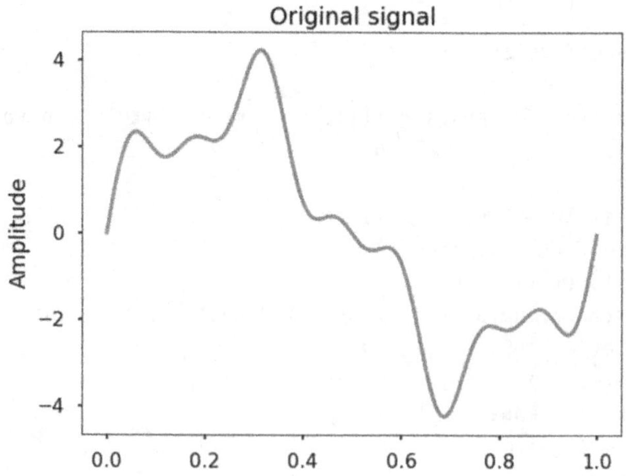


```
In [14]: # FFT the signal
         sig_fft = fft(x)
         # copy the FFT results
         sig_fft_filtered = sig_fft.copy()

         # obtain the frequencies using SciPy function
         freq = fftfreq(len(x), d=1./2000)

         # define the cut-off frequency
         cut_off = 6

         # high-pass filter by assign zeros to the
         # FFT amplitudes where the absolute
         # frequencies smaller than the cut-off
         sig_fft_filtered[np.abs(freq) < cut_off] = 0

         # get the filtered signal in time domain
         filtered = ifft(sig_fft_filtered)

         # plot the filtered signal
         plt.figure(figsize = (12, 6))
         plt.plot(t, filtered)
         plt.xlabel("Time (s)")
         plt.ylabel("Amplitude")
         plt.show()
```

```
# plot the FFT amplitude before and after
plt.figure(figsize = (12, 6))
plt.subplot(121)
plt.stem(freq, np.abs(sig_fft), "b", markerfmt=" ", basefmt="-b")
plt.title("Before filtering")
plt.xlim(0, 10)
plt.xlabel("Frequency (Hz)")
plt.ylabel("FFT Amplitude")
plt.subplot(122)
plt.stem(freq, np.abs(sig_fft_filtered), "b", markerfmt=" ", basefmt="-b")
plt.title("After filtering")
plt.xlim(0, 10)
plt.xlabel("Frequency (Hz)")
plt.ylabel("FFT Amplitude")
plt.tight_layout()
plt.show()
```

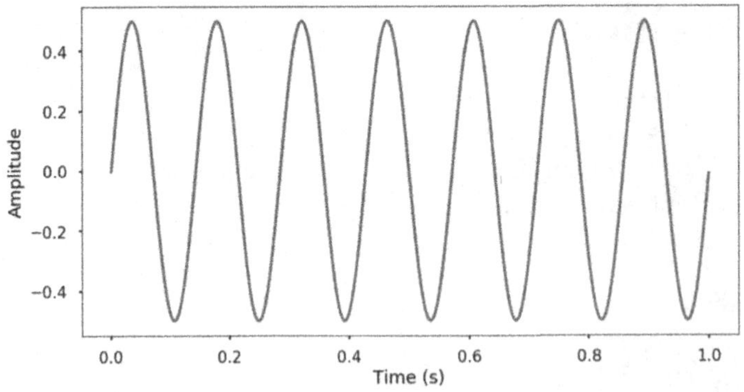

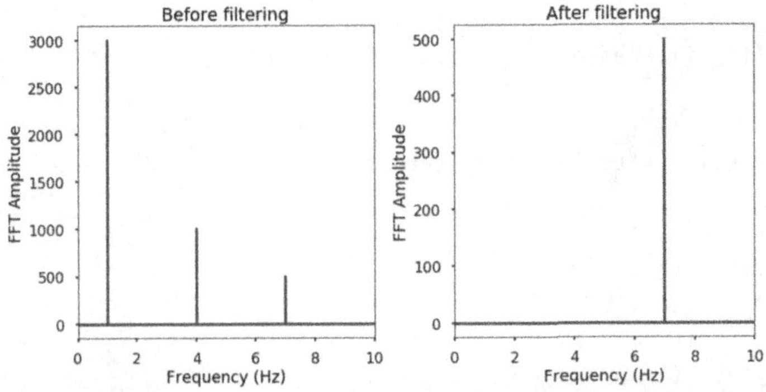

In the above example, we assigned any absolute frequencies of the FFT amplitude to zero,and returned back to time domain signal; we achieved a very basic high-pass filter in a few steps. Therefore, FFT can help us get the signal we are interested in and remove the ones that are unwanted.

24.5 SUMMARY AND PROBLEMS

24.5.1 SUMMARY

1. We learned the basics of the waves: frequency, period, amplitude, and wavelength are characteristics of the waves.
2. The Discrete Fourier Transform (DFT) is a way to transform a signal from the time domain to the frequency domain using the sum of a sequence of sine waves.
3. The Fast Fourier Transform (FFT) is an algorithm used to calculate the DFTs efficiently by taking advantage of the symmetry properties in DFT.

24.5.2 PROBLEMS

1. You are asked to measure the temperature of the room for 30 days. Every day at noon you measure the temperature and record the value. What's the frequency of the temperature signal you get?

2. What is the relationship between the frequency and the period of a wave?

3. What is the difference between period and wavelength? What are the similarities between them?

4. What are the time domain and frequency domain representation of a signal?

5. Generate two signals: signal 1 is a sine wave with 5 Hz, amplitude 3, and phase shift 3, and signal 2 is a sine wave with 2 Hz, amplitude 2, and phase shift −2. Plot the signal for 2 s.
   Test cases:

```
In [1]: # sampling rate
        sr = 100
        # sampling interval
        ts = 1.0/sr
        t = np.arange(0,2,ts)

        freq = 5.
        x = 3*np.sin(2*np.pi*freq*t + 3)

        freq = 2
        x += 2*np.sin(2*np.pi*freq*t - 2)


        plt.figure(figsize = (8, 6))
        plt.plot(t, x, "r")
        plt.ylabel("Amplitude")
        plt.xlabel("Time (s)")
        plt.show()
```

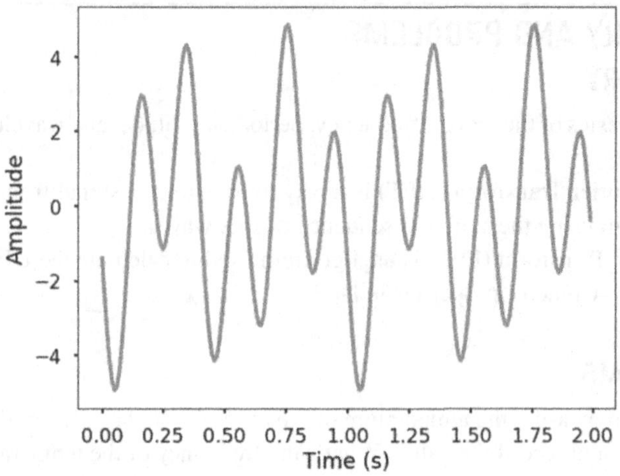

6. Sample the signal you generated in Problem 5 using a sampling rate 5, 10, 20, 50, and 100 Hz, and see the differences between different sampling rates.
7. Given a signal `t = [0,1,2,3]`, and `y = [0,3,2,0]`, find the real DFT of X. Write the expression for the inverse DFT. Do not use Python to find the results. Write out the equations and calculate the values.
8. What are the amplitude and phase of the DFT values for a signal?
9. We implemented the DFT previously. Can you implement the inverse DFT in Python similarly?
10. Use the DFT function and inverse DFT we implemented, and generate the amplitude spectrum for the signal you generated in Problem 5. Normalize the DFT amplitude to get the correct corresponding time domain amplitude.
11. Can you describe the tricks used in FFT to make the computation faster?
12. Use the `fft` and `ifft` function from scipy to repeat Problem 10.
13. Add a random normal distribution noise into the signal in Problem 5 using `NumPy` and plot the FFT amplitude spectrum. What do you see? The signal with noise will be shown in the following test case.
    Test case:

```
In [2]: np.random.seed(10)
        x_noise = x + np.random.normal(0, 2, size = len(x))

        plt.figure(figsize = (8, 6))
        plt.plot(t, x_noise, "r")
        plt.ylabel("Amplitude")
        plt.xlabel("Time (s)")
        plt.show()
```

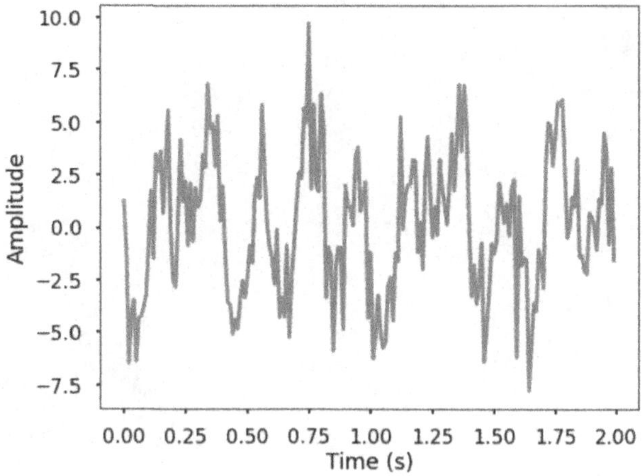
10.0
7.5
5.0
2.5
0.0
−2.5
−5.0
−7.5
Amplitude
0.00 0.25 0.50 0.75 1.00 1.25 1.50 1.75 2.00
Time (s)

APPENDIX

GETTING STARTED WITH PYTHON IN WINDOWS

A

A.1 GETTING STARTED WITH PYTHON IN WINDOWS

A.1.1 SETTING UP YOUR WORKING ENVIRONMENT IN WINDOWS

Before we start to use Python, we need to set up our Python working environment on the computer. In this section, we will introduce the processes to get it started.

There are different ways to install Python and related packages, here we recommend to use Anaconda[1] or Miniconda[2] to install and manage your packages. Depending on the *operating system* (OS) you are using, i.e., Windows, Mac OS X, or Linux, you need to download a specific installer for your machine. Both Anaconda and Miniconda are aiming to provide easy ways to manage Python work environment in scientific computing and data science.

Here we will use Windows as an example to show you the installation processes. For Mac and Linux users, please read Chapter 1.1 for all the processes. The main differences between Anaconda and Miniconda are:

- **Anaconda** is a complete distribution framework that includes the Python interpreter, package manager, as well as the commonly used packages in scientific computing.
- **Miniconda** is a light version of Anaconda that does not include the common packages, therefore, you need to install all the different packages by yourself. But it does have the Python interpreter and package manager.

The option we choose here is to use Miniconda to manage our installation of the packages. This way we can only install those we need.

The Miniconda install process is described below:

Step 1. Download the Miniconda installer from the website[3] as shown in Fig. A.1. Here you can choose a different installer based on your OS. We choose the Windows installer and Python 3.7 as an example.

Step 2. Run the installer by double clicking the installer. After you run the installer, follow the guide and you will successfully install it (Fig. A.2). One thing to note is that you can change the installation location by giving it an alternative location on your machine, we will use the default path here (Fig. A.3).

[1] https://www.anaconda.com/download/.
[2] https://conda.io/miniconda.html.
[3] https://conda.io/miniconda.html.

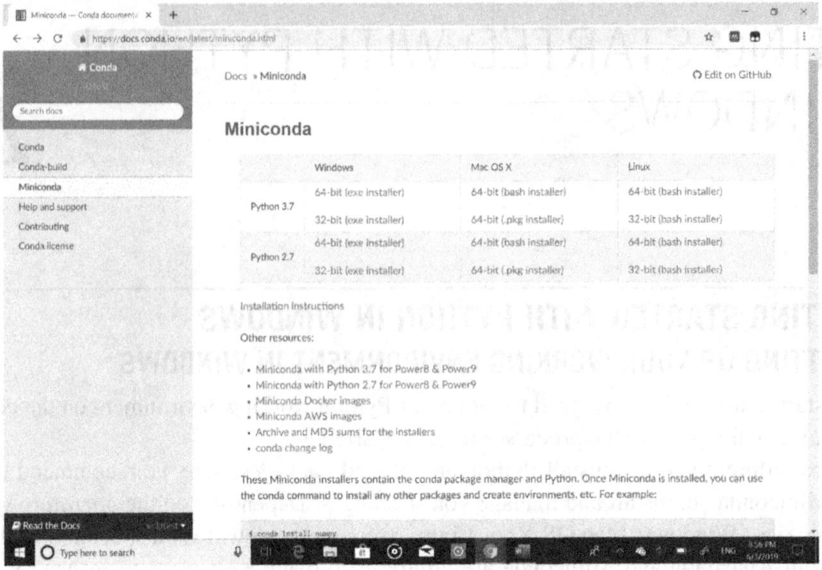

FIGURE A.1

The Miniconda download page, choose the installer based on your operating system.

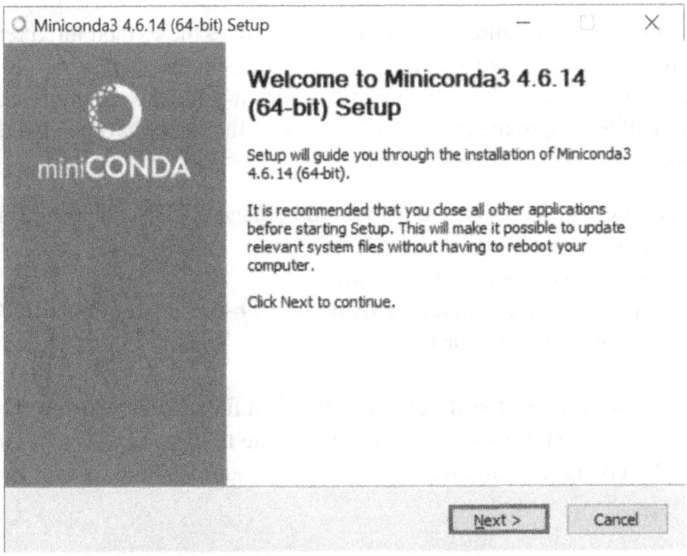

FIGURE A.2

Screen shot of running the installer in Anaconda prompt.

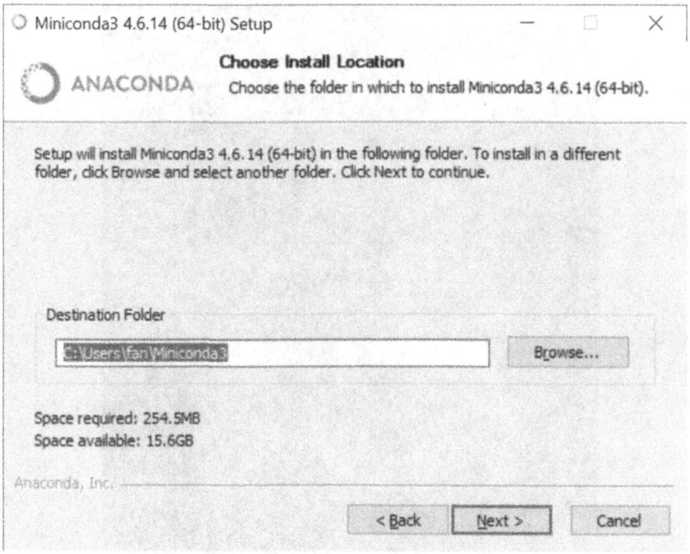

FIGURE A.3

The default installation location of your file system.

After installation, we can open the **Anaconda prompt** (the equivalent terminal on Mac or Linux) from the start menu as shown in Fig. A.4. Then we can check whether installation is success or not by typing the following commands shown in Fig. A.5.

Step 3. Following Fig. A.6, install the basic packages that are used in this book. Let us first install some packages for our book – `IPython`, `NumPy`, `SciPy`, `pandas`, `matplotlib` and `Jupyter notebook`. We will talk more about the management of the packages using `pip` and `conda` later.

A.1.2 THREE WAYS TO RUN PYTHON CODE

There are different ways to run Python code, they all have different usages. In this section, we will quickly introduce the three different ways to get you started.

Using Python or IPython shell

The easiest way to run Python code is through the Python or IPython shell (which stands for Interactive Python). The IPython shell is richer than Python shell due to, e.g., tab autocompletion, color-highlighted error messages, basic UNIX shell integration, and so on. Since we just installed IPython, let us try to run the "hello world" example with it. The way we launch either Python or IPython shell is by typing it in the Anaconda prompt (see the figure below). Then we can run Python command by typing it into the shell, by pressing `Enter`, we immediately see the results from the command. For example, we can print out "Hello World" by typing `print("Hello World")` as shown in Fig. A.7.

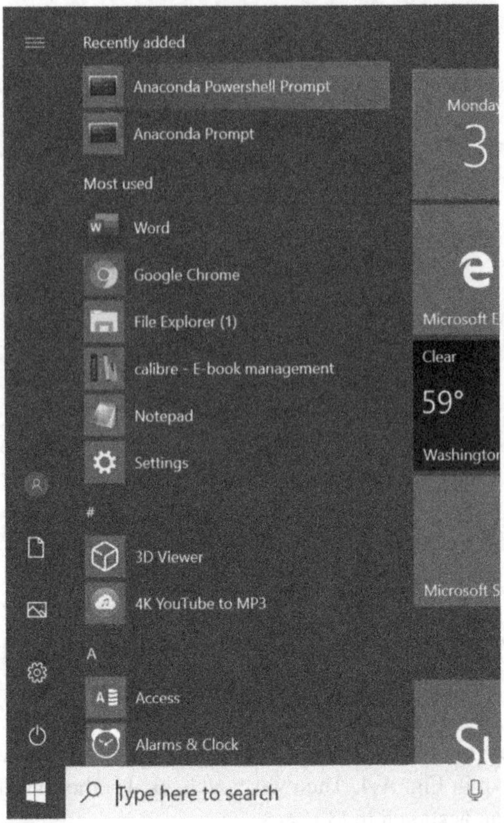

FIGURE A.4

Open the Anaconda prompt from the start menu.

In the above command, the `print()` is a function in Python, and "Hello World" is a string data type that we will introduce in the book.

Run Python script/file from command line

The second way to run Python code is to put all the commands into a file and save it as a file with extension `.py` (the extension of the file could be anything, but by convention, it is usually `.py`). For example, use your favorite text editor (Shown here is the Visual Studio Code[4]), put the command in a file called *hello_world.py* as shown in Fig. A.8. Then just run it from the prompt (Fig. A.9).

Using Jupyter notebook

The third way to run Python is through **Jupyter notebook**. It is a very powerful browser-based Python environment, we will talk more about it in details later in this chapter. Here we just quickly see

4 https://code.visualstudio.com.

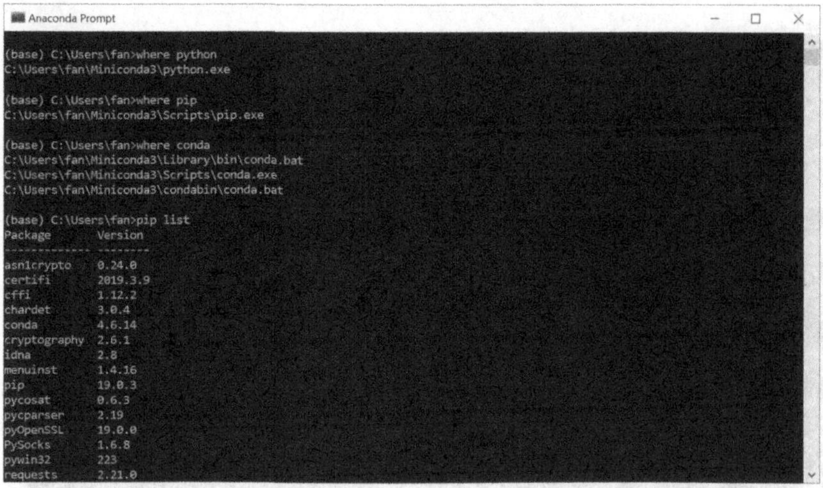

FIGURE A.5

A quick way to check if your installed Miniconda runs properly.

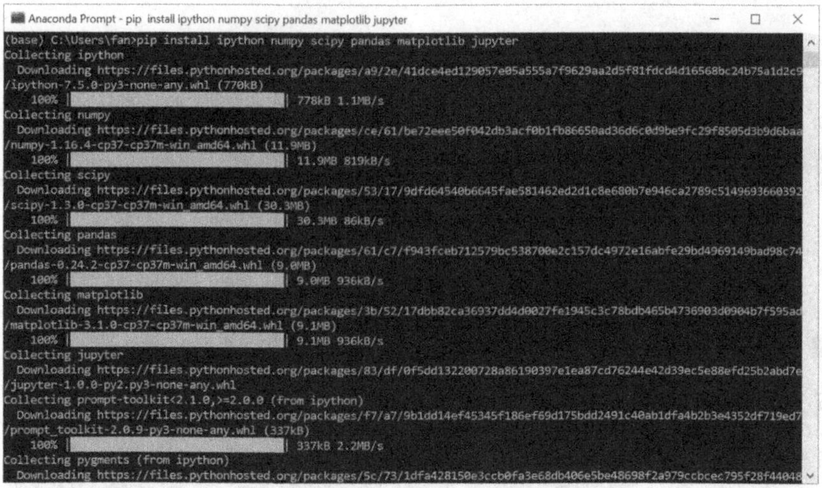

FIGURE A.6

Installation process for the packages that will be used in the rest of the book.

how we could run the code from a Jupyter notebook. Type the `jupyter notebook` in the bash command line:

```
jupyter notebook
```

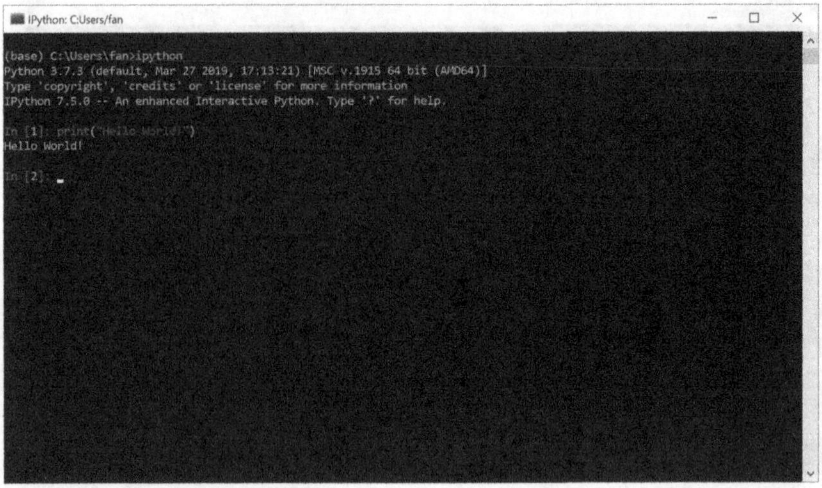

FIGURE A.7

Run "Hello World" in IPython shell by typing the command, "`print`" is a function that we will learn to print out anything within the parentheses.

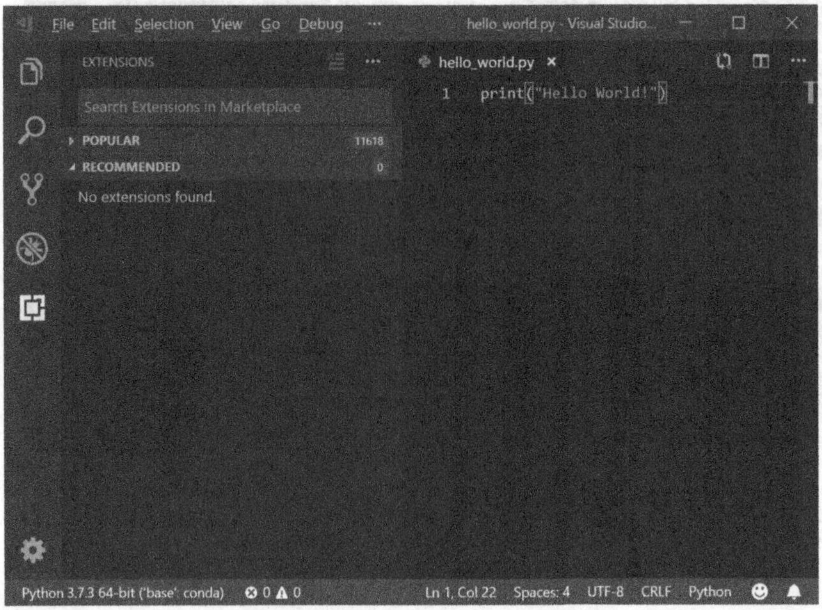

FIGURE A.8

A Python script file example using Visual Studio Code. You type in the commands you want to execute and save the file with a proper name.

FIGURE A.9

To run the Python script from command line, we can type “python hello_world.py”. This line tells Python that we will execute the commands that were saved in this file.

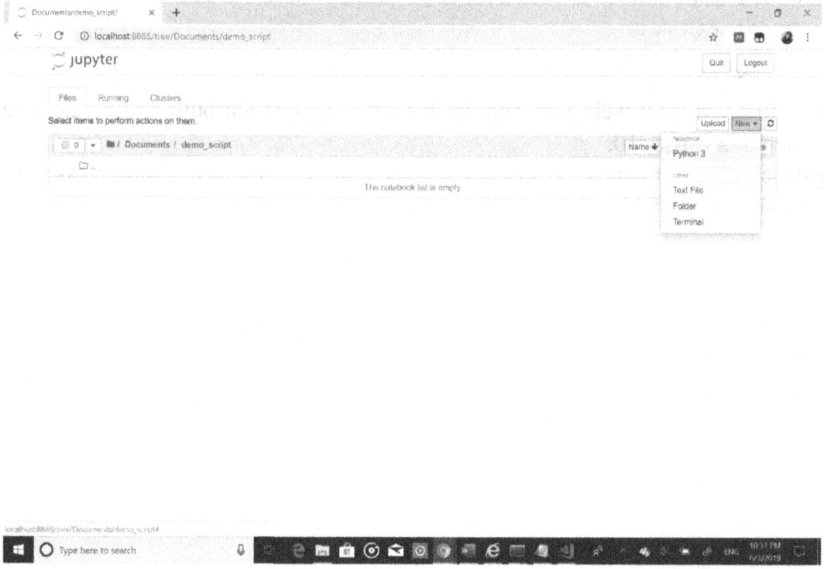

FIGURE A.10

To launch a `Jupyter notebook` server, type `jupyter notebook` in the command line, which will open a browser page as shown here. Click “New” button at the top right corner, and choose “Python3” which will create a Python notebook to run Python code.

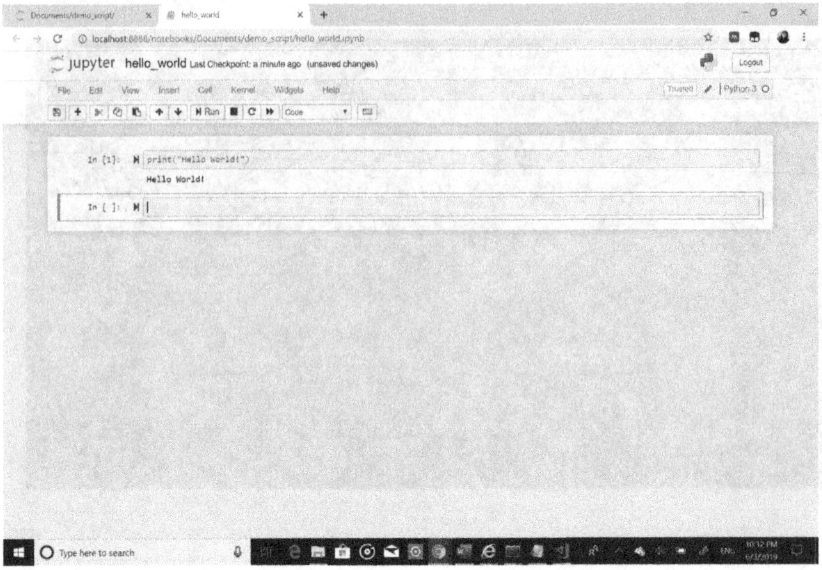

FIGURE A.11

Run the Hello World example within `Jupyter notebook`. Type the command in the code cell (the grey boxes) and press `Shift + Enter` to execute it.

Then you will see a local web page pop up, from the upper right button to create a new Python3 notebook as shown in Fig. A.10.

Running code in Jupyter notebook is easy, you type your code in the cell, and press `Shift + Enter` to run the cell, the results will be shown below the code (Fig. A.11).

Index

Printed in the United States
By Bookmasters